Let's Review Regents:

Living Environment Revised Edition

Gregory Scott Hunter, M.A.
has served New York State public education as

Superintendent of Schools
Chatham Central School District
Chatham, New York

Superintendent of Schools
Mexico Academy and Central School District
Mexico, New York

School Business Administrator and
Biology and General Science Teacher
Schodack Central School District
Castleton-On-Hudson, New York

Consultant
New York State Education Department
Bureaus of Science Education, Educational Testing,
and Curriculum Development
Albany, New York

Kaplan North America, LLC d/b/a Barron's Educational Series
1515 West Cypress Creek Road
Fort Lauderdale, Florida 33309
www.barronseduc.com

ISBN: 978-1-5062-6478-3

10 9 8

Kaplan North America, LLC d/b/a Barron's Educational Series print books are available at special quantity discounts to use for sales promotions, employee premiums, or educational purposes. For more information or to purchase books, please call the Simon & Schuster special sales department at 866-506-1949.

Contents

Preface

TO THE STUDENT

For Which Course Can This Book Be Used?

This book is designed to be used as a review text for the New York State Regents course in Living Environment. The material presented illustrates and complements the Core Curriculum for this course. Don't let the words "Core Curriculum" scare you. The Core Curriculum simply lists and describes what topics you must learn to complete this course. Because this book is comprehensive, it can be used to supplement any college-preparatory course in biology taught anywhere in the United States.

What Special Features Does This Book Have?

The topics in this book parallel those of the New York State Core Curriculum for Living Environment. The Core Curriculum is reproduced in its entirety within the text of this book. All Key Ideas, Performance Indicators, and Major Understandings for Standards 1 and 4 of the Core Curriculum appear within the text. In addition, this edition contains information about the laboratory experiences required of all New York State biology students.

Each section contains the material considered to be most relevant to a Major Understanding. However, the contents of this book do not represent all the information you will be required to learn for this course. Requirements will vary among teachers and school districts. This book illustrates and explains the concepts contained in the Major Understandings. Students are expected to do more than simply memorize facts. You are also required to demonstrate understanding of scientific concepts. Your teacher may supplement material in the basic curriculum due to interest expressed by a class or a particular student. Feel free to ask your teacher about the areas of biology that interest you.

Practice question sets that mirror the style and content of the Regents examination appear in this book. The multiple-choice questions usually require that you use your knowledge of biology to make judgments and to select an appropriate choice from among a number of possible responses. Free-response questions ask you to construct single-word, complete-sentence, or essay answers that are understandable as well as scientifically accurate. Graphical analysis questions challenge you to organize, represent in graphic form, and draw inferences from experimental data.

Answers Explained for the practice multiple-choice questions provide an extensive analysis of each question, not simply an answer key. The explanations give you information about why choices are correct or incorrect. Sample free-response and essay answers guide you toward the standard required to perform successfully on a Regents examination in biology. Graphs and charts provide guidance for construction of responses that will meet new Regents standards.

An extensive glossary provides an alphabetical listing of terms commonly used in biology courses together with simple, easy-to-understand definitions. Many of these terms will appear in Regents examination questions or in your reading assignments.

Finally, two full-length Regents examinations are included. They will give you an opportunity to test your knowledge of biology and to practice your question-answering skills before taking the Regents examination.

Who Should Use This Book?

Any student enrolled in the New York State course on Living Environment will find this book to be a valuable supplement to the regular class textbook. In addition, this book can be used to prepare for tests throughout the school year and for the year-end Regents examination on Living Environment. It will help you do your best on the Regents examination by increasing your self-confidence in your test-taking abilities.

Students in any secondary-level biology course anywhere in the United States can use this book in much the same way. Its clear, concise style helps to clarify concepts. The absence of extraneous material enables you to concentrate on the learning standards without being bogged down in unnecessary details.

NEW YORK STATE LEARNING STANDARDS

Graduation Requirements

Several graduation or commencement standards are required of students in New York State public schools regarding their performance in math, science, and technology. Living Environment Core curriculum addresses two of these standards as follows:

Standard 1: Students will use mathematical analysis, scientific inquiry, and engineering design, as appropriate, to pose questions, seek answers, and develop solutions.

Standard 4: Students will understand and apply scientific concepts, principles, and theories pertaining to the physical setting and living environment and recognize the historical development of ideas in science.

The Core Curriculum for Living Environment was devised from these two commencement standards. The Core Curriculum is not a detailed outline, or syllabus, for biology. It does not prescribe what must be taught and learned in any particular classroom. Instead, it defines the skills and major understandings that you must master to achieve graduation standards for life science.

Key Ideas, Performance Indicators, and Major Understandings

Each topic within the graduation standards is broken down into a number of Key Ideas. Key Ideas are broad, general statements about what you as a student will be expected to know. Within Standard 1, three Key Ideas are concerned with laboratory investigation and data analysis. Within Standard 4, seven Key Ideas present a set of concepts that are central to the science of biology. The unifying principles develop your understanding of the essential characteristics of living things and how these characteristics allow them to be successful in diverse habitats.

Each Key Idea contains Performance Indicators. The Performance Indicators spell out what skills you should be able to demonstrate when you have mastered the Key Idea. Performance Indicators help you understand what is expected of you.

Performance Indicators are further subdivided into Major Understandings. The Major Understandings contain specific concepts that you must master in order to demonstrate the skills required by a Performance Indicator. The Major Understandings contain the material that will appear on your Regents exam.

Laboratory Component

A meaningful laboratory experience is essential to the success of any science course. You're expected to develop a good sense of how scientific inquiry is carried out by a professional scientist and how these same techniques can assist in the full understanding of scientific concepts. The Regents requires 1,200 minutes of laboratory experience coupled with satisfactory written reports of your findings.

Living Environment Regents: Format and Scoring

The format of the Living Environment Regents is as follows:

Part A: 35 multiple-choice questions that test your knowledge of specific factual information. All questions in Part A must be answered.

Part B: Variable number of questions representing a mixture of multiple-choice and free-response items. Questions may be based on your direct knowledge of biology, interpretation of experimental data, analysis of readings in science, and the ability to deal with representations of biological phenomena. All questions must be answered in Part B.

Part C: Variable number of free-response questions. Questions may be based on your direct knowledge of biology, interpretation of experimental data, analysis of readings in science, and ability to deal with representations of biological phenomena. All questions must be answered in Part C.

Part D: Laboratory component of the Regents examination in which four required laboratory experiences are tested. See pages xvii and xviii for additional information on the Part D requirement. All questions must be answered in Part D.

HOW TO STUDY

You've spent the school year learning many different facts and concepts—far more than you could hope to remember from one review session. Your teacher has drilled you on these facts and concepts. You've done homework, taken quizzes and tests, and reviewed the material at intervals throughout the year. Now it's time to put it all together. The Regents exam may be only a few weeks away. If you and your teacher have planned properly, you will have finished the course material about a month before the exam. You now have to make efficient use of the days and weeks ahead to review all that you have learned and demonstrate your mastery of the material on the Regents.

It may seem like an impossible task, but you shouldn't be discouraged. You've actually retained much more of the year's material than you realize. The review process should be one that helps you to recall the many facts and concepts you have stored away in your memory. This book will help you to review this material as efficiently as possible.

You also have to get yourself in the right frame of mind. Being nervous and stressed during the review process won't help you one bit. The best way to avoid being stressed during an exam is to be well rested, prepared, and confident. This book is here to help you get prepared and build your confidence. So, now is the time to get started on the road to a successful exam experience. To begin, carefully read and follow the following steps:

1. *Get started!* Start your review early; don't wait until the last minute. Allow at least two weeks to prepare for the Regents exam. Set aside an hour or two a day over the next two weeks for your review. Less than an hour a day is not enough time for you to concentrate on the

material in a meaningful way; more than two hours a day will give diminishing returns on your investment of time.

2. *Get lost!* Find a quiet, comfortable place to study. You should seat yourself at a well-lit work surface free of clutter and in a room without distractions of any kind. (You may enjoy watching TV or listening to music curled up in a nice, soft chair, but these and other distractions should be avoided when studying.)

3. *Get prepared!* Make sure you have the tools you need to work, including this book, a pen and pencil, and some scratch paper for taking notes and doing calculations. Keep your class notebook at hand for looking up information between test-taking sessions. Have this book available for quick and efficient review of important concepts.

4. *Get focused!* Concentrate on the study material in the question sets in this book. Think about the questions that you review. Read carefully and thoughtfully. Try to make sense out of the questions. Think carefully about the answers that you choose. See the section "Using This Book for Study" for additional information about question-answering techniques.

5. *Get help!* Use available resources, including a dictionary and the glossary of this book, to look up the meanings of unfamiliar words in the practice questions. Remember that these same terms can appear on the Regents exam, so take the opportunity to learn them now.

6. *Go get 'em!* Remember—study requires time and effort. Your investment in study now will pay off when you take the Regents exam.

Using This Book for Study

This book will be an invaluable tool for you if used properly. Answer all of the questions in the question sets and practice exams even though the exam directions may allow you to choose which questions to answer. The more you study and practice, the better your chances of increasing your knowledge about biology and of obtaining a high grade on the Regents exam. To maximize your chances, use this book as follows.

1. Answer all of the questions in each question set. Check and correct your responses by using Answers Explained and Wrong Choices Explained at the back of the book. Follow this procedure for each question set. Study your areas of weakness. Then redo the questions you missed on the first round. Make sure you now understand fully what the question is asking and what makes one answer choice the correct answer.

2. When you've completed the questions in all the question sets, go on to the examination section. Take the exam under test conditions. Plan to answer all test questions; allow yourself no choices at this stage of the study process.

3. Test conditions should include the following:
   - Be well rested—get a good night's sleep before attempting any exam
   - Find a quiet, comfortable room in which to work
   - Allow no distractions of any kind
   - Have a pen, pencil, and some scratch paper handy
   - Set an alarm clock or watch for the three-hour exam limit

4. This step is very important. Take a deep breath, close your eyes for a moment, and RELAX! Tell yourself that you know this material. You have lots of time to take the Regents exam; use it to your advantage by reducing your stress level. Forget about your plans for later—your first priority right now is to do your best on the Regents exam, whether it is a practice exam in this book or the real thing.

5. Read all test directions carefully. Be sure you know how many questions you must answer to complete each part of the exam. If test questions relate to a reading passage, diagram, chart, or graph, be sure you fully understand the supplemental information before you attempt to answer the questions that relate to it.

6. When answering the multiple-choice questions on the Regents exam, TAKE YOUR TIME! Be sure to read the stem of the question very carefully; read it over several times. These questions are painstakingly written by the test preparers, and every word is chosen to convey a specific meaning. If you read the questions *carelessly*, you may answer a question that was never asked. Read each of the four choices carefully, using a pencil to mark the test booklet next to the answer you feel is correct.

7. Remember that three of the choices are incorrect; these incorrect choices are called distracters because they are selected by the test preparers to seem like plausible answers to poorly prepared or careless students. Make sure that you think clearly, using everything you have learned about biology since the beginning of the year, to eliminate each of the distracters as correct answers. This elimination process is just as important to your success on the Living Environment Regents exam as knowing the correct answer. If more than one answer seems to be correct, reread the question to find the words that will help you to distinguish between the correct answer and the distracters. When you have made your best judgment about the correct answer, circle its number in pencil in your test booklet.

8. Free-response questions appear in a number of different forms. Students may be asked to select a term from a list, write the term on the answer sheet, and define the term. Students may be asked to describe some biological phenomenon or state a true fact in biology using a complete sentence. Students may be asked to read a value from a diagram of a measuring instrument and write that value in a blank on the answer sheet. Exercise care when answering this type of question—follow directions precisely. A complete sentence must contain a noun and a verb, must be punctuated, and must be written in an understandable way in addition to accurately answering the scientific part of the question. Values must be written clearly and accurately. They must include a unit of measure, if appropriate. Failure to follow the directions for a question may result in a loss of credit for that question.

9. The essay or paragraph question is a special type of free-response question. Typically, essay or paragraph questions provide an opportunity to earn multiple credits for answering the question correctly. As in the free-response questions described above, you must follow the directions for a question if you hope to earn the maximum number of credits for the question. Typically, the question will outline exactly what must be included in your essay to gain full credit. Follow these directions step by step, double-checking to be certain that all question components are addressed in your answer. In addition, your essay or paragraph should follow the rules of good grammar and good communication so that it is readable and understandable. Of course, it should contain correct information and answer all parts of the question asked.

10. Graphs and charts are used in a special type of question that requires the student to organize and represent data in graphic format. Typically, for such questions you are expected to place unorganized data into ascending order in a data chart or table. You may also be asked to plot unorganized data on a graph grid, connect the plotted points, and label the graph axes appropriately. Finally, questions regarding data trends and extrapolated projections may be asked, requiring you to analyze the data in the graph and draw inferences from it. As with all examination questions, always follow all directions for the question. Credit can be granted only for correctly followed directions and accurate interpretation of the data.

11. When you've completed the exam, relax for a moment. Check your time; have you used the entire three hours? Probably not. Resist the urge to quit. Go back to the beginning of the exam and, in the time remaining, **retake the exam in its entirety**. Try to ignore the penciled notations you made the first time through the exam. If you come up with a different answer the second time through, *stop* and read over the question with extreme care before deciding which is the correct

response. Once you've decided on the correct response, finalize the answer in ink in the answer booklet.

12. Score the exam using the answer key at the end of the exam. Review the Answers Explained section of *Barron's Regents Exams and Answers—Living Environment* for each question to aid your understanding of the exam and material. Remember that understanding why an answer is incorrect is just as important as understanding why an answer is correct.

13. Finally, focus your between-exam study on your areas of weakness in order to improve your performance on the next practice exam. Complete all the practice exams using the preceding techniques.

Test-Taking Techniques

The following is a summary of test-taking techniques you should follow when taking the actual exam.

1. Complete your study and review at least one day before taking this examination. Last minute cramming may actually hurt, rather than help, your performance on the exam.

2. Be well rested the day of the exam. Get a good night's sleep before taking any examination.

3. Bring two pens, two pencils, and an eraser to the exam. If your school requires it, bring some form of identification with you, as well. Wear a watch, or sit where you see a clock. Before entering the room, remember that you will remain for the entire three-hour examination period.

4. Be familiar with the format of the examination. You must answer all questions in Part A (multiple-choice questions), all questions in Part B (mixed-format questions), all questions in Part C (free-response questions), and all questions in Part D (mixed-format questions).

5. Before beginning the exam, take a deep breath, close your eyes for a moment, and RELAX. Use this technique any time you feel yourself tensing up during the exam.

6. Read all exam directions carefully. Be sure you fully understand supplemental information (reading passages, charts, diagrams, graphs) before you attempt to answer the questions that relate to it.

7. When answering questions on the Regents exam, TAKE YOUR TIME. Be sure to read the stem of multiple-choice questions very carefully. Read each of the four answer choices carefully, as well.

With a pencil, make a mark in the test booklet next to the answer you feel is correct. If you're temporarily stumped by a question, put a check mark next to it and go on to the next question. Come back to the question later, when your mind is clear.

8. Remember that three of the multiple-choice answers are incorrect (known as distracters). If more than one answer seems to be correct, reread the question to find the words that will help you distinguish between the correct answer and the distracters. When you have made your best judgment about the correct answer, circle its number in pencil on your answer sheet.

9. Be certain to follow directions when completing free-response questions. Use complete sentences whenever they're required.

10. Complete charts and graphs according to instructions. If you're instructed to circle some part of an answer, be sure to circle it. If the directions call for you to connect points or list items in ascending order, then be certain to do so.

11. Answer essay or paragraph questions according to instructions. Be sure to include correct information that addresses every required point. Doing so will help to ensure that you receive full credit for your answers.

12. When you have completed the exam, relax for a moment. Go back to the beginning and, in the time remaining, **retake the exam in its entirety**. Once you have decided on the correct response, finalize the answer by marking an X in *ink* through the penciled circle on the answer sheet for multiple-choice questions. Be sure to finalize all of the answers for all parts of the exam.

13. Be certain to sign the declaration on your answer sheet. Unless this declaration is signed, your exam cannot be scored.

TIPS FOR TEACHERS

For teachers and administrators in New York State public schools, this book will provide an excellent source of primary instructional material for locally developed curricula for Living Environment. The examples given and the factual material presented represent a reasonable body of knowledge to complement the commencement standards for life science. In addition, it will provide a ready source of review material to prepare students for the New York State Regents examination on Living Environment.

All teachers will be able to use this book with their students as a companion to their regular textbooks and will find that students will gain considerable self-confidence and facility in test taking through its consistent use. Some school systems may wish to use this book as the primary text for their courses in college-preparatory biology; others may wish to employ it as the review text for test preparation.

Local Curricula

Teachers and administrators need to develop local curricula that complement the Core Curriculum, promulgated by the State in 1999. It is up to the teacher/administrator to decide what examples and factual knowledge will best illustrate the concepts presented in the Core Curriculum, what concepts need to be reinforced and enhanced, what experiences will add measurably to the students' understanding of science, and what examples of local interest should be included. The teacher will immediately recognize the need to go beyond this level in the classroom, with examples, specific content, and laboratory experiences that complement and illuminate these Major Understandings. It is on this level that the locally developed curriculum is essential. Each school system is challenged to develop an articulated K-12 curriculum in mathematics, science, and technology that will position students to achieve a passing standard at the elementary and intermediate levels, such that success is maximized at the commencement level.

The addition of factual content must be accomplished without contradicting the central philosophy of the learning standards. If local curricula merely revert to the fact-filled syllabi of the past, then little will have been accomplished in the standards movement other than to add yet another layer of content and requirements on the heads of students. A balance must be struck between the desire to build students' ability to think and analyze and the desire to add to the content they are expected to master.

ESSENTIAL QUESTIONS

A key component of a thoughtfully developed local curriculum is the development of Essential Questions. Essential Questions state, in interrogative form, the key concepts we expect our students to master. Posed correctly by the teacher, they provide a pathway for students to follow as they pursue their study of the subject matter, in this case, the subject matter of biology. Essential Questions also help students answer the questions "Why do I have to know this material? What will I be able to do with what I learn? How will knowing this help me to achieve my life goals? How will having this skill contribute to my growth as a person and a citizen?"

Teachers should be skilled in the art of developing Essential Questions, which serve as a guide not only to students but also to teachers. Essential Questions should be used by teachers in the preparation of class activities and of student performance assessments.

A feature of this book is the inclusion of Essential Questions at key points throughout the text. These Essential Questions organize and frame the concepts to be understood and the skills to be mastered through the student's study of biology. Likewise, they focus the concepts to better enable the teacher to develop class activities and assessments that truly add to the student's experience in measurable ways. They are meant to stimulate the teacher to develop his or her own Essential Questions so as to make the local curriculum come alive and provide an opportunity for student engagement in the learning process.

Included below is a listing of Essential Questions included in this book as samples to be considered by the teacher in delivering the Living Environment local curriculum. These questions are meant to be samples only, to help stimulate the teachers' thinking as they develop their own Essential Questions to assist in student learning.

Essential Questions:

Standard 1:

- Why is science important?

- How is good science done?

Key Idea 1.1

- How do models help scientists understand natural phenomena?

   Performance Indicators:

   1.1.1 How does expressing an idea in visual or mathematical terms assist in our understanding?

   1.1.2 What can we learn from studying the work of others?

   1.1.3 How can we tell the difference between good information and bad information?

   1.1.4 How can data be effectively represented in a manner that assists in our understanding of science?

Key Idea 1.2

- What are the steps in the scientific method?

- How do these steps ensure that science is valid?

   Performance Indicators:

   1.2.1 How do questions and observations help scientists to conduct good experiments?

1.2.2 How do scientists use the work of other scientists to help them in their own work?

1.2.3 What is a hypothesis and how does it help the scientist to focus an experiment?

1.2.4 What techniques and tools should be selected when designing an experiment?

Key Idea 1.3

- What are the different ways that scientists analyze scientific data?

Performance Indicators:

1.3.1 How can tables, charts, and graphs of data help to illustrate the findings of a scientific inquiry?

1.3.2 How do scientists measure the accuracy of their own experimental results?

1.3.3 How does a scientist draw inferences and conclusions based on data?

1.3.4 How does one experiment lead to another once results are known?

1.3.5 What are the appropriate ways to present the findings of a scientific investigation?

Standard 4:

- What are the principles that guide scientists in their study of the natural world?

- How is modern science built on the work of scientists of the past?

Key Idea 4.1

- How are living things similar to and different from other living things and nonliving things?

Performance Indicators

4.1.1 How does the great variety of living species impact environmental stability?

4.1.2 How are humans adapted to survive in their environment?

4.1.3 How are unicellular life forms able to carry out the life processes needed to sustain life?

Key Idea 4.2

- How does genetic inheritance help ensure the continuity of life?

Performance Indicators

4.2.1 How does DNA control the inheritance of genetic traits in living things?

4.2.2 What effect will genetic engineering have on the future of genetic inheritance?

What cautions should be considered in the use of genetic engineering?

Key Idea 4.3

- What has caused the wide variety of living things that have inhabited Earth over the past 3.5 billion years?

Performance Indicator

4.3.1 How do new species arise from previously existing species?

Key Idea 4.4

- What mechanisms are at work to create and perpetuate life on Earth?

Performance Indicator

4.4.1 How do living things produce more living things of the same kind?

Key Idea 4.5

- How do living things create a balanced steady state that maintains the living condition?

Performance Indicators

4.5.1 What biochemical reactions are common in living cells and promote the living condition?

4.5.2 How does the body defend itself against disease?

What happens when a living organism gets sick?

4.5.3 How do the activities that sustain the life of a cell contribute to the survival of a multicellular organism?

Key Idea 4.6

- How do species of different kinds depend on each other and contribute to each other's survival?
- How do conditions in the nonliving environment contribute to the survival of plant/animal communities?

Performance Indicators

4.6.1 What forces of nature manage the balance of species populations?

4.6.2 Why is it important to protect endangered/threatened species and habitats?

4.6.3 How do Earth's environments affect, and how are they affected by, the living community?

Key Idea 4.7

- What is the proper role of the human species in the ecological community?
- What responsibility does the human species have to protect the natural ecosystem?
- What trade-offs must be considered when making economic and political decisions that affect the environment?

Performance Indicators

4.7.1 How do humans positively and negatively affect the ecosystem?

4.7.2 How has human technology and the growth of the human population affected the quality of life for other species?

4.7.3 How can humans take positive action to reverse the degradation of the natural environment caused by human technology?

Lab Needs

A positive benefit of the reduction of factual detail in the Core Curriculum is that it should allow more in-depth treatment of laboratory investigations to be planned and carried out than was possible under the previous syllabus. Laboratory experiences should be designed to address Standard 1 (inquiry techniques) but should also take into account Standards 2 (information systems), 6 (interconnectedness of content), and 7 (problem-solving approaches). They should also address the laboratory skills listed in Appendix A of the Core Curriculum and of this book.

It is important to recognize that this and other assessment tools based on the learning standards may undergo transition in the first few years of implementation. Teachers and students should remain alert to the possibility of changes in the format and scoring of the Regents examination.

New York State is a participant in a multi-state effort to develop "Common Core Standards" in biology. These standards are scheduled to be implemented in the near future. When implemented, these standards will require some changes in the way that biology is taught and tested.

UNIT ONE _____

Scientific Inquiry

STANDARD 1

Students will use mathematical analysis, scientific inquiry, and engineering design, as appropriate, to pose questions, seek answers, and develop solutions.

Essential Questions:
- *Why is science important?*
- *How is good science done?*

Good science is a logical, yet creative, process that applies human intelligence to discovering how the world and the universe work. It depends as much on the published reports of past scientific work as on the ongoing experimental method used to create new knowledge. Scientists continually question the meaning of scientific discoveries and develop new theories based on combining old knowledge with new knowledge. In this way, there is a continual turnover of knowledge that serves the best interests of all people everywhere.

Discovering Truths

Understanding scientific inquiry and discovery is important because the truths that emerge from this discovery can and should be used in everyday social and ethical decision making. These decisions often involve issues of personal and community health, economic considerations, and technological claims. To fail to develop basic scientific literacy is to place oneself and one's community at great potential risk since proposed changes may carry significant negative consequences to any or all of these items.

I. EXPLANATIONS OF NATURAL PHENOMENA

> ## KEY IDEA 1—PURPOSE OF SCIENTIFIC INQUIRY
> The central purpose of scientific inquiry is to develop explanations of natural phenomena in a continuing and creative process.

Essential Question:

- *How do models help scientists understand natural phenomena?*

Scientific inquiry is used as a tool to aid scientists in the study of the natural world. Scientists ask questions, pose hypotheses, design experiments to test these hypotheses, collect and organize data, analyze collected data, draw inferences, and develop conclusions. Major scientific principles are developed from the combined knowledge that results from such experimentation. Scientific inquiry is a creative process that has been ongoing since humans have been on the Earth and will continue long into the future. We are only beginning to unlock the many secrets of the natural world.

Following Procedures

Scientific inquiry depends on an objective application of certain procedures. Some of these include modeling, research, reconciliation of opposing viewpoints, and presentation of explanations. The following sections will discuss each of these processes.

Modeling

Essential Question:

- *How does expressing an idea in visual or mathematical terms assist in our understanding?*

Performance Indicator 1.1.1 *The student should be able to elaborate on basic scientific and personal explanations of natural phenomena and develop extended visual models and mathematical formulations to represent one's thinking.*

Modeling is a way of putting complex scientific data into understandable terms. These models may be mathematical models, schematic models, or physical models. Models frequently use easy references to help you understand complex principles and relationships.

Scientific explanations are built by combining evidence that can be observed with what people already know about the world. Scientific explanations of natural phenomena come from many different sources, including scientific and personal observations. The modern scientist builds on the work of many earlier scientists who asked questions, developed hypotheses, designed experiments, made observations, analyzed experimental results, and drew inferences and conclusions. These scientists also wrote about their experiments and results, creating vast amounts of scientific literature. Scientists often compare their results with the results of other scientists performing the same or similar experiments.

In many cases, these scientists developed models to better understand and to help others understand their thinking. A biochemist, for example, may develop a model of a biochemical pathway in the form of a series of formulas that represent different steps in that pathway. An ecologist may model a complex set of nutritional relationships in the form of a food web or pyramid. A geneticist may use three-dimensional models to illustrate how the shape of a molecule of DNA helps to predict how it will carry out its genetic control. A cytologist may find that a physical model of a cell helps to explain the movement of materials through the cell membrane. In each of these examples, a model was developed to help explain an abstract scientific concept. During your study of biology, you will be expected to be able to develop models to illustrate what you learn.

Decision Making

An important way to enlarge one's scientific knowledge is to use it to make sound decisions. Science provides knowledge, but values are also essential to making effective and ethical decisions about the application of scientific knowledge. You need to understand scientific knowledge and its implications for social action. Technological advance can impact the lives of individuals, communities, and civilizations. Therefore, the decision must be made whether or not an advance should be introduced, controlled, or set aside.

As a responsible individual, you have an obligation to use your knowledge of scientific principles to make such judgments and act accordingly. You have an ethical and moral responsibility to exercise sound judgment based on a solid understanding of scientific fact.

Research

Essential Question:

- *What can we learn from studying the work of others?*

Performance Indicator 1.1.2 *The student should be able to hone ideas through reasoning, library research, and discussion with others, including experts. Humans sharpen their intellect by reasoning, reading the ideas of* *others, and discussing topics with people whose ideas may differ from their own.*

By starting with the knowledge base of past generations, we develop ever greater understanding of the world around us. We constantly seek to understand natural phenomena in greater and greater detail. This is possible only by applying the scientific method of inquiry.

Inquiry involves asking questions and locating, interpreting, and processing information from a variety of sources. Scientists do not rely solely on the results of their own experiments when developing scientific principles. To broaden their knowledge, they depend heavily on a large body of **scientific literature** created from centuries of experimentation, thinking, and discussion. Each new generation of scientists adds to this body of knowledge and moves forward with the discovery of scientific truths.

Reconciliation of Opposing Viewpoints

Essential Question:

- *How can we tell the difference between good information and bad information?*

Performance Indicator 1.1.3 *The student should be able to work toward reconciling competing explanations and clarify points of agreement and disagreement.*

Inquiry involves making judgments about the reliability of the source and relevance of information. Not all material written about science is reliable or accurate. In some cases, such material may be misleading or biased and may unfairly represent the data in order to support a particular point of view. Good scientific writing involves complete explanation of the experimental method used, a clear presentation of the data, and an objective analysis of the data with respect to other, prior experimentation.

In order to be useful to science, the results of these reported experiments have to be **reproducible** by other scientists using the same experimental methods. Experiments that are not reproducible are considered unreliable and are usually not considered in this process. A good scientist also needs to be able to sort through scientific literature and discard material that does not appear to be relevant to the inquiry at hand. Relevant material should be reviewed and summarized as an introduction to the new inquiry, even though it may suggest conclusions that differ from the scientist's hypothesis.

Scientific explanations are accepted when they are consistent with experimental and observational evidence and when they lead to accurate predictions. A scientific principle, if it is to gain widespread acceptance, must enable scientists to make **accurate predictions** of how the natural world operates under a given set of conditions. The conclusions drawn from experimental results must accurately reflect an objective analysis of the data gathered from such experiments. These conclusions must predict the outcome of similar experimentation with reasonable accuracy in order to be considered valid. When they do not do so, conclusions are thrown into question until further investigation is conducted to determine their validity.

All **scientific explanations are tentative** and subject to change or improvement. Each new bit of evidence can create more questions than it answers. This increasingly leads to better understanding of how things work in the living world. Scientists are always at work to add to the body of knowledge we call science. This new knowledge often supports, but sometimes contradicts, previously held understandings. When contradictions occur, further scientific work is needed to determine where the truth lies.

Scientists do not always agree on the interpretation of scientific data. Divergent ideas and attitudes, as well as diametrically opposed conclusions and theories, also exist. Far from impeding the development of scientific thinking, these divergent concepts are used to stimulate new thinking and experimentation in a never-ending quest for the truths behind the theories. Ultimately, some theories are proven incorrect. However, the process of continual questioning and inquiry is responsible for moving science in a positive direction.

Presentation of Explanations

Essential Question:

- *How can data be effectively represented in a manner that assists in our understanding of science?*

Performance Indicator 1.1.4 *The student should be able to coordinate explanations at different levels of scale, points of focus, and degrees of complexity and specificity, and recognize the need for such alternative representations of the natural world.*

The information drawn from scientific investigation is discussed and analyzed from different points of view. By changing the way they look at the data, scientists begin to see patterns and trends not immediately apparent from a one-dimensional analysis.

When data collected from an experiment is represented graphically, the **scale** used to represent the dependent variable may be large or small. A large scale includes a wide range of possible data points. A small scale uses fewer possible data points. When a large scale is used, small variations in the data may not be easily seen. By decreasing the scale of the graph, the range of data represented is reduced so that small variations can be seen more easily. Analytical methods may involve decreasing the scale of the data to amplify small changes, or increasing it to examine general trends.

Looking at a broad range of data or focusing on specific components of the data may isolate the specific change being measured. The **focus area** may be the point at which the experimental variable has the most significant effect on the dependent variable. By focusing on this narrow range of data, the researcher can learn much about the changes that occur within this range. The scientist may then design a new experiment that breaks the experimental variable into smaller increments in order to determine more precisely how small changes in the experimental variable work to affect the dependent variable.

Changing the way data is viewed may involve **describing the data** in new ways to highlight a researcher's interpretation of the results of one or more scientific studies. Many concepts in the life sciences are extremely complex, with multiple variables operating at once to produce conditions that effect the changes noted by the researcher. It is often the job of the researcher to reduce this complexity for the reader by increasing the specificity of the study. By dealing with a **single variable** at a time, the researcher determines the effect of each variable independent of the others. When the isolated effect of each variable is understood, the reader can then comprehend how multiple environmental conditions affect the dependent variable in complex ways.

Well-accepted theories are the ones that are supported by different kinds of scientific investigations often involving the contributions of individuals from different disciplines. These varying methods of interpretation help to illuminate the data in new ways and to stimulate still more ideas about its meaning. Well-accepted theories are those **supported by investigations** involving different scientific disciplines (for example, biochemistry, cytology, genetics). As more and more evidence mounts supporting a theory, the more credibility it gains with the scientific community. Each discipline reports its findings independently of the others so that independent corroboration is achieved. In this way, an inference becomes a hypothesis, which then may develop into an accepted theory, and finally it becomes established scientific principle.

II. TESTING OF PROPOSED EXPLANATIONS

KEY IDEA 2—METHODS OF SCIENTIFIC INQUIRY

Beyond the use of reasoning and consensus, scientific inquiry involves the testing of proposed explanations involving the use of conventional techniques and procedures that usually require considerable ingenuity.

Essential Questions:

- *What are the steps in the scientific method?*
- *How do these steps help to ensure that science is valid?*

Scientific inquiry involves several well-defined procedural steps known collectively as the **scientific method**. These steps include several elements, the first five of which are discussed in the following section:

1. Pose a scientific question based on observation of the natural world.
2. Review available scientific literature relating to the scientific question.
3. Develop a hypothesis about the proposed answer to the scientific question.
4. Design an appropriate experimental procedure to test the hypothesis.
5. Collect and record data resulting from the experimental procedure.

By following this design, the researcher defines his/her goals prior to commencing an experiment. Doing so helps the researcher to focus on the problem at hand. The researcher can then conduct the experiment in a manner that will generate useful data. This data can then be analyzed to yield new knowledge about how the natural world operates.

Question and Observation

Essential Question:

- *How do questions and observations help scientists to conduct good experiments?*

Performance Indicator 1.2.1 *The student should be able to devise ways of making observations to test proposed explanations.*

Scientific inquiry, as its name implies, always begins with the fundamental step of asking a question. A nonscientist may be curious about something he/she observes in nature and ask a question something like "How do leaves turn color in the fall?" or "Why do fruit flies have red eyes?" While questions such as these provide a starting point for inquiry, they do not lend themselves to easy and understandable answers. They do not point to aspects of the observed phenomena that can be measured. They are focused more on reasons for the phenomena than on their observable and measurable characteristics.

By contrast, a **scientific question** is always asked in terms of observable qualities and measurable quantities. Scientific questions might ask "What pigments are found in the leaves of sugar maple trees?" or "What is the inheritance pattern of red eyes in fruit flies?" Questions such as these lead the researcher toward established laboratory methods. For instance, chromatography (separating pigments of differing molecular characteristics) and genetic crosses (studying the inheritance of genetic traits) provide clues to the broader questions posed above.

These questions, even though less general, can be broken down still further into even more specific, measurable questions. For example, "What pigment is most abundant in sugar maple leaves?" "How does temperature affect the production of chlorophyll in sugar maple leaves?" "What is the relative rate of destruction of the leaf pigments chlorophyll and anthocyanin under constant, intense white light?" and "At what temperature does chlorophyll function most effectively? Least effectively?" The researcher uses these, more focused, questions to help point the way toward research studies. These studies should produce data that lead to new knowledge about natural phenomena.

The simplest method for testing is direct **field observation** of the experimental subject. Such observations may be conducted informally but by taking careful notes. Later, more formal and **controlled observations** may be made as a part of an experimental procedure. Such observations are always made objectively, accurately, and in great detail.

Conventional **laboratory techniques** may take on various characteristics depending on what idea is being tested and what variables are being measured. For example, an investigation into the anatomic characteristics of an earthworm should be conducted using dissection tools. An experiment designed to discover the different pigments present in a maple leaf should employ techniques of chromatography. An inquiry into the genetics of the fruit fly should include the tools and techniques necessary to isolate, mate, and count the offspring of fruit flies displaying specific genetic characteristics.

Review of Literature

Essential Question:

- *How do scientists use the work of other scientists to help them in their own work?*

Performance Indicator 1.2.2   *The student should be able to refine research ideas through library investigations, including electronic information retrieval and reviews of the literature, and by peer feedback obtained from review and discussion.*

A vast body of **scientific information** is available on a wide range of scientific topics. This information is available in books, professional journal articles, research databases, and other reference sources. This literature may contain the findings of other scientists who have conducted similar studies in the past. It may provide good clues to the researcher as to the likely outcome of his/her investigation. Pertinent sources in the scientific literature are normally cited in the introductory material of a report on the experiment at hand.

Increasingly, these resources are available though networked computer databases. You should possess the skills needed to search the Internet to locate reliable research information and to **discriminate among sources** based on their authenticity. This may require focusing on the archived databases of universities and research laboratories while excluding data sources that do not carry the endorsement of such institutions. You should recognize that posting information on the Internet does not require the editorial scrutiny of competent scientists that is necessary to publish in a professional journal. For this reason, exercise care when using the Internet for conducting literature reviews since the data and analyses presented may be inaccurate, misleading, or biased.

Development of a **research plan** involves researching background information and understanding the major concepts in the area being investigated. Recommendations for methodologies, use of technologies, proper equipment, and safety precautions should also be included. Such sources provide information about the appropriate methods and techniques to be used in conducting research in a particular area of scientific inquiry. They also inform the reader about what equipment and **safety precautions** should be used in such investigations. Having this information available in the literature saves valuable time since these techniques and precautions do not have to be discovered by each research scientist conducting a study in a similar area of inquiry. Having this information available also assists the researcher in developing a formal hypothesis.

Finally, being able to review and summarize information gained from such sources and to discuss it with other researchers engaged in the same sorts of investigations is an important skill. **Peer feedback** can benefit all participants since the shared information and ideas provide everyone with a better understanding of the phenomena being studied. Collaborative sharing is also an important skill that you, as a student, should learn.

Hypothesis and Research Plan

Essential Question:

- *What is a hypothesis and how does it help the scientist to focus an experiment?*

Performance Indicator 1.2.3 *The student should be able to develop and present proposals including formal hypotheses to test explanations, that is, predict what should be observed under specific conditions if the explanation is true.*

Hypotheses are predictions based upon both research and observation. A **hypothesis** is an educated guess made by a researcher about the likely outcome of an investigation. It is a prediction of what should be observed in an experiment run under a specific set of conditions. Hypotheses are based upon both research and observation.

Hypotheses are widely used in science for determining what data to collect and as a guide for interpreting the data. The hypothesis serves as a guide. It assists the researcher in designing the experiment and interpreting the data derived from that experiment. For example, a hypothesis concerning the effect of temperature on bacterial growth might state, "*Escherichia coli* bacteria will grow best at human body temperature." This hypothesis suggests that a series of identical experimental setups be created that provide a nutrient medium to a single species of bacterium. It further suggests that these setups be placed under a range of temperatures including temperatures below, at, and above human body temperature. Finally, it suggests that bacterial growth be measured in some way that will give the researcher a clear, objective understanding of differential growth rates.

If the researcher has interpreted the literature appropriately, the hypothesis will often be correct, and the results of the experiment will be as predicted. However, the experimental results may not match the hypothesis. When this occurs, it is *not* viewed as a failure by the researcher but as a means to suggest more probing questions that get at the real relationships among the variables in the experiment.

Experimentation

Essential Question:

- *What tools and techniques should be selected when designing an experiment?*

Performance Indicator 1.2.4 *The student should be able to carry out a research plan for testing explanations, including selecting and developing techniques, acquiring and building apparatus, and recording observations as necessary.*

Carrying out a research plan requires a **procedure,** or set of steps. These steps are laid out within the plan and are specific to the experimental question being asked.

Development of a **research plan** for testing a hypothesis requires planning to avoid bias (for example, repeated trials, large sample size, and objective data-collection techniques). A research plan is developed in such a way as to provide objective data that will help to answer the experimental question without introducing bias into the experiment. **Bias** is the tendency to prejudice the results of an experiment with one's own assumptions about the likelihood of its outcome. While developing a hypothesis is an important step in scientific research, the researcher must not let that hypothesis color his/her objective interpretation of the data. To do otherwise would be to introduce bias into the experiment. To ensure objective results, researchers use techniques such as repeated trials, large sample size, adequate controls, and objective data collection techniques.

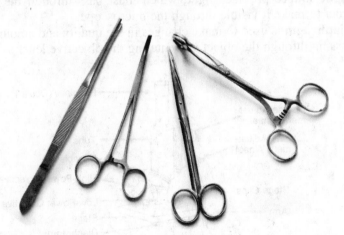

The **equipment** being used, the **method** being followed, and the means of **recording and analyzing data** must be appropriate to answering that question. For example, an experiment to determine the structural similarities of five species of worm would probably require dissection equipment and a dissecting microscope. It would involve studying and collecting data about the external and internal structures of a reasonable number of representatives of each species. An experiment on the effects of auxins on geranium stem growth, on the other hand, would require several geranium plants, various concentrations of auxins, a constant light source, and instruments to measure stem elongation and rate of bending. Each experiment has its own unique setup and design. Following is a description of some of the commonly used techniques and tools used by biologists when performing scientific investigations.

The Compound Light Microscope

The compound light microscope, developed hundreds of years ago by early scientists such as Anton von Leeuwenhoek and Robert Hooke, still represents a major tool of cell study used by biologists all over the world. Compound light microscopes of the type used in most school laboratories produce magnifications of 50× to 500× and are suitable for viewing whole cells and large organelles such as nuclei and chloroplasts. Since this tool is used commonly in the biology classroom, it is important that you understand its parts and workings thoroughly. The critical parts of the microscope include:

- **Eyepiece lens** (ocular) is the lens closest to the eye during study.
- **Objective lens** is the lens closest to the object of study.
- **Coarse-adjustment knob** is used to make large changes in the focus of the microscope; fine adjustment knob is used for small changes.
- **Light source** provides light, which must pass through the object in order to make it visible through the microscope.
- **Diaphragm** is used to make changes in the quality and amount of light passing through the object and entering the objective lens.

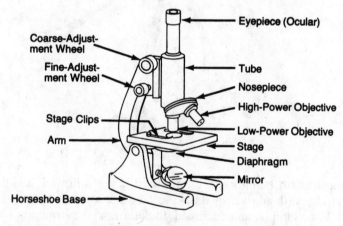

Compound Light Microscope

The compound light microscope provides two major advantages to the scientist studying the cell structure of living things:

- **Magnification** is the quality of the compound light microscope that makes the image of an object appear larger than the object itself. The magnification of a compound light microscope is determined by multiplying the power of the ocular by the power of the objective lens (for example, 10× ocular × 40× objective = 400× total magnification).
- **Resolution** is the quality of the compound light microscope that makes it possible to see two or more objects that are very close together as separate objects.

Use of the microscope is an important skill best learned in the laboratory. Practical laboratory experience is the best way to master the skills involved in preparing wet mount slides, focusing the microscope, and making observations of biological specimens. Students may wish to attempt to answer the practice questions below in conjunction with such practical laboratory experience. These questions are provided here and in the Laboratory Skills section to illustrate the kind of question that may be asked on the Regents examination. See the Laboratory Skills section for additional information.

Scientific Measurement

The researcher must also have a good understanding of the various units of measurement and means of data collection used in each type of research study. The **metric system** is used universally in scientific measurement. Knowledge of the relative sizes of various metric measures is important to accurate data collection. For example, the units of linear measure include the **meter** (m), the **centimeter** (cm—1/100 of a meter), **millimeter** (mm—1/1,000 of a meter), and **micrometer** (μm—1/1,000,000 of a meter). The units of volume include the **liter** (l) and **milliliter** (mL). The metric unit of temperature is the **degree** (°C). Metric units of mass include the **gram** (g), **kilogram** (kg), and **milligram** (mg).

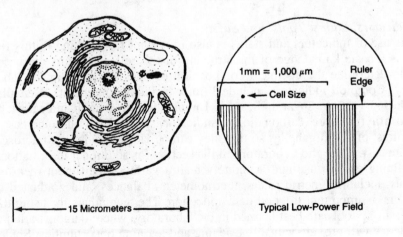

Cell Measurement

The Micrometer

Understanding the use of the compound microscope and the units of measurement for microscopic objects is very important for conducting research in the biological sciences. The unit of linear measurement used to measure cells and their organelles is the **micrometer (μm)**. Micrometers are so small that 1,000 of them can fit inside 1 millimeter, and a million of them are required to equal 1 meter. Cells typically are found to have diameters of between 10 and 50 micrometers. If the diameter of the microscope field of view is known, it can be used to estimate the size of objects under it. The low-power field can be measured directly by placing a metric ruler under the low-power field of view. Most low-power fields have diameters of between 1,200 and 1,600 micrometers (1.2 and 1.6 millimeters). The diameter of a high-power field can be determined mathematically using a simple proportion as follows:

$$\frac{\text{Low-Power Magnification}}{\text{High-Power Magnification}} = \frac{\text{High-Power Field Diameter}}{\text{Low-Power Field Diameter}}$$

To illustrate the use of this proportion, assume you are using a microscope with a 100× low-power magnification and a 400× high-power magnification. Direct measurement shows that the low-power field diameter is 1,200 micrometers (1.2 mm). The diameter of the high-power field is determined by substituting the known values into the formula and solving algebraically for the unknown value as follows.

$$\frac{100\times}{400\times} = \frac{\text{High-Power Field Diameter}}{1,200\ \mu\text{m}}$$

(400×) (High-Power Field Diameter) = (100×) (1,200 ×m)
High-Power Field Diameter = 120,000/400 ×m
High-Power Field Diameter = 300 ×m

Indicators, Stains, Tools, and Safety

The use of indicators and stains is also important for working in this field, as is a basic knowledge of the structure of the cell. Indicators are used to help determine the chemical composition of materials used in the laboratory. Common indicators include **litmus** (acid/base indicator), **Benedict's solution** (simple sugar indicator), **Lugol's iodine** (starch indicator), and **bromthymol blue** (carbon dioxide indicator). Stains are used to increase the contrast of cell organelles so that they can be viewed under the microscope more easily. See the Laboratory Skills section for additional information.

Being able to use and accurately read a variety of different measuring tools, including metric rulers, thermometers, balances, and graduated cylinders, is important to good data collection. The techniques for using these measuring tools are best learned in the laboratory, where demonstration and practice will provide valuable teaching and learning opportunities. Students may wish to attempt to answer the practice questions that follow in conjunction with such practical laboratory experience. See the Laboratory Skills section for additional information.

Finally, a good understanding of **laboratory safety** is essential for conduction laboratory work in any area of study. Like the use of measuring tools, laboratory safety is something that is best learned in practice. A few simple safety rules should always be followed, however. A sampling is shown below. See the Laboratory Skills section for additional information.

- Always follow instructions given to you by your instructor and by your laboratory manual.
- Never use laboratory equipment that is damaged or with which you are unfamiliar.
- Never endanger yourself or others with dangerous horseplay in the laboratory.
- Never touch, taste, or smell objects, chemicals, or solutions unless you know what they are and know they are safe.
- Always wear protective eyewear when working with chemicals, hot liquids, or sharp tools.

QUESTION SET 1.1—EXPLANATIONS OF NATURAL PHENOMENA (ANSWERS EXPLAINED, P. 251)

1. A biologist plans to spend a year investigating the mating behavior of a certain species of frog. To make meaningful observations, the biologist should observe
   (1) a small number of frogs in their natural habitat
   (2) a large number of frogs in their natural habitat
   (3) several groups of frogs maintained in different temperatures in the laboratory
   (4) several groups of frogs maintained on different diets in the laboratory

2. In the diagram below, letter A represents the starting volume of liquid in a graduated cylinder. Letter B represents the volume after 8 milliliters of this liquid was removed.

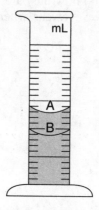

   This information indicates that the scale on this graduated cylinder is in milliliter increments of
   (1) 1   (3) 8
   (2) 2   (4) 4

3. Which sentence represents a hypothesis?
   (1) Environmental conditions affect germination.
   (2) Boil 100 milliliters of water, let it cool, and then add 10 seeds to the water.
   (3) Is water depth in a lake related to available light in the water?
   (4) A lamp, two beakers, and elodea plants are selected for the investigation.

4. Zebra finches are small black-and-white birds that lay eggs about the size of a bean seed. Which unit of measurement is best for accurately measuring the length of these eggs?
   (1) millimeters
   (2) micrometers
   (3) feet
   (4) meters

5. Each division of the metric ruler shown in the diagram below equals 1 millimeter.

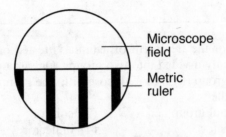

The diameter of the field of vision is approximately
   (1) 2,800 μm
   (2) 3,700 μm
   (3) 4,400 μm
   (4) 4,700 μm

6. How does a control setup in an experiment differ from the other setups in the same experiment?
   (1) It tests a different hypothesis.
   (2) It has more variables.
   (3) It differs in the one variable being tested.
   (4) It utilizes a different method of data collection.

7. When a test tube of water containing elodea (an aquatic plant) is placed near a bright light, the plant gives off gas bubbles. When the light is placed at different distances from the plant, the rate of bubbling is affected. The experimental variable in this demonstration is the
   (1) concentration of gas in the water
   (2) type of aquatic plant in the test tube
   (3) amount of water in the test tube
   (4) distance of the plant from the light

8. Which procedure would be part of a laboratory investigation designed to determine if a specific nutrient is present in a food?
   (1) Test a moist sample of the food with pH paper.
   (2) Add Lugol's iodine solution to a sample of the food.
   (3) Place a sample of the food into a test tube containing methylene blue.
   (4) Add bromthymol blue to a sample of the food.

9. What is the first step of a scientific investigation?
   (1) Perform the experiment.
   (2) Analyze the experimental data.
   (3) Formulate a hypothesis.
   (4) State the problem.

10. By using one or more complete sentences, state one safety procedure that a student should follow when dissecting a preserved frog.

_____

11. A new drug for the treatment of asthma is tested on 100 people. The people are evenly divided into two groups. One group is given the drug and the other group is given a glucose pill. The group given the glucose pill serves as the
   (1) experimental group   (3) control
   (2) limiting factor   (4) indicator

12. A scientific study showed that the depth at which algae were found in a lake varied from day to day. On clear days, the algae were found as much as 6 meters below the surface of the water but were only 1 meter below the surface on cloudy days. Which hypothesis best explains these observations?
   (1) Light intensity affects the growth of algae.
   (2) Wind currents affect the growth of algae.
   (3) Nitrogen concentration affects the growth of algae.
   (4) Precipitation affects the growth of algae.

13. The diagram below represents a hydra as viewed with a compound light microscope.

If the hydra moves toward the right side of the slide preparation, which diagram best represents what will be observed through the microscope?

   (1)   (2)   (3)   (4)

14. The diagram below represents two cells next to a metric measuring device as seen under low power with a compound light microscope.

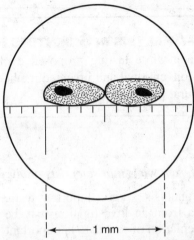

1 mm

What is the length of a nucleus of one of these cells?
(1) 100 μm
(2) 500 μm
(3) 1,000 μm
(4) 1,500 μm

15. The diagram below represents measurements of two leaves.

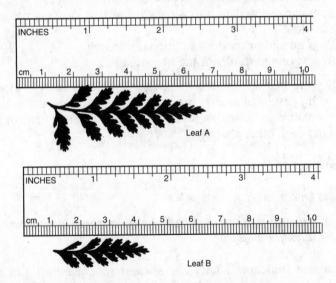

The difference in length between leaves A and B is closest to
(1) 20 mm
(2) 20 cm
(3) 0.65 m
(4) 1.6 μm

19

III. ANALYSIS OF TESTING RESULTS

> **KEY IDEA 3—ANALYSIS IN SCIENTIFIC INQUIRY** The observations made while testing proposed explanations, when analyzed using conventional and invented methods, provide new insights into natural phenomena.

Essential Question:

- *What are the different ways that scientists analyze scientific data?*

Good, objective data analysis is very important to the scientific method of inquiry. The raw data from an investigation may be extensive, so proper organization and presentation of that data is important. Analysis of this data allows the researcher to compare experimental results against predicted results, to perform additional research, and finally to report confidently on the results of the research. The following section discusses the last five steps involved in data analysis.

6. Organize data in a manner that facilitates analysis.
7. Analyze data to identify trends and points of focus.
8. Summarize the data into a conclusion that answers the experimental question.
9. Propose and/or conduct additional research.
10. Report on the results of the investigation.

By following this design, the researcher is able to analyze effectively and report on the results of an experiment. This organization and analysis follows certain conventions, allowing researchers from different parts of the world to understand each other's work.

Organization of Data

Essential Question:

- *How can tables, charts, and graphs of data help to illustrate the findings of a scientific inquiry?*

Performance Indicator 1.3.1 *The student should be able to use various methods of representing and organizing observations (for example, diagrams, tables, charts, graphs, equations, matrices) and insightfully interpret the organized data.*

Diagrams of observational data can be important for certain kinds of investigations, such as those carried out with the help of the compound microscope. The researcher focuses on the components being studied and sketches his/her interpretations of what is seen. The researcher emphasizes the details pertaining to the study at hand. Diagrams must be accurately labeled to be useful in research. An important skill to be used with this method is estimating the sizes in micrometers (μm) of microscopic objects when seen in the low-power and high-power fields.

Charts or tables are appropriate for presenting many kinds of data. These grids have the experimental variable presented in ascending order on the left side of the chart or table and the dependent variable labeled at the top. The dependent variable values are placed next to the corresponding experimental variable data. A data matrix operates in much the same way. It organizes data by category for ease of comparison.

When data elements relate to each other in complex ways, representing the data as **mathematical equations** is sometimes preferred. Such equations are derived only after extensive field and laboratory study have supplied extensive data that supports the mathematical model.

Often, a **graphic representation** of the data will aid in its interpretation, especially where trends are clearly identifiable. Selecting the kind of graph that will best illustrate the data being analyzed is important. A line graph is useful when both experimental variable and dependent variable are numerical. A bar graph is used when the experimental variable cannot be expressed as a number but the dependent variable can be. Pie graphs are useful for depicting data that adds up to 100 percent of some value. Appropriately title the graph in each graph style. Label and scale the axes in line and bar graphs. Accurately plot data points in line and bar graphs or determine percentage sections in pie graphs. Connect points in line graphs, complete bars in bar graphs or complete sections in pie graphs.

From these depictions of the data, seeing broad trends that would not be obvious if the data were not organized in chart or graphic formats becomes possible. By studying the trends in these data elements, researchers are able to develop new and more accurate hypotheses. These hypotheses are then tested using the scientific method of inquiry. With further data to review, researchers can make generalizations, draw inferences and conclusions, and formulate explanations of natural phenomena. At this point, the research is considered to have added new knowledge to science.

Statistical Analysis of Data

Essential Question:

- *How do scientists measure the accuracy of the results of their own experiments?*

Performance Indicator 1.3.2 *The student should be able to apply statistical analysis techniques when appropriate to test if chance alone explains results.*

One of the simplest statistical methods used in science is calculating the **percentage of error**. In this method, the value obtained in the laboratory investigation is compared with the result expected through a mathematical formula. The difference between these values is divided by the expected value, and the result multiplied by 100 to obtain the percentage of error in the experiment. The researcher should always speculate on the probable causes of the percentage of error. These causes generally fall into three categories: (1) experimenter error (the researcher fails to follow proper procedure), (2) instrument error (the laboratory equipment introduces error by being inaccurate), or (3) calculation error (calculating the values used in the experiment is inaccurate or rounded inappropriately).

A slightly more complex method used to determine whether the data is valid within statistical norms is the **test of statistical significance**. The accuracy of any investigation depends on the size of the experimental sample being taken. Small samples often yield unreliable results, while large samples tend to yield more reliable results when probabilities are involved. The test of statistical significance compares what is expected according to probability against results obtained in the laboratory in the light of the size of the sample taken. Specific examples of tests of statistical significance are beyond the scope of this course of study, but may be implemented at the local level as necessary.

Summary and Conclusion

Essential Question:

- *How does a scientist draw inferences and conclusions based on data?*

Performance Indicator 1.3.3 *The student should be able to assess correspondence between the predicted result contained in the hypothesis and the actual result.*

The student should reach a **conclusion** as to whether the explanation on which the prediction was based is supported. After conducting a well-designed experiment with appropriate controls; having collected, organized, and objectively analyzed data; and determining the reliability of this data, the researcher is then ready to summarize his/her findings in the form of a conclusion. In its simplest form, the conclusion restates the experimental question and hypothesis. Then it objectively assesses the accuracy of that hypothesis in light of the data presented. If the hypothesis and the experimental results closely correspond, the conclusion should then go on to examine whether the results were produced in a manner that follows the logic used by the researcher in developing the hypothesis. In a well-researched investigation, both the hypothesis and the logic behind it should be reasonably accurate.

When a hypothesis is not supported by the experimental evidence, the conclusion should state that fact. The scientist should examine the literature review, experimental technique, and data analysis to try to detect flaws in the methodology used. Some fundamental omission in the literature review may have lead to the incorrect hypothesis. The techniques used may have been flawed, producing inaccurate data. The organization and analysis of the data may have been insufficient to permit trends to be appropriately identified. Far from being viewed as a failure, this situation is used by professional scientists to stimulate new questioning, additional research, and honed data analysis. Eventually, better results are obtained.

Additional Research

Essential Question:

- *How does one experiment lead to another once results are known?*

Performance Indicator 1.3.4   *The student should be able to revise the explanation and contemplate additional research based on the results of the test and through public discussion.*

Investigations in science do not always work out as predicted. Frequently, the results vary in some way from those anticipated in the hypothesis. When this happens, a competent researcher will use the opportunity to examine the experiment, make improvements, and conduct further research. Performing additional research is basic to the pursuit of scientific knowledge.

Hypotheses, even if found to be untrue, are valuable because they help to stimulate further investigation. As a part of the investigation, the researcher makes an objective judgment as to the accuracy of his/her hypothesis. If it is substantially correct, the hypothesis may add measurably to the information available to answer the question asked. Instead, it may take a partially answered question to new levels of understanding. If incorrect, the hypothesis leads the researcher to question the premise upon which the hypothesis was based, reexamine the literature for possible misinterpretation, and review the methods used to conduct the experiment. After performing this review, the researcher may well enter into a new round of experimentation to address any deficiencies found.

Claims should be questioned if the data are based on samples that are very small, biased, or inadequately controlled or if the conclusions are based on the faulty, incomplete, or misleading use of numbers. Claims should be questioned if the experimental design is not consistent with generally accepted experimental procedure. If samples are small or unrepresentative, the data collected may not fairly represent the population as a whole, leading to skewed results. If the data are biased or slanted to match the researcher's hypothesis, the objectivity of the entire investigation is called into question. Since science is based on objective interpretation of data, bias invalidates the inferences drawn from such experiments.

If the experimental method is flawed (uses inappropriate tools or techniques), the data will be meaningless and not able to answer the question. An experiment that lacks appropriate controls (allows multiple variables to operate) does not allow the researcher to determine which variable is at work in producing the results reported. If data analysis is not accurate and objective, but rather is faulty, incomplete, or misleading, then trends in the data may be difficult or impossible to identify and project beyond the limited scope of the experiment. Conclusions drawn from flawed data or flawed data analysis are invalid since the basis of all science is objective data.

Claims should be questioned if fact and opinion are intermingled, if adequate evidence is not cited, or if the conclusions do not follow logically from the evidence given. Claims should be questioned if the data and its analysis are not objective in nature. If fact is mingled with opinion when reporting results, if adequate evidence is not cited, or if the conclusions do not follow logically from the evidence provided, be skeptical of the validity of the experiment. These flaws call into question the objectivity and completeness of the data collection and data analysis procedures. Good science is not based on opinion but on factual evidence. This evidence must be collected in sufficient quantity so it can provide a firm foundation for the conclusions drawn analyzing it. Conclusions must come from an objective analysis of the data collected and presented. Those that seem to ignore or make peripheral use of the data are immediately suspect and should be discarded.

Report of Experimental Results

Essential Question:

- *What are the appropriate ways to present the findings of a scientific investigation?*

Performance Indicator 1.3.5 *The student should be able to develop a written report for public scrutiny that describes the proposed explanation, including a literature review, the research carried out, its result, and suggestions for further research.*

When scientific investigation results in valid conclusions, the researcher may choose to summarize his/her findings in a written report. If the findings are of sufficient significance, the report may be published in a professional journal or other medium for review by other researchers and the general public.

Repeatability

One assumption of science is that other individuals could arrive at the same explanation if they had access to similar evidence. Scientists make the results of their investigations public; they should describe the investigations in ways that enable others to **repeat the investigations**. A major assumption of professional science is that other researchers, by following the explicit descriptions found in such a report, could arrive at the same results as those reached in the original study.

A **scientific report** should accurately and clearly summarize the literature examined when formulating the hypothesis, describe the methods used in designing and carrying out the investigation, present the data and its analysis in an organized fashion, state the conclusions drawn from this analysis, and suggest additional research to carry the investigation further. These items must be described in sufficient detail to allow others to duplicate the experiment in their own laboratories. If this is done and the same results are obtained, the additional research helps to validate the original investigation. If this is done but different results are obtained, doubt is cast on the original investigation and its conclusions.

Peer Review

Scientists use **peer review** to evaluate the results of scientific investigations and the explanations proposed by other scientists. They analyze the experimental procedures, examine the evidence, identify faulty reasoning, point out statements that go beyond the evidence, and suggest alternative explanations for the same observations. Peer review either lends support to, or helps to discredit, the reported investigation. It also guides the researcher in the investigation's design, execution, analysis, and reporting.

Peer review may be able to offer alternative explanations not presented by the researcher. Such peer review is meant to strengthen the quality of scientific inquiry and reporting. As this quality improves, so does the body of useful knowledge available to the public.

QUESTION SET 1.2—ANALYSIS OF TESTING RESULTS (ANSWERS EXPLAINED, P. 255)

1–4. Base your answers to questions 1 through 4 on the information and data table below and on your knowledge of biology. The table shows the average systolic and diastolic blood pressure measured in millimeters of mercury (Hg) for humans between the ages of 2 and 14 years.

DATA TABLE

	Average Blood Pressure (mm of Hg)	
Age	Systolic	Diastolic
2	100	60
6.	101	64
10	110	72
14	119	76

Directions (1–4): By using the information in the data table, construct a line graph on the grid provided, following the directions below.

1. Mark an appropriate scale on each labeled axis.

2. Plot the data for systolic blood pressure on your graph. Surround each point with a small triangle, and connect the points.

   Example: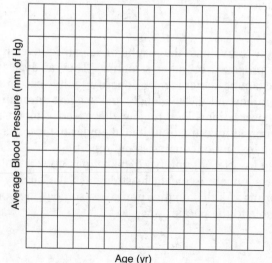

3. Plot the data for diastolic blood pressure on your graph. Surround each point with a small circle, and connect the points.

   Example:

The Effect of Age on Human Blood Pressure

Average Blood Pressure (mm of Hg)

Age (yr)

Key
△ Systolic blood pressure
⊙ Diastolic blood pressure

26

4. By using one or more complete sentences, state one conclusion that compares systolic blood pressure with diastolic blood pressure in humans between the ages of 2 and 14 years.

_____

5. The graph below shows the results of an experiment.

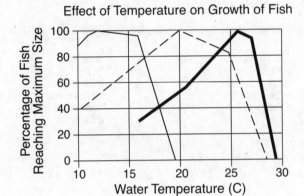

Effect of Temperature on Growth of Fish

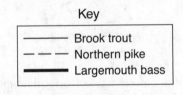

Key

—————— Brook trout
– – – – Northern pike
━━━━━ Largemouth bass

At 16°C, what percentage of the brook trout reached maximum size?
(1) 30%   (3) 75%
(2) 55%   (4) 95%

6. An experiment is represented in the diagram below.

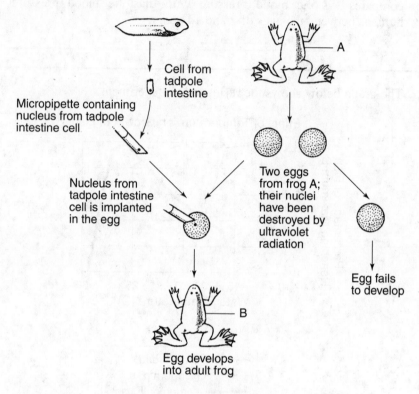

An inference that can be made from this experiment is that

(1) adult frog B will have the same genetic traits as the tadpole
(2) adult frog A can develop from only an egg and a sperm
(3) fertilization must occur in order for frog eggs to develop into adult frogs
(4) the nucleus of a body cell fails to function when transferred to other cell types

7. The charts below show the relationship of recommended weight to height in men and women age 25–29.

HEIGHT–WEIGHT CHARTS

MEN Age 25–29 Weight (lb)					WOMEN Age 25–29 Weight (lb)				
Height		Small Frame	Medium Frame	Large Frame	**Height**		Small Frame	Medium Frame	Large Frame
Feet	Inches				Feet	Inches			
5	2	128–134	131–141	138–150	4	10	102–111	109–121	118–131
5	3	130–136	133–143	140–153	4	11	103–113	111–123	120–134
5	4	132–138	135–145	142–156	5	0	104–115	113–126	122–137
5	5	134–140	137–148	144–160	5	1	106–118	115–129	125–140
5	6	136–142	139–151	146–164	5	2	108–121	118–132	128–143
5	7	138–145	142–154	149–168	5	3	111–124	121–135	131–147
5	8	140–148	145–157	152–172	5	4	114–127	124–138	134–151
5	9	142–151	148–160	155–176	5	5	117–130	127–141	137–155
5	10	144–154	151–163	158–180	5	6	120–133	130–144	140–159
5	11	146–157	154–166	161–184	5	7	123–136	133–147	143–163
6	0	149–160	157–170	164–188	5	8	126–139	136–150	146–167
6	1	152–164	160–174	168–192	5	9	129–142	139–153	149–170
6	2	155–168	164–178	172–197	5	10	132–145	142–156	152–173
6	3	158–172	167–182	176–202	5	11	135–148	145–159	155–176
6	4	162–176	171–187	181–207	6	0	138–151	148–162	158–179

The recommended weight for a 6'0" man with a small frame is closest to that of a
(1) 5'10" man with a medium frame
(2) 5'9" woman with a large frame
(3) 6'0" man with a medium frame
(4) 6'0" woman with a medium frame

8–11. Base your answers to questions 8 through 11 on the information below and on your knowledge of biology.

A group of biology students extracted the photosynthetic pigments from spinach leaves using the solvent acetone. A spectrophotometer was used to measure the percent absorption of six different wavelengths of light by the extracted pigments. The wavelengths of light were measured in units known as nanometers (nm). One nanometer is equal to one-billionth of a meter. The following data were collected:

> yellow light (585 nm)—25.8% absorption
> blue light (457 nm)—49.8% absorption
> orange light (616 nm)—32.1% absorption
> violet light (412 nm)—49.8% absorption
> red light (674 nm)—41.0% absorption
> green light (533 nm)—17.8% absorption

8. Complete all three columns in the data table below so that the wavelength of light either increases or decreases from the top to the bottom of the data table.

Color of Light	Wavelength of Light (nm)	Percent Absorption by Spinach Extract

Directions (9–10): By using the information in the data table, construct a line graph on the grid provided, following the directions below.

9. Mark an appropriate scale on the axis labeled "Percent Absorption."

10. Plot the data from the data table on your graph. Surround each point with a small circle, and connect the points.

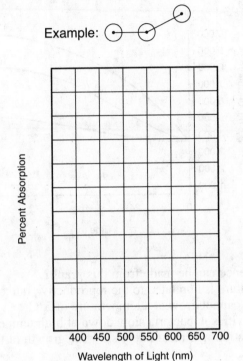

11. Which statement is a valid conclusion that can be drawn from the data obtained in this investigation?
    (1) Photosynthetic pigments in spinach plants absorb blue and violet light more efficiently than red light.
    (2) The data would be the same for all pigments in spinach plants.
    (3) Green and yellow light are not absorbed by spinach plants.
    (4) All plants are efficient at absorbing violet and red light.

12. The graph below represents the results of an investigation of the growth of three identical bacterial cultures incubated at different temperatures.

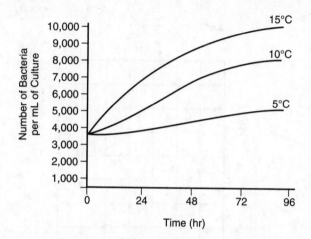

Which inference can be made from this graph?
(1) Temperature is unrelated to the reproductive rate of bacteria.
(2) Bacteria cannot grow at a temperature of 5°C.
(3) Life activities in bacteria slow down at high temperatures.
(4) Refrigeration will most likely slow the growth of these bacteria.

13. A study was conducted using two groups of ten plants of the same species. During the study, the plants were placed in identical environmental conditions. The plants in one group were given a growth solution every 3 days. The heights of the plants in both groups were recorded at the beginning of the study and at the end of a 3-week period. The data showed that the plants given the growth solution grew faster than those not given the solution.

When other researchers conduct this study to test the accuracy of the results, they should
(1) give growth solution to both groups
(2) make sure that the conditions are identical to those in the first study
(3) give an increased amount of light to both groups of plants
(4) double the amount of growth solution given to the first group

14. Worker bees acting as scouts are able to communicate the distance of a food supply from the hive by performing a waggle dance. The graph below shows the relationship between the distance of a food supply from the hive and the number of turns in the waggle dance every 15 seconds.

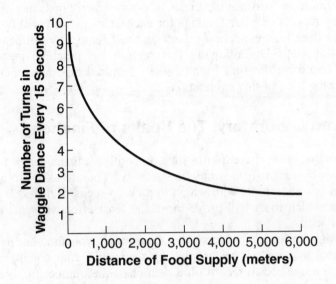

By using one or more complete sentences, state the relationship between the distance of the food supply from the hive and the number of turns the bee performs in the waggle dance every 15 seconds.

_____

15. Based on experimental results, a biologist in a laboratory reports a new discovery. If the experimental results are valid, biologists in other laboratories should be able to perform
    (1) an experiment with a different variable and obtain the same results
    (2) the same experiment and obtain different results
    (3) the same experiment and obtain the same results
    (4) an experiment under different conditions and obtain the same results

IV. LABORATORY REQUIREMENTS FOR PART D

New York State's biology students are required to complete four state-developed laboratory activities as part of their laboratory experience.

In addition to performing these laboratory activities and completing laboratory reports on your findings for each of them, you will be expected to answer questions concerning them on the Living Environment Regents examination, Part D. Following this section are practice questions and answers based on the first four of these required laboratory activities as known at the time of this publication.

Required Laboratory: The Beaks of Finches

This laboratory exercise explores the competitive advantages of beak sizes and shapes by means of a simulation activity. You are asked to examine a variety of common tools with which to pick up seeds of different sizes and to predict which tools will prove most and least effective in picking up a particular seed.

In the data collection portion of the experiment, you and your partner will use the tool assigned to you to pick up seeds over four equal-length time trials. You are asked to record on a data chart the number of seeds picked up during each trial.

In a variation of the data-collection process described above, you and your partner will join other lab teams to compare the effectiveness of your assigned tool against those of the other teams.

By completing this laboratory activity, you will be able to:

- Describe how structural differences in beak shape may affect the survival rate of members of a species;
- Analyze the role of competition for limited resources in the survival of members of a species with varying beak shapes;
- Comprehend how environmental conditions act to select favorable variations within a species;
- Draw inferences and conclusions concerning the accuracy of your hypothesis.

Required Laboratory: Relationships and Biodiversity

This laboratory exercise helps you to explore similarities and differences among different plant species. You will carry out observations of the gross and microscopic anatomical features, chemical characteristics, and genetic makeup of four simulated plant species.

Based on your observations of anatomical features, you will be expected to develop a hypothesis concerning the species that is most similar to a hypothetical species, *Botana curus*. You and your lab partner will then conduct a number of tests, including chromatography, use of indicators, gel electrophoresis, and genetic sequencing, to test your hypothesis.

As a result of completing this activity you should be able to:

- Accurately predict the relative degree of relatedness of two or more plant species based on their anatomical features;
- Analyze the results of several laboratory tests to determine the biochemical relatedness of two or more plant species;
- Draw inferences and conclusions concerning the accuracy of your hypothesis.

Required Laboratory: Making Connections

In this laboratory activity, you and your lab partner will explore the reactions of the human body to fatigue. An important skill needed for this activity is the ability to accurately take another person's pulse. Once you have mastered this skill, you will be asked to use it as one of your primary data-collection techniques.

You are asked first to determine your average resting pulse rate. This data will become the "baseline" data for your experiment. You will then be expected to pool this data with that of the rest of your class and to represent class data graphically on a histogram.

In the next segment of the laboratory, you are asked to gather data about the effect of fatigue on muscle performance. Squeezing a clothespin repeatedly is the measurable activity.

Finally, you will be presented with hypothetical conflicting claims concerning the relationship between exercise and the rate at which a clothespin can be squeezed. You and your lab partner are asked to develop a hypothesis concerning which of these claims you believe to be true. You will then be expected to design a scientific experiment to test your hypothesis.

As a result of completing this experiment you will be able to:

- Accurately determine the pulse rate of another person;
- Collect, organize, and analyze class data on average pulse rate;
- Develop a hypothesis concerning the effect of exercise on muscle performance;
- Design and carry out an experiment to test your hypothesis;
- Draw inferences and conclusions concerning the accuracy of your hypothesis;
- Synthesize a report of your findings and present it to the class.

Required Laboratory: Diffusion Through a Membrane

This laboratory experience involves the creation of a simulated "cell" from dialysis tubing and some simple mixtures of common substances. The cell you create will behave in much the same way as a real cell. Like a real cell, it will contain a mixture of complex organic molecules suspended or dissolved in water. Like a cell membrane, the dialysis tubing will be a semipermeable membrane, allowing some substances to pass through and others not.

Once constructed, the cell will be subjected to two chemical tests to determine the nature of its semipermeability. These tests will involve the use of indicator solutions that indicate the presence of either glucose (blue-colored Benedict's solution) or starch (tan-colored Lugol's solution).

In the second part of the experiment, you will be asked to investigate the nature of a special kind of diffusion known as osmosis (the diffusion of water). You and your lab partner will place living cells in a solution containing dissolved salt. Your observations will help you to understand the nature of osmosis in living tissues.

By the time you complete this activity you should be able to:

- Demonstrate the use of chemical indicators to identify substances dissolved or suspended in water;
- Explain diffusion and osmosis as they relate to semipermeable membranes;
- Describe how the various chemical substances used in this experiment diffused in or out of the semipermeable cell constructed from dialysis tubing.

Required Laboratory: Adaptations for Reproductive Success in Flowering Plants

No information was available about the details of this laboratory activity at the time of this revision. Presumably, this laboratory will deal with the structure and function of the flower as a reproductive structure found in many plant species.

Required Laboratory: DNA Technology

No information was available about the details of this laboratory activity at the time of this revision. Presumably, this laboratory will deal with the various techniques used to analyze the structure and function of DNA.

Required Laboratory: Environmental Conditions and Seed Germination

No information was available about the details of this laboratory activity at the time of this revision. Presumably, this laboratory will deal with how environmental conditions such as temperature, moisture, and oxygen affect the germination of seeds of various types.

QUESTION SET 1.3—PART D REQUIRED LABORATORY EXPERIENCES (ANSWERS EXPLAINED, P. 259)

Base your answers to questions 1 through 3 on the diagram below which represents various finch species on the Galapagos Islands, your specific knowledge of the laboratory entitled "The Beaks of Finches," and your general knowledge of biology.

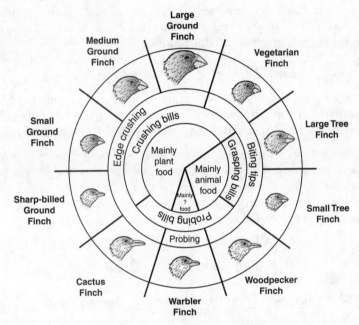

— from *Galapagos: A Natural History Guide*

1. An isolated island in the Galapagos supports four native finch species. These species are the large ground finch, the large tree finch, the small ground finch, and the small tree finch. A new species (the sharp-billed ground finch) is introduced to the island. Which of the native species will be most strongly affected by the introduction of the sharp-billed ground finch, and how will it be affected? In your answer, be sure to include:

   • the native finch species affected [1]
   • what will likely occur that will affect this native species [1]

   _____

   _____

2. A major environmental change occurs on this island that eliminates most of the plants that produce small seeds, leaving only plants that produce large seeds with thick, hard coverings. In terms of food gathering, the finch species having the greatest adaptive advantage under this changed set of conditions is most probably the
   (1) large ground finch
   (2) small ground finch
   (3) large tree finch
   (4) small tree finch

3. When one finch species tries to gain an advantage in obtaining the specific type of food it needs for survival to the disadvantage of other species in the same habitat, this represents
   (1) variation within a species
   (2) environmental change
   (3) interspecies competition
   (4) mutagenic agents

Base your answers to questions 4 through 6 on the partial genetic code chart and information below, your specific knowledge of the laboratory entitled "Relationships and Biodiversity," and your general knowledge of biology.

A plant species with the scientific name *S. hunta* is known to produce a protein substance with valuable medicinal qualities. Its genetic makeup is analyzed, and the gene for this protein is located at a specific gene locus on a certain chromosome. This gene is found to contain the following sequence of DNA codons:

GAA-CAA-TGA-CTT-GTA-GTA-GGG-CAA

Partial Genetic Code Chart

DNA Codons	mRNA Codons	Amino Acids
CTT	GAA	Glu
CTC	GAG	
GTA	CAU	His
GTG	CAC	
GAA	CUU	Leu
GAC	CUG	
GGA	CCU	Pro
GGG	CCC	
TGA	ACU	Thr
TGG	ACC	
CAA	GUU	Val
CAC	GUG	

4. The messenger RNA (mRNA) codon sequence that will result from this DNA sequence is
   (1) CUU-GUU-ACU-GAA-CAU-CAU-CCC-GUU
   (2) GAA-CAA-UGA-CUU-GUA-GUA-GGG-CAA
   (3) CTT-GTT-ACT-GAA-CAT-CAT-CCC-GTT
   (4) UGG-AGG-CUG-ACC-UCG-UCG-UUU-AGG

5. The amino acid sequence in a protein segment that will result from this DNA codon sequence during protein synthesis is
   (1) Leu-Glu-His-Pro-Val-Val-Pro-Val
   (2) Leu-Val-Thr-Leu-Glu-Glu-His-Pro
   (3) Leu-Pro-His-Val-Thr-Thr-Glu-Pro
   (4) Leu-Val-Thr-Glu-His-His-Pro-Val

6. Four new plants (species W, X, Y, and Z) are discovered that appear to be related to *S. hunta*. Upon analysis of their genetic makeup, the following DNA sequences were found at the same gene locus on the same chromosome as was found in *S. hunta*:
   Species W: GAC-CAC-TGA-CTT-CAA-GAA-TGA-CAA
   Species X: GAA-CAA-TGG-CTT-GTA-GTG-GGA-CAA
   Species Y: GAC-GGG-GTG-GTA-CAC-GTA-GGG-CAC
   Species Z: GAA-CAA-GAA-GGG-CTT-CTC-TGG-CAC

   Which species produces a protein segment most similar to that produced by *S. hunta*?
   (1) Species W   (3) Species Y
   (2) Species X   (4) Species Z

Base your answers to questions 7 through 9 on the experimental data below, your specific knowledge of the laboratory entitled "Making Connections," and your general knowledge of biology.

Data on Resting Pulse Rate (Beats per Minute)

Student (M/F)*	Average Pulse Rate (3 Trials)	Student (M/F)*	Average Pulse Rate (3 Trials)
Juan (M)	58	Jessica (F)	66
Sue (F)	81	Sean (M)	75
Tae Moon (M)	67	Mai Le (F)	86
Louise (F)	69	Cesar (M)	80
Sam (M)	72	Rita (F)	63
Erika (F)	60	Guiseppe (M)	71
Jose (M)	85	Meredith (F)	68
Tatiana (F)	65	Bill (M)	88
Brad (M)	91	Concetta (F)	74
Madelena (F)	76		
Greg (M)	62	AVERAGE	72.85

*M/F = male/female.

7. Represent the data in the chart in the form of a histogram (bar chart) in the grid below. Shade the histogram bars that result from your data analysis. [1]

Number of Students							
10							
9							
8							
7							
6							
5							
4							
3							
2							
1							
	<51	51–60	61–70	71–80	81–90	>90	

Average Pulse Rate Range (beats per minute)

8. In a complete sentence, describe one pattern that is evident in the data that would help someone else to understand how pulse rate is distributed among these 20 students. [1]

_____

9. State a question that someone might ask about pulse rate that could be answered by further analyzing the data above or by collecting additional data about the same 20 students. [1]

_____

Base your answers to questions 10 through 12 on the experimental data below, your specific knowledge of the laboratory entitled "Diffusion Through a Membrane," and your general knowledge of biology.

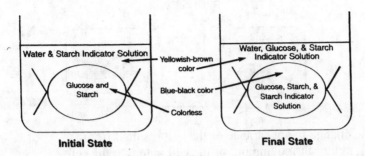

| Initial State | Final State |

10. The following chart shows data that was collected by a student laboratory group as a result of performing the laboratory entitled "Diffusion Through a Membrane." It shows the results of chemical tests that were done on various fluids in the initial and final setups shown above.

Indicator Fluid Tested	Initial Setup		Final Setup	
	Lugol's Solution	Benedict's Solution	Lugol's Solution	Benedict's Solution
Beaker fluid	Tan	Blue	Tan	Red
Dialysis tube fluid	Blue-black	Red	Blue-black	Red

The difference between the fluids in the initial and final setups is that
(1) Lugol's solution has been converted to Benedict's solution
(2) starch has been converted to glucose by the indicator solutions
(3) starch has diffused from the fluid inside the dialysis tubing, across the membrane, and into the beaker fluid
(4) glucose has diffused from the fluid inside the dialysis tubing, across the membrane, and into the beaker fluid

11. A biologist observed a plant cell in a drop of water as shown in diagram A. The biologist then added a 10% salt solution to the slide and observed the cell as shown in diagram B.

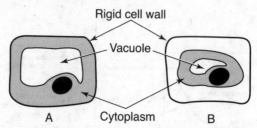

The change in appearance of the cell resulted from
(1) more salt moving out of the cell than into the cell
(2) more salt moving into the cell than out of the cell
(3) more water moving into the cell than out of the cell
(4) more water moving out of the cell than into the cell

12. A farmer irrigating his corn fields irrigates field A with normal well water. In preparing to irrigate field B, he makes an error and loads a 10% saltwater solution (instead of well water) into his irrigation tanker. He then proceeds to irrigate field B with this 10% saltwater solution. Within 24 hours, he notes that all the corn plants in field A are standing straight and healthy, whereas all the corn plants in field B have wilted. Fortunately, a rainstorm the next day restores the corn plants in field B to a healthy condition.

Write a brief paragraph that describes the biological explanation for the farmer's observations of the corn plants in field B. In your answer, be sure to:

• define the terms *osmosis* and *equilibrium* [2]
• describe the *movement of water* into and out of the corn plants in field B during *each of the two days* of this observation [2]

V. LABORATORY SKILLS AND CURRENT EVENTS IN BIOLOGY

The New York State Regents Examination on Living Environment tests the student's knowledge of both content and practical skills. In general, the skills required include basic laboratory technique, analysis of readings in science, written expression of biological concepts, and general awareness of current events in biology.

LABORATORY SKILLS

The Laboratory Checklist (see pages 47 and 48) mentions a set of specific laboratory skills that should be mastered in preparation for the year-end Regents examination. Students should be able to do the following.

1. Formulate a question or define a problem and develop a hypothesis to be tested in an investigation.
2. When given a laboratory problem, select suitable lab materials, safety equipment, and appropriate observation methods.
3. Distinguish between controls and variables in an experiment.
4. Identify parts of a light microscope and their functions, and focus in low and high power.
5. Determine the size of microscopic specimens in micrometers.
6. Prepare wet mounts of plant and animal cells, and apply staining techniques using iodine and methylene blue.
7. Identify cell parts under the compound microscope, such as the nucleus, cytoplasm, chloroplast, and cell wall.
8. Use and interpret indicators, such as pH paper, Benedict's (Fehling's) solution, iodine (Lugol's) solution, and bromthymol blue.
9. Use and read measurement instruments, such as metric rulers, Celsius thermometers, and graduated cylinders.
10. Dissect plant and animal specimens for the purpose of exposing major structures for suitable examination. Suggestions of specimens include seeds, flowers, earthworms, and grasshoppers.
11. Demonstrate safety skills involved in heating materials in test tubes or beakers, using chemicals, and handling dissection instruments.
12. Collect, organize, and graph data.
13. Make inferences and predictions based on data collected and observed.
14. Formulate generalizations or conclusions based on the investigation.

15. Assess the limitations and assumptions of the experiment.
16. Determine the accuracy and repeatability of the experimental data and observations.

These laboratory skills may be subdivided into three broad areas:

A. The methods by which a laboratory experiment is designed.
B. The techniques used in conducting a laboratory experiment.
C. The skills related to interpreting the results of a laboratory experiment.

A. Methods Used in Designing a Laboratory Experiment (Skills 1–3)

Skill 1   Students should be able to express an experimental problem as a statement or as an experimental **question** to be answered. Such questions should be written so as to indicate a quantity to be measured in a laboratory experiment. Students also should be able to state the expected results of such an experiment in the form of a **hypothesis**.

Skill 2   When presented with a hypothetical experiment to perform, students should be able to **select the group of tools** that would be most appropriate to use for conducting that experiment. For example, an experiment involving comparative anatomy would probably be carried out most effectively with dissecting instruments; one involving the chemical nature of an unknown food would be performed most effectively using chemical indicators; one involving plant growth might use a variety of measuring instruments.

Skill 3   When given an outline of an experiment, students should be able to identify the **variables**, which are the (changing) quantities being measured in the experiment. Certain variables are manipulated by the investigator (independent variables), whereas others vary as a result of experimental factors (dependent variables). Students should also be able to identify the aspects of the experiment, known as **controls**, designed to exclude possible interference by unwanted variables.

B. Techniques Used in Conducting Laboratory Experiments (Skills 4–11)

Skill 4   Students should have a basic familiarity with the **compound light microscope**—its parts, the function of each part, its use as a tool for measuring small objects, the procedure for determining its magnification, and the appearance of objects within its visual field. Students should be prepared to describe each of these aspects in sentence form.

Skill 5   Students should be able to express the sizes of microscopic objects in metric units. An understanding of the **micrometer** (μm), a metric unit of linear measure, is required. Students should be able to convert measurements in micrometers to millimeters (or centimeters) and from larger units to micro-meters (1 μm = 0.001 mm = 0.0001 cm = 0.000001 m).

Skill 6 Students should be able to describe, in sentence form, how to prepare a wet-mount slide for examination under the compound light microscope. A basic familiarity with the application of biological stains and the stains **iodine** and **methylene blue** is also required.

Skill 7 Students should be able to recognize and label the major **organelles** of typical plant, animal, and protozoan cells as shown in photomicrographs (photographs taken through a microscope). Students should also be prepared to describe the major functions of these organelles in sentence form as well as to judge their sizes in micrometers.

Skill 8 Students should be familiar with the use of **indicators** used to determine the chemical characteristics of food samples and solutions. Students should be prepared to describe, in sentence form, the use of each of the following indicators.

- **pH paper** is used to determine the relative acidity (pH) of a solution. A pH paper containing litmus will turn red in acid solution and blue in basic solution.
- **Bromthymol blue** turns yellow under acid conditions and remains blue under basic conditions. It may be used to detect the presence of carbon dioxide in solution since carbon dioxide forms a weak acid when dissolved in water.
- **Benedict's (Fehling's) solution** is used to detect the presence of simple sugars in food samples or solutions. Benedict's solution is blue when first prepared. When heated in the presence of simple sugar, it turns color, ranging from yellow to brick red.
- **Iodine (Lugol's) solution**, normally light tan, turns blue-black when applied to a food sample or solution containing starch.

Skill 9 Students should be able to determine **quantities and dimensions**, using a variety of metric measuring instruments. The dimensions of common objects should be determinable in millimeters and centimeters, using metric rulers and scales. Students should be able to read correctly the volume of liquid in a graduated cylinder by sighting the bottom of the meniscus. Students should be able to read the temperature indicated on a Celsius thermometer.

Skill 10 Students should be able to recognize, identify, and label the major **organs and organ systems** of common dissection specimens, including earthworms, grasshoppers, seeds, and flowers.

Skill 11 Students should be able to describe the proper methods of dealing with a variety of **laboratory situations**. This includes the correct means of handling chemicals so as not to cause harm to oneself or others. Also included is the safe use of dissecting tools, such as scalpels and other sharp instruments. Students should also be able to describe the approved techniques for heating liquids and for handling hot objects. Students should be prepared to describe these methods in sentence form.

C. Skills Used in Interpreting Experimental Results (Skills 12–16)

Skill 12 When given unorganized raw data, students should be able to **collect and organize** these data in chart form according to increasing values of the independent variable. Students should also be familiar with the proper techniques to use in representing data in graph form. Knowledge of the correct methods for constructing both bar charts and line graphs is required. Students should be able to label and increment graph axes properly, appropriately title graphs, and correctly plot graphic data points.

Skill 13 Students should be able to analyze the data that result from a laboratory experiment and to draw **inferences** (conclusions based on facts) that help to solve the experimental problem or answer the experimental question. Students should also be able to predict the outcome of experiments that broaden the range of the independent variable. In order to do this, students should understand how to interpret data organized in either chart or graph form. Students should be prepared to describe their inferences and predictions in sentence form.

Skill 14 When given an experiment whose data have been properly organized and analyzed, students should be able to develop generalizations (broad **conclusions**) concerning the effect of the test variable on the experimental question. It should then be possible to project these generalizations onto situations outside the laboratory where similar variables interact. Students should be prepared to express such generalizations in sentence form.

Skill 15 Students should recognize the **limitations** of their experimental methods in the context of professional science. This requires an understanding of the limits of accuracy of measuring equipment, the errors that may be introduced through incompletely developed laboratory skills, the ambiguities that result from inadequate experimental controls, and other limitations affecting experimental results.

Skill 16 Once a laboratory experiment is completed, students should be able to **determine the accuracy of the experimental results** through a review of the experimental methods. Where appropriate, calculating the percent error may assist in determining experimental accuracy. The experimental methods should allow for repeatability of the experiment to assist in the verification of experimental accuracy.

LABORATORY CHECKLIST

In addition to demonstrating performance indicators relating to scientific inquiry described in Standard 1, biology students need to develop proficiency in certain laboratory or technical skills in order to successfully conduct investigations in biological science. During the school year, students must develop the capacity to successfully perform each of the laboratory skills listed below. Proficiency in performing these laboratory skills may also be evaluated by items found on certain parts of the New York State Living Environment assessment, including the new Part D.

Follows safety rules in the laboratory

Selects and uses correct instruments

- Uses graduated cylinders to measure volume
- Uses metric ruler to measure length
- Uses thermometer to measure temperature
- Uses triple-beam or electronic balance to measure mass

Uses a compound microscope/stereoscope effectively to see specimens clearly, using different magnifications

- Identifies and compares parts of a variety of cells
- Compares relative sizes of cells and organelles
- Prepares wet-mount slides and uses appropriate staining techniques
- Uses other laboratory skills
- Designs and uses dichotomous keys to identify specimens
- Makes observations of biological processes
- Dissects plant and/or animal specimens to expose and identify internal structures
- Follows directions to correctly use and interpret chemical indicators
- Uses chromatography and/or electrophoresis to separate molecules
- Demonstrates ability to design, carry out, and report the results of simple biological experiments
- Designs and carries out a controlled, scientific experiment based on biological processes
- States an appropriate hypothesis
- Differentiates between independent and dependent variables
- Identifies the control group and/or controlled variables
- Collects, organizes, and analyzes data, using a computer and/or other laboratory equipment
- Organizes data through the use of data tables and graphs
- Analyzes results from observations/expressed data

- Formulates an appropriate conclusion or generalization from the results of an experiment
- Recognizes assumptions and limitations of the experiment

ANALYSIS OF READINGS IN SCIENCE

Part B and C of the New York State Regents examination may include items that require the analysis of short readings in science. These readings may deal with New York State Core Curriculum understandings or with concepts related to, but outside of, the curriculum. Students are expected to comprehend the meanings of technical terms that appear in the curriculum (see the glossary of this book). If other technical terms are used, they will be defined within the passage.

Students are expected to be able to read through the passage and answer questions based on it. Students should also be able to draw on their knowledge of Major Understandings to answer some questions about these reading passages.

WRITTEN EXPRESSION OF BIOLOGICAL CONCEPTS

Students in the Regents Living Environment course are expected to be able to express themselves in complete sentences concerning biological principles. Although this requirement is not meant to be a test of the grammatical skills of the student, it is a test of the student's ability to express himself/herself clearly in scientific terms. Students should be prepared to answer Part B and C questions in sentence form.

CURRENT EVENTS IN BIOLOGY

Students should maintain an awareness of current events in biology that have reached statewide, national, or international prominence. Students will not be tested directly on current events but rather through use of reading comprehension and analysis of graphs and other data representations. Topics that may be selected for inclusion on Parts B and C include environmental situations (for example, acid precipitation, toxic waste disposal), advances in genetic research (for example, genetic engineering), aspects of biomedical research (for example, immunology, AIDS research), and others of a similar nature.

Following are typical Part B and C questions, grouped by the skills tested, that have appeared on actual Regents examinations in recent years. Students can find additional practice questions in the Regents examinations at the end of this book.

QUESTION SET 1.4—LABORATORY SKILLS QUESTION SET (SKILLS 1–14) (ANSWERS EXPLAINED, P. 264)

Skill 1

1. A student reported that a wilted stalk of celery became crisp when placed in a container of ice water. The student then suggested that water entered the stalk and made it crisp. This suggestion is considered to be
(1) a control
(2) a hypothesis
(3) an observation
(4) a variable

Skill 2

Base your answers to questions 2 through 4 on the four sets of laboratory materials listed below and on your knowledge of biology.

Set A
Light source
Colored filters
Beaker
Test tubes
Test tube stand

Set C
Scalpel
Forceps
Scissors
Pan with wax bottom
Pins
Stereomicroscope
Goggles

Set B
Droppers
Benedict's solution
Iodine
Test tubes
Starch solution
Sugar solution
Test tube holder
Test tube rack
Heat source
Goggles

Set D
Compound light microscope
Glass slides
Water
Forceps

2. Which set should a student select in order to test for the presence of a carbohydrate in food?

3. Which set should a student select to determine the location of the aortic arches in the earthworm?

4. Which set should a student use to observe chloroplasts in elodea (a green water plant)?

49

Skill 3

5. Some scientists have concluded that stressful situations cause a decrease in the normal operation of the immune system in human beings. In a recent study, people who were under severe stress were examined to measure how well their immune systems were functioning. These people showed poorer immune system function during times of severe stress than when they were under less stress.

   If the experimental group studied consisted of truck drivers who drove daily for 8 hours in very heavy traffic, a corresponding control group would most likely consist of truck drivers who drove
   (1) daily for 12 hours in very heavy traffic
   (2) every *third* day for 8 hours in very heavy traffic
   (3) every other day for 12 hours in very light traffic
   (4) daily for 8 hours in very light traffic

6. A student is studying the effect of temperature on the hydrolytic action of the enzyme gastric protease, which is contained in gastric fluid. An investigation is set up using five identical test tubes, each containing 40 milliliters of gastric fluid and 20 millimeters of glass tubing filled with cooked egg white. The five test tubes are each placed in a different temperature-controlled environment at 0°C, 10°C, 20°C, 30°C, and 40°C. After 48 hours, the amount of egg white hydrolyzed in each tube is measured. Which is a variable in this investigation?
   (1) gastric fluid
   (2) length of glass tubing
   (3) temperature
   (4) time

Skill 4

7. The diagram below represents a compound light microscope. Choose *one* of the numbered parts. *In a complete sentence,* name the part selected and describe its function.

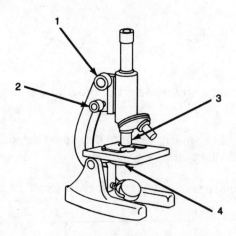

8. To view cells under the high power of a compound microscope, a student places a slide of the cells on the stage and moves the stage clips over to secure the slide. She then moves the high-power objective into place and focuses on the slide with the coarse adjustment.

   Two steps in this procedure are incorrect. For this procedure to be correct, she should have focused under
   (1) low power using the coarse and fine adjustments, and then under high power using only the fine adjustment
   (2) high power first, and then under low power using only the fine adjustment
   (3) low power using the coarse and fine adjustments, and then under high power using the coarse and fine adjustments
   (4) low power using the fine adjustment, and then under high power using only the fine adjustment

Skill 5
Base your answers to questions 9 and 10 on the information following and on your knowledge of biology.

A student was using a microscope with a 10× eyepiece and 10× and 40× objective lenses. He viewed the edge of a metric ruler under low power and observed the following field of vision.

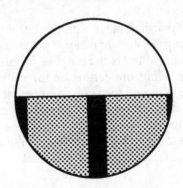

9. What is the diameter, in micrometers, of the low-power field of vision?
   (1) 1   (3) 1,000
   (2) 2   (4) 2,000

10. The diameter of the high-power field of vision of the same microscope would be closest to
   (1) 0.05 mm   (3) 5 mm
   (2) 0.5 mm   (4) 500 mm

Skill 6

11. Which substance, when added to a wet mount containing starch grains, would react with the starch grains and make them more visible?
    (1) litmus solution
    (2) iodine solution
    (3) distilled water
    (4) bromthymol blue

Base your answers to questions 12 and 13 on the diagrams below and on your knowledge of biology. The diagrams show wet-mount microscope slides of fresh potato tissue.

Slide A   Slide B

12. The formation of air bubbles on slide *A* could have been prevented by
    (1) using a thicker piece of potato and less water
    (2) using a longer piece of potato and a coverslip with holes in it
    (3) holding the coverslip parallel to the slide and dropping it directly onto the potato
    (4) bringing one edge of the coverslip into contact with the water and lowering the opposite edge slowly

13. A drop of stain is put in contact with the left edge of the coverslip on slide *B*, and a piece of absorbent paper is placed in contact with the right edge of the coverslip. What is the purpose of this procedure?
    (1) It prevents the stain from getting on the ocular of the microscope.
    (2) It prevents the water on the slide from penetrating the potato tissue.
    (3) It allows the stain to penetrate the potato tissue without the removal of the coverslip.
    (4) It helps increase the osmotic pressure of the solution.

Skill 7

Base your answers to questions 14 and 15 on the drawing below, which shows a piece of tissue stained with iodine solution, as viewed with a microscope under high power.

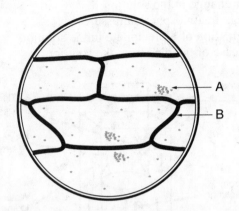

14. The tissue represented in the drawing is most likely made up of
    (1) onion epidermal cells
    (2) ciliated protists
    (3) cardiac muscle cells
    (4) blue-green algae

15. The organelle labeled *B* in the drawing is most likely a
    (1) mitochondrion   (3) lysosome
    (2) centriole   (4) cell wall

16. Diagram *A* represents the appearance of a wet mount of plant tissue as seen through a compound light microscope. Diagram *B* represents the appearance of the same field of view after the fine adjustment knob is turned. What is the best conclusion to be made from these observations?

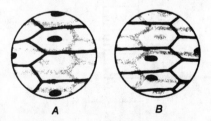

A   *B*

    (1) The tissue is composed of more than one layer of cells.
    (2) The tissue is composed of multinucleated cells.
    (3) The cells are undergoing mitotic cell division.
    (4) The cells are undergoing photosynthesis.

Skill 8

17. A student was testing the composition of exhaled air by exhaling through a straw into a solution of bromthymol blue. The presence of carbon dioxide in the exhaled air would be indicated by
    (1) a color change in the solution
    (2) a change in atmospheric pressure
    (3) the formation of a precipitate in the solution
    (4) the release of bubbles from the solution

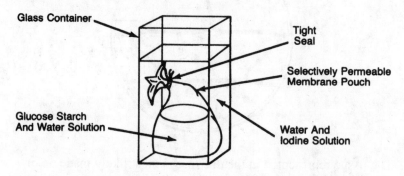

18. A student tested a sample of the fluid in the glass container for glucose 30 minutes after the apparatus had been set up. Which indicator should be used for this test?
    (1) iodine solution
    (2) bromthymol blue
    (3) Benedict's solution
    (4) pH paper

Skill 9

19. What is the total volume of water indicated in the graduated cylinder illustrated below?

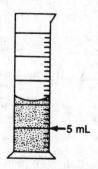

(1) 10 mL   (3) 12 mL
(2) 11 mL   (4) 13 mL

20. The diagram below represents a segment of a metric ruler and part of an earthworm. What is the length of the part of the earthworm shown? You must include the correct units in your answer.

21. Which group of measurement units is correctly arranged in order of increasing size?
    (1) micrometer, millimeter, centimeter, meter
    (2) millimeter, micrometer, centimeter, meter
    (3) meter, micrometer, centimeter, millimeter
    (4) micrometer, centimeter, millimeter, meter

Skill 10
22. In earthworms and grasshoppers, which structure is ventral to the esophagus?
    (1) gizzard
    (2) brain
    (3) intestine
    (4) nerve cord

Base your answers to questions 23 and 24 on the illustration below of the flower of an amaryllis plant.

23. Name the circled part of the stamen.

24. Using a complete sentence, state a process carried out within the circled structure.

Skill 11

25. A student performing an experiment noticed that the beaker of water she was heating had a slight crack in the glass, but was not leaking. What should the student do?
    (1) Discontinue heating and attempt to seal the crack.
    (2) Discontinue heating and report the defect to the instructor.
    (3) Discontinue heating and immediately take the beaker to the instructor.
    (4) Continue heating as long as fluid does not seep from the crack.

26. Which would be the proper laboratory procedure to follow if some laboratory chemical splashed into a student's eyes?
    (1) Send someone to find the school nurse.
    (2) Rinse the eyes with water and do not tell the teacher because he or she might become upset.
    (3) Rinse the eyes with water; then notify the teacher and ask further advice.
    (4) Assume that the chemical is not harmful and no action is required.

27. While a student is heating a liquid in a test tube, the mouth of the tube should always be
    (1) corked with a rubber stopper
    (2) pointed toward the student
    (3) allowed to cool
    (4) aimed away from everybody

Skill 12

A student was investigating the relationship between different concentrations of substance X and the height of bean plants. He started with six groups, each of which contained the same number of bean plants with identical heights. Conditions were kept the same except that each group was watered with a different concentration of substance X for a period of 2 weeks. Then the concentration of substance X used in watering each group of plants and the average height for each group of plants were recorded by the student as follows:

Group A—6%, 32.3 cm   Group D—8%, 37.1 cm

Group B—0%, 28.7 cm   Group E—4%, 31.5 cm

Group C—2%, 29.4 cm   Group F—10%, 30.7 cm

For questions 28 and 29, organize the above data by filling in the Data Table on page 57, following the directions given in the questions.

28. Label column III with an appropriate heading. [*Include the proper unit of measurement.*]

29. Complete all three columns in the Data Table so that the concentrations of substance X are increasing from the top to the bottom of the Data Table.

DATA TABLE

I	II	III
Group	**Concentration of Substance X (%)**	

The Data Table shows the wolf and moose populations recorded at the end of June from 1970 to 1980 on an isolated island national park that serves as a natural refuge for wildlife. Before the arrival of wolves on the island (1965), the moose population had increased to more than 300 members. Wolves have been observed many times on this island hunting cooperatively to prey upon moose.

DATA TABLE

	Number of Members	
Year	**Wolf Population**	**Moose Population**
1970	10	90
1972	12	115
1974	20	145
1976	25	105
1978	18	95
1980	18	98

For questions 30 through 33, use the information in the Data Table to construct a line graph on the grid, following the directions given in the questions.

30. Mark an appropriate scale on the axis labeled "Number of Members of Each Population."

31. Mark an appropriate scale on the axis labeled "Year."

32. Plot the data for the wolf population on the graph. Surround each point with a small triangle and connect the points.

    Example

33. Plot the data for the moose population on the graph. Surround each point with a small circle and connect the points.

    Example

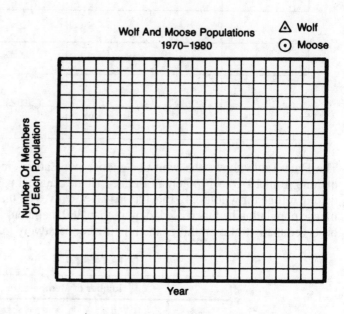

Wolf And Moose Populations
1970–1980

△ Wolf
⊙ Moose

Number Of Members
Of Each Population

Year

Skill 13

Base your answers to questions 34 through 36 on the following graph representing survival rates of fish species at various pH levels.

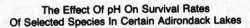

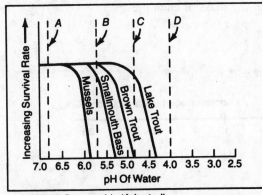

The Effect Of pH On Survival Rates Of Selected Species In Certain Adirondack Lakes

Key:
A - pH Of A Certain Group of Adirondack Lakes, 1880
B - pH Of Rainfall, 1880
C - pH Of The Same Group Of Adirondack Lakes, 1980
D - pH Of Rainfall, 1980

— National Geographic (Adapted)

34. Which species can tolerate the highest level of acidity in its water environment?
    (1) mussels   (3) brown trout
    (2) smallmouth bass   (4) lake trout

35. In the years between 1880 and 1980, which species would most likely have been eliminated first because of the gradual acidification of Adirondack lakes?
    (1) mussels   (3) brown trout
    (2) smallmouth bass   (4) lake trout

36. What is the total change in the pH value of rainwater from 1880 to 1980?
    (1) 1.3   (3) 5.3
    (2) 1.7   (4) 9.7

Base your answers to questions 37 and 38 on the information and the graph given below, which shows the effect of sugar concentration on osmotic balance.

Four pieces of apple were cut so that all were the same mass and shape. The pieces were placed in four different concentrations of sugar water. After 24 hours, the pieces were removed and their masses determined. The graph below indicates the change in the mass of each piece.

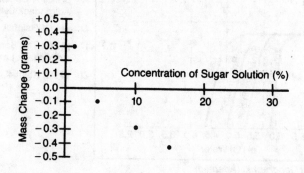

37. What was the change in mass of the apple piece in the 10% sugar solution?
    (1) a decrease of 0.45 gram
    (2) an increase of 0.30 gram
    (3) a decrease of 0.30 gram
    (4) an increase of 0.10 gram

38. At approximately what sugar concentration should the pieces neither lose nor gain weight?
    (1) 6%   (3) 3%
    (2) 10%   (4) 20%

Skill 14

39. The graph below shows the average growth rate for 38 pairs of newborn rats. One of each pair was injected with anterior pituitary extract. The other member of each pair served as a control.

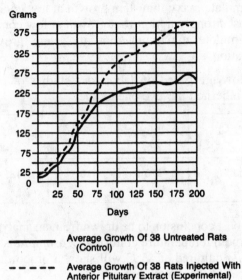

Grams

Days

———— Average Growth Of 38 Untreated Rats
(Control)

– – – – – Average Growth Of 38 Rats Injected With
Anterior Pituitary Extract (Experimental)

Based on the graph, it can be correctly concluded that the pituitary extract
(1) is essential for life
(2) determines when a rat will be born
(3) affects the growth of rats
(4) affects the growth of all animals

40. Plant hormones are chemical regulators that stimulate or inhibit growth depending on their concentration and the type of tissue in which they are found.

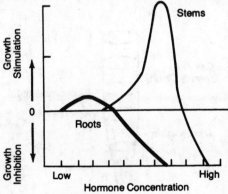

Based on the information in the preceding graph, which is a correct conclusion about plant hormones?
  (1) They stimulate maximum root growth and stem growth at the same concentration.
  (2) They stimulate maximum stem growth at low concentrations.
  (3) They most strongly inhibit root growth at low concentrations.
  (4) They stimulate maximum root and stem growth at different concentrations.

41. The data below are based on laboratory studies of male *Drosophila,* showing the inherited bar-eye phenotype.

Culture Temperature (°C) During Development	15	20	25	30
Number of Compound Eye Sections	270	161	121	74

Which is the best conclusion to be drawn from an analysis of these data?
  (1) The optimum temperature culturing *Drosophila* is 15°C
  (2) *Drosophila* cultured at 45°C will show a proportionate increase in the number of compound eye sections.
  (3) Temperature determines eye shape in *Drosophila.*
  (4) As temperature increases from 15°C to 30°C, the number of compound eye sections in male *Drosophila* with bar-eyes decreases.

42. The diagram below represents the result of spinning a suspension of broken cells in an ultracentrifuge. Which is a correct conclusion?
  (1) Ribosomes are more dense than mitochondria.
  (2) Nuclei are more dense than mitochondria.
  (3) Mitochondria and ribosomes are equal in density.
  (4) The cell consists of only solid components.

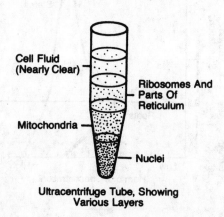

Ultracentrifuge Tube, Showing
Various Layers

Base your answers to questions 1 through 5 on the passage below.

Gene Splicing

Recent advances in cell technology and gene transplanting have allowed scientists to perform some interesting experiments. Some of these experiments have included splicing a human gene into the genetic material of bacteria. The altered bacteria express the added genetic material.

Bacteria reproduce rapidly under certain conditions. This means that bacteria with the gene for human insulin could multiply rapidly, resulting in a large bacterial population which could produce large quantities of human insulin.

The traditional source of insulin has been the pancreases of slaughtered animals. Continued use of this insulin can trigger allergic reactions in some humans. The new bacteria-produced insulin does not appear to produce these side effects.

The bacteria used for these experiments are *E. coli,* bacteria common to the digestive systems of many humans. Some scientists question these experiments and are concerned that the altered *E. coli* may accidentally get into water supplies.

For each statement below, write the number 1 if the statement is true according to the paragraph, the number 2 if the statement is false according to the paragraph, or the number 3 if not enough information is given in the paragraph.

1. Transplanting genetic material into bacteria is a simple task.
2. Under certain conditions bacteria reproduce at a rapid rate.
3. Continued use of insulin from other animals may cause harmful side effects in some people.
4. The bacteria used in these experiments are normally found only in the nerve tissue of humans.
5. Bacteria other than *E. coli* are unable to produce insulin.

Base your answers to questions 6 through 8 on the reading passage below. Write your answers in complete sentences.

Time Frame for Speciation

Evolution is the process of change through time. Theories of evolution attempt to explain the diversification of species existing today. The essentials of Darwin's theory of natural selection serve as a basis for our present understanding of the evolution of species. Recently, some scientists have suggested two possible explanations for the time frame in which the evolution of species occurs.

Gradualism proposes that evolutionary change is continuous and slow, occurring over many millions of years. New species evolve through the accumulation of many small changes. Gradualism is supported in the fossil record by the presence of transitional forms in some evolutionary pathways.

Punctuated equilibrium is another possible explanation for the diversity of species. This theory proposes that species exist unchanged for long geological periods of stability, typically several million years. Then, during geologically brief periods of time, significant changes occur and new species may evolve. Some scientists use the apparent lack of transitional forms in the fossil record in many evolutionary pathways to support punctuated equilibrium.

6. Identify one major difference between gradualism and punctuated equilibrium. [1]
7. According to the theory of gradualism, what may result from the accumulation of small variations? [1]
8. What fossil evidence indicates that evolutionary change may have occurred within a time frame known as gradualism? [1]

---

9. An organism contains many structures that enable it to survive in a particular environment. The human body has many such structures that have adaptive value. Several of these structures are listed below.

- sweat gland
- pancreas
- liver
- epiglottis
- capillary
- villus
- kidney
- platelet

Choose *three* (3) of the structures listed above. For *each* one chosen, write the name of the structure and then, using a complete sentence, describe one of its adaptive values to the human body. [3]

Base your answers to questions 10 through 14 on the information below and on your knowledge of biology.

Acid rain is a serious environmental problem in large areas of Canada and the northeastern United States, including New York State. It is partly created as rain "washes out" sulfur and nitrogen pollutants from the air. Acid rain alters the fundamental chemistry of sensitive freshwater environments and results in the death of many freshwater species. The principal sources of this pollution have been identified as smokestack gases released by coal-burning facilities located mainly in the midwestern United States.

"Unpolluted" rain normally has a pH of 5.6. Acid rain, however, has been measured at pH values as low as 1.5, which is more than 10,000 times more acidic than normal. Commonly, acid rain has a pH range of 3 to 5, which changes the acidity level of the freshwater environment into which it falls. The effect of the acid rain depends on the environment's ability to neutralize it. Evidence is accumulating, however, that many environments are adversely affected by the acid rain. As a result, the living things within lakes and streams that cannot tolerate the increasing acidity gradually die off.

There are many environmental problems that result from acid rain. Most of these problems center around the food web upon which all living things, including humans, depend. If freshwater plants, animals, and protists are destroyed by the acid conditions, then terrestrial predators and scavengers dependent on these organisms for food are forced to migrate or starve. These changes in a food web can eventually affect the human level of food consumption.

10. The accompanying scale shows the pH values of four common household substances. Acid rain has a pH closest to that of which of these substances?

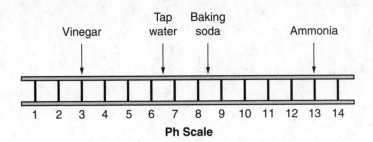

(1) ammonia
(2) tap water
(3) baking soda
(4) vinegar

11. What is most likely the source of acid rain in New York State?
(1) far western United States
(2) midwestern United States
(3) far eastern Canada
(4) far western Europe

12. Which food chain includes organisms that would most immediately be affected by acid rain?
(1) grass → rabbit → fox → decay bacteria
(2) algae → aquatic insect → trout → otter
(3) shrub → mouse → snake → hawk
(4) tree → caterpillar → bird → lynx

13. Acid rain is generally considered a negative aspect of human involvement with the ecosystem. As such, it would most correctly be classified as a type of
    (1) biological control
    (2) conservation of resources
    (3) technological oversight
    (4) land-use management

14. A strain of fish that could survive under conditions of increased acidity could best be obtained by
    (1) binary fission
    (2) vegetative propagation
    (3) selective breeding
    (4) budding

Understanding and Applying Scientific Concepts

STANDARD 4

Students will understand and apply scientific concepts, principles, and theories pertaining to the physical setting and living environment and will recognize the historical development of ideas in science.

Essential Questions:

- *What are the principles that guide scientists in their study of the natural world?*
- *How is modern science built on the work of scientists of the past?*

I. SIMILARITY AND DIVERSITY AMONG LIVING THINGS—LIFE PROCESSES

> **KEY IDEA 1—APPLICATION OF SCIENTIFIC PRINCIPLES**   Living things are both similar to and different from each other and from nonliving things.

Essential Question:

- *How are living things similar to and different from other living things and nonliving things?*

Life Processes

Living things rely on many of the same processes to stay alive. Yet the ways that these processes and interactions are carried out are diverse. The **life processes** carried out by living things include nutrition, transport, respiration, regulation, reproduction, and (in animals) locomotion. These life processes provide mechanisms for living things to obtain and process foods (nutrition), circulate essential materials around their bodies (transport), convert food energy to cell energy (respiration), coordinate life activities (regulation), produce more members of their kind (reproduction), and move from place to place in their environment (locomotion). Although almost all living things perform these same life functions, they are performed in many different ways in different kinds of living organisms. These different mechanisms are known as adaptations.

Nearly all living things are composed of cells. Cells are small living units that serve as the structural and functional building blocks of most living organisms. Cells carry on dynamic chemical activities that together constitute the cell's metabolism (**metabolic activity**). By carefully balancing this metabolic activity, cells are able to maintain a balanced internal stability known as **homeostasis**. Nonliving things differ from living things in that nonliving things lack the ability to perform these life processes.

The components of living systems must work together to maintain homeostatic balance. This is true whether the systems are represented by cells, multicelled organisms, or ecosystems. Each level of complexity brings a new degree of variation to the theme of homeostasis. On the cellular level, life functions are performed by **organelles**, or small organs, found within the cell. Multicellular organisms, depending on their complexity, are made up of specialized organ systems that perform different aspects necessary to carry out separate, but interrelated, life functions. On the ecosystem level, species populations of all kinds interact to help achieve homeostatic balance in the living environment.

Diversity and Ecosystem Stability

Essential Question:

- *How does the great variety of living species impact environmental stability?*

Performance Indicator 4.1.1 *The student should be able to explain how diversity of populations within ecosystems relates to the stability of ecosystems.*

Recognizing the interrelatedness of all living things, including humans, is important. It is also important to understand the unique, yet interdependent, roles that these organisms play in maintaining the stability of life on Earth. It is the richness of **species diversity** that provides for this stability. When even one species in an ecosystem is eliminated, several other species are adversely affected, interrupting stability. See material under "Interdependence of Living Things" for a complete discussion of ecology (pages 200–224).

Life Functions in Humans

Human Physiology

Essential Question:

- *How are humans adapted to survive in their environment?*

Performance Indicator 4.1.2 *The student should be able to describe and explain the structures and functions of the human body at different organizational levels (for example, systems, tissues, cells, organelles).*

Humans, like all living things, must perform basic life functions in order to survive. These life functions include nutrition, transport, respiration, excretion, regulation, and locomotion. The bodies of human beings are organized to provide efficient and effective means of performing these life functions. Each life function in a human is carried out by an **organ system** that contains **organs** and **tissues** specialized for the tasks involved. In turn, these structures are composed of cells whose metabolic processes are specialized for assisting these life functions. In addition, the activities of each organ and organ system must be coordinated so as to provide an integrated, homeostatic balance that promotes the maintenance of life. When any part of this complex system breaks down or becomes inefficient, this balance may be disrupted. **Diseases** may result from such disturbances.

Important levels of organization for structure and function include **organelles**, **cells**, **tissues**, **organs**, **organ systems**, and whole **organisms**. Several specific adaptations are found in humans (and in many other living things) to carry out these life functions. The simplest of these adaptations is the **cell**, which is the unit of structure and function of all living things. Cells contain subcomponents, known as **organelles**, that are specialized to perform aspects of these life functions. Some important organelles and their respective functions are discussed on pages 106–107.

Above the cellular level of organization, the bodies of living things contain groupings of cells that are similar in structure and function. Such groupings of similar cells are known as **tissues**. For example, muscle tissues in humans are composed of spindle-shaped cells rich in contractile fibers. These cells are specialized to act together and assist in body movement. Because of their similar structure and function, muscle cells comprise a tissue. The bodies of complex organisms such as humans contain hundreds of distinct tissues.

An **organ** is a structure, composed of several different tissues, that plays a major role in the performance of a life function. Examples of organs in the human include the heart, the stomach, the liver, and the kidney. A series of organs that function together to assist in the performance of a life function is known collectively as an **organ system**. For example, the digestive system of the human is made up of the esophagus, stomach, small intestine, and large intestine, as well as several accessory organs. Each organ in the system has a specific role to play in the performance of the life function. Organ systems of humans are described in detail below.

Human Organ Systems

Humans are complex organisms. They require multiple systems for digestion, respiration, circulation, excretion, movement, coordination, and immunity. The systems interact to perform the life functions. Humans are complex organisms whose existence depends on the coordinated functioning of several integrated organ systems. Each organ system is specialized for the tasks involved in carrying out a separate and distinct life function. By utilizing these systems, humans are able to:

- take in and process complex foods to produce simpler subunits;
- move essential materials to all parts of the body;
- obtain respiratory gases and metabolize these foods to produce cellular energy;
- remove potentially harmful wastes from the body;
- coordinate all life functions to produce an efficient, integrated system;
- move from place to place in the environment; and
- provide immunity from disease.

Perhaps the most essential feature of this process is the coordination of the separate systems into an integrated whole, promoting the maintenance of life in each individual human being. Without this integration, complex functional living units would probably not exist on Earth.

Nutrition is the life function by which human beings obtain materials needed for energy, growth and repair, and other life functions. As part of this process, these materials are converted to a simplified form that can be used by the cell. Nutritional requirements of humans are similar to those of other animals. A diet that includes the proper balance of carbohydrate, protein, lipid, roughage, vitamins, and minerals is essential to the physical health of the human body. These requirements are known to vary, however, with the age, sex, and physical activity of the individual.

The digestive system enables the human to carry out the life function of nutrition. In general, the human digestive tract resembles that of simpler organisms whose body plan is that of a one-way tube within a tube. The outer tube is the body exterior; the inner tube is the digestive tract. Within the digestive tract, food materials are progressively converted into molecular end products. The components of this system are as follows.

- **Oral cavity**—The human oral cavity (mouth) is used to ingest food. The teeth and tongue help to manipulate and break the food down mechanically. This process increases the surface area of the food, thereby aiding the process of chemical digestion by enzymes in saliva. Saliva is produced by salivary glands and secreted into the mouth cavity. The enzymes in saliva are responsible for the partial digestion of complex carbohydrates, such as starch, into double sugars, such as maltose.
- **Esophagus**—The esophagus is a short tube that connects the oral cavity to the stomach. The swallowing action initiated at the back of the oral cavity continues in the esophagus as a wave of muscular contraction known as **peristalsis**. This peristaltic action moves the chewed food to the stomach for further digestive action.
- **Stomach**—The stomach is a muscular organ whose main function is to liquefy and further digest food materials. The lining of the stomach contains digestive glands that secrete the digestive enzymes and hydrochloric acid that make up gastric fluid. Stomach enzymes are specifically designed to digest proteins. Unlike most of the body's enzymes, gastric enzymes work best in an acidic condition. This acidic condition is provided by the hydrochloric acid secreted by the stomach's digestive glands.
- **Small intestine**—Liquefied and partially digested food enters the small intestine from the stomach. There, enzymes secreted by intestinal glands complete the digestive process. These enzymes include protease (to digest proteins), lipase (to digest fats and oils), and enzymes such as maltase and sucrase (to digest maltose and sucrose sugars). In addition to the digestive process carried on within the small intestine, the lining of the small intestine acts as the principal surface for the absorption of the molecular end products of digestion. To facilitate this absorption, the lining of the small intestine contains millions of microscopic projections known as **villi**. The villi contain microscopic blood vessels, known as **capillaries**, and extensions of the lymphatic system, known as **lacteals**, that receive the dissolved nutrients and conduct them throughout the body.
- **Large intestine**—The large intestine receives food materials that have passed through the entire digestive tract but have not been digested. These materials are normally in a liquid state when they pass from the small to the large intestine. The large intestine reabsorbs much of the water from the waste matter and condenses it into semisolid waste known as **feces**. The feces are stored in the lower end of the large intestine in an area known as the **rectum**. The feces pass out of the body via the anus by means of strong muscular (peristaltic) contractions. This action constitutes the process of egestion.

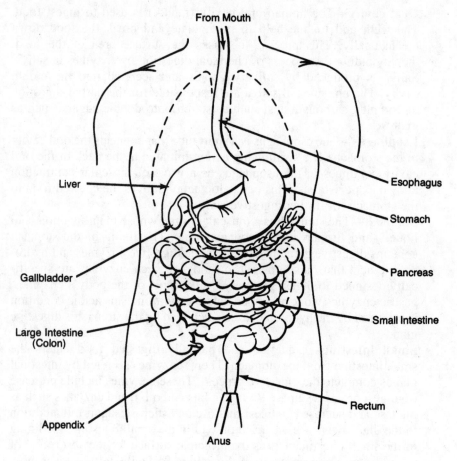

Digestive Tract

- **Accessory organs**—Accessory digestive organs include the salivary glands, liver, gallbladder, and pancreas. The term accessory refers to the fact that these organs are not a part of the food tube. The liver produces bile, which is important in the digestion of fats. Bile acts by breaking up (emulsifying) fat into small droplets. The gallbladder stores bile for release into the small intestine. The pancreas manufactures and stores several digestive enzymes important for the complete digestion of several kinds of complex molecules into simple, soluble end products. These end products include glucose, amino acids, fatty acids, and glycerol.

The underlying mechanism of digestion is the chemical process known as **hydrolysis**. This term literally means "splitting by adding water." In this reaction, large, insoluble molecules are converted to small, soluble molecules. To accomplish this, the cell uses enzymes to add the atoms that constitute water (hydrogen and oxygen) to the structure of the complex molecule. This reaction breaks the complex molecule at specific locations, producing smaller subunit molecules that can be absorbed by the cell.

This enzyme-catalyzed reaction may be accomplished with each of the major types of food and is illustrated by the following equation:

$$\text{maltose} + \text{water} \xrightarrow{\text{maltase}} 2 \text{ glucose}$$

$$C_{12}H_{22}O_{11} + H_2O \xrightarrow{\text{maltase}} 2\ C_6H_{12}O_6$$

In this reaction, the double sugar maltose is hydrolyzed to two molecules of the simple sugar glucose by the addition of a single molecule of water. The enzyme maltase catalyses this reaction. In similar reactions, starch molecules are hydrolyzed to double sugars, protein molecules are hydrolyzed to amino acid molecules, and fat molecules are hydrolyzed to fatty acid and glycerol molecules.

Transport is the life function by which human beings absorb and distribute the materials necessary to maintain life. The human circulatory system is specially adapted to move essential materials through the body to all cells. At the same time, waste materials resulting from cellular metabolism are carried to areas where they can be released to the environment away from living cells. **Immunity** from disease is also provided by transport mechanisms. Specialized tissues and organs assist in the transport function.

Transport media include blood and lymph. The blood is a fluid tissue suspended in liquid plasma. Plasma is made up of water containing dissolved salts, nutrients, gases, and molecular wastes. Also found in the plasma are hormones and a large variety of manufactured proteins such as antibodies, clotting proteins, and enzymes. The plasma also suspends the cellular fraction of the blood.

The cellular fraction of the blood contains several different cell types. **Red blood cells** are the most abundant cell type in the blood. These small (about 8 micrometers in diameter), dish-shaped cells lack nuclei and cannot reproduce. Red blood cells contain **hemoglobin**, a red oxygen-carrying pigment that makes the blood an efficient medium for transporting oxygen to all body tissues. **Phagocytes**, a type of white blood cell, engulf and destroy bacteria that enter the bloodstream though breaks in the skin surface. **Lymphocytes**, another type of white blood cell, produce **antibodies** specifically designed to recognize and attack particular types of proteins (**antigens**) that may enter the blood by various routes. **Platelets** are small, noncellular components of the blood that contain chemicals important to the clotting process. The body's immune reactions are discussed in detail in the section "Disease as a Failure of Homeostasis."

Intercellular fluid (ICF) and lymph are abundant in human tissues. All cells in the body are bathed in ICF, which is rich in salts and various components important to the homeostatic balance of the cell. ICF drains from the tissues within lymphatic vessels, where it is known as lymph.

The **circulatory system** includes a series of structures (vessels and organs) designed to move the transport fluids throughout the body efficiently. These transport vessels and organs include the following.

- **Arteries**—Arteries are relatively thick-walled blood vessels that contain cardiac muscle tissues. These muscles enable the artery to maintain blood flow via rhythmic contractions known as the **pulse**. Arteries are always involved with conducting the blood away from the heart toward the body's tissues.
- **Veins**—Veins are relatively thin-walled blood vessels that lack muscular tissues. Veins contain one-way valves that aid the forward movement of blood by preventing backflow within the vein. Veins are always involved with conducting the blood back toward the heart and away from the body tissues.
- **Capillaries**—Capillaries are microscopic blood vessels whose walls are only one cell in thickness. Capillaries branch from the ends of small arteries and carry blood rich in oxygen and nutrients to all tissues in the body. Dissolved materials are readily exchanged by diffusion between the blood and the body tissues through the thin walls of the capillaries.
- **Lymph vessels**—Lymph vessels form a branching series of microscopic vessels containing lymph. The lymph vessels carry lymph to and from the body tissues where it bathes the cells. To aid the movement of the lymph, the lymph vessels contain valves similar to those found in veins. Lymph vessels become enlarged and are gathered in masses known as lymph nodes at specific parts of the body. These nodes contain phagocytic white blood cells that attack and destroy bacteria in the lymph.
- **Heart**—The structure of the heart permits efficient movement of blood throughout the body within blood vessels. The heart is a muscular pump with four chambers. Two of these chambers, the **atria**, receive blood from veins leading from body organs. Two other chambers, the **ventricles**, have thick, muscular walls that contract to force blood out under great pressure through arteries to other organs.

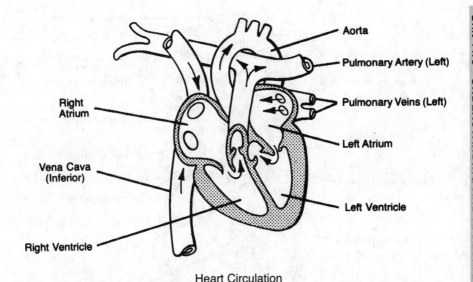

Heart Circulation

The circulatory function can be traced from the entry of deoxygenated blood (blood whose hemoglobin has given up its oxygen) into the **right atrium** of the heart via the **vena cava**. The vena cava is a large vein that collects blood from smaller veins. This blood then passes through a one-way valve to the **right ventricle**. Strong muscular contractions in this ventricle force the blood out through the **pulmonary arteries** and into the lungs. A valve in the pulmonary artery prevents the backflow of blood. In the lungs, the blood passes through **capillaries**, where gas exchange occurs. This process adds oxygen to the blood and removes carbon dioxide from the blood. The oxygenated blood then returns via the **pulmonary veins** to the heart, entering through the **left atrium**. From the left atrium, the blood passes through another one-way valve on its way to the **left ventricle**. Contraction of the left ventricle sends the oxygenated blood out of the heart to body organs by way of the **aorta**. As in the pulmonary artery, a valve prevents the blood from flowing backward into the heart. The aorta branches into a series of smaller arteries. They eventually terminate in capillary networks within body tissues. In the capillaries, oxygen and nutrients are absorbed from the blood and carbon dioxide and metabolic wastes are absorbed into the blood. The capillaries then carry the blood to veins, which eventually lead back to the vena cava and into the heart.

The circulation of blood through the lungs is known as **pulmonary circulation**. The circulation of blood through the body organs is known as the **systemic circulation**. The movement of blood through the blood vessels serving the heart muscle is known as the **coronary circulation**.

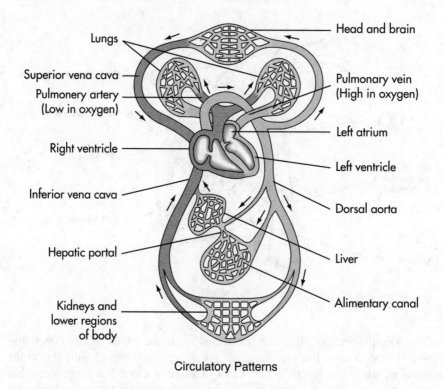

Circulatory Patterns

Blood pressure from both the pumping action of the heart and the contractions of the muscular artery walls maintains the flow of blood through the arteries and capillaries. During blood pressure testing, the higher pressure (**systole**) is registered when the ventricles contract; the lower pressure (**diastole**) is registered when the ventricles relax.

QUESTION SET 2.1—HUMAN PHYSIOLOGY 1
(ANSWERS EXPLAINED, P. 282)

1. Some characteristics of digestive systems are listed below.

   A Food is moved along by peristalsis.
   B Food is moved along by involuntary muscles.
   C Accessory organs are present.

   Which characteristics best describe the human digestive system?
   (1) A, B, and C
   (2) A only
   (3) B only
   (4) B and C only

2. Which statement most accurately describes the human heart?
   (1) It has two atria and one ventricle, and it pumps blood directly into veins.
   (2) It has one atrium and one ventricle, and it is composed of cardiac muscle.
   (3) It has one atrium and two ventricles, and it is composed of visceral muscle.
   (4) It has two atria and two ventricles, and it pumps blood directly into arteries.

3. A malfunction of the lymph node would most likely interfere with the
   (1) release of carbon dioxide into the lymph
   (2) filtering of glucose from the lymph
   (3) release of oxygen into the lymph
   (4) filtering of bacteria from the lymph

4. A heart attack may be due to all of the following except
   (1) an increase in arterial blood pressure
   (2) oxygen deprivation of cardiac muscle
   (3) narrowing of the arteries transporting blood to the heart muscle
   (4) decreased consumption of complex carbohydrates

5–7. Base your answers to questions 5 through 7 on the diagram below of the human digestive system and on your knowledge of biology.

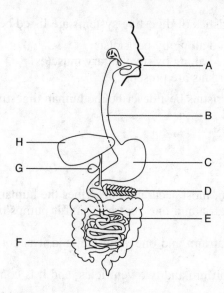

5. In which structure does the initial hydrolysis of carbohydrates occur?
   (1) A   (3) C
   (2) E   (4) D

6. From which structure are glucose and amino acids normally absorbed into the circulatory system?
   (1) F   (3) C
   (2) H   (4) E

7. In which structure does extracellular chemical digestion of protein begin?
   (1) G   (3) C
   (2) B   (4) E

---

8. A pulse can be detected most easily in
   (1) an artery   (3) a capillary
   (2) a vein   (4) a lacteal

9–10. Base your answers to questions 9 and 10 on the diagram below of the human heart and on your knowledge of biology.

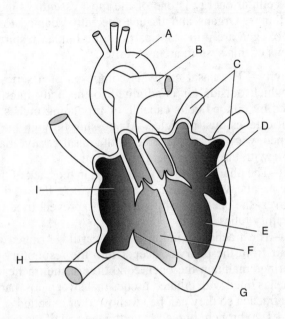

9. Which structures are most closely associated with the transport of deoxygenated blood?
   (1) A, B, and C
   (2) B, F, and I
   (3) C, D, and E
   (4) D, H, and I

10. A structure that prevents the backflow of blood into an atrium is indicated by letter
    (1) G
    (2) B
    (3) C
    (4) H

11. Which transport structures have specialized regions for filtering out bacteria and dead cells?
    (1) arteries
    (2) capillaries
    (3) veins
    (4) lymph vessels

Respiration is the life function by which human beings convert the chemical energy stored in foods to a form that the cells can use more easily. The production of cellular energy is one of the most essential functions carried on by living things. Organs and tissues specially adapted to promote the absorption of oxygen are important parts of the human respiratory system. The components of this system are as follows.

- **Nasal cavity**—The nasal cavity is composed of a series of channels through which outside air is admitted to the body interior. The cavity is lined by ciliated cells capable of producing mucus. Hair is also present; it catches dust particles. Air passing through the nasal cavity is moistened, warmed, and filtered in the nasal cavity, making it more compatible with the environment of the lung.
- **Pharynx**—The pharynx (throat) is the area in the back of the oral cavity where the nasal cavity joins in. In the pharynx, a flap of tissue, the epiglottis, covers the open end of the trachea to prevent food from entering the respiratory tubes.
- **Trachea**—The trachea (windpipe) is a cartilage-ringed tube used to conduct air from the pharynx deeper into the respiratory system. The cartilage rings maintain the open condition of the trachea at all times. The trachea is lined with ciliated tissues that sweep dust particles up and out of the trachea so they can be swallowed or expelled.
- **Bronchi**—Two bronchi branch from the the end of the trachea and lead to the two lungs. Like the trachea, the bronchi are ringed with cartilage and lined with ciliated mucous membrane.
- **Bronchioles**—The bronchioles are highly branched tubules that subdivide from the ends of the bronchi and become progressively smaller as they pass deeper into the lungs. The bronchioles lack cartilage rings.
- **Alveoli**—The alveoli, tiny air sacs, are found at the ends of each of the bronchioles. The alveoli are lined with cells that constitute the actual respiratory surface of the lung. Hence, the alveoli are the functional unit of the lung. Capillaries surround the alveoli. They carry away the oxygen absorbed in the moist lining so it can be transported to body tissues.
- **Lungs**—The lungs are composed of the bronchi, bronchioles, alveoli, and their supporting tissues. The functional unit of the lung is the alveolus (alveoli, plural), whose function is described above.

Breathing is a mechanical process used to move air into the lungs as efficiently as possible. This process involves muscular movements of the **diaphragm** and rib cage, which raise and lower pressures within the chest cavity. As the diaphragm is lowered, pressure in the chest cavity decreases and air containing oxygen is forced into the lungs by normal atmospheric pressures. As the diaphragm is raised, pressure in the chest cavity increases and air containing carbon dioxide and water vapor is expelled.

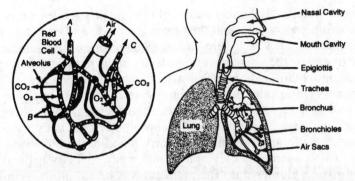

Deoxygenated blood enters through capillaries *(A)* branching from the pulmonary artery. As the blood passes through the capillary network *(B)* surrounding the alveoli, oxygen diffuses into the blood and carbon dioxide diffuses out of the blood. Oxygenated blood then returns to the heart via capillaries *(C)* leading to the pulmonary veins.

Adaptations for Human Respiration

Gas exchange in the alveoli is accomplished by simple diffusion. Oxygen in the bloodstream combines with hemoglobin to form **oxyhemoglobin**. In a reversal of the process that forms oxyhemoglobin, oxygen is released into the cells, where it is used in the chemical process of **aerobic respiration**. At the cells, carbon dioxide formed as a waste product of cellular respiration diffuses into the blood and is carried in the plasma in the form of the **bicarbonate ion**. Carbon dioxide and water vapor are released from the blood into the alveoli for removal from the body.

At the cellular level, food molecules are broken down chemically into energy and certain waste products. When this process occurs in the absence of molecular oxygen, the process is known as **anaerobic respiration**. Both anaerobic and aerobic respiration are discussed in detail on pages 178–180.

Excretion is how human beings remove metabolic wastes from their cells and release them into the environment. The by-products of metabolic activity in humans may be harmful to the human body if they are not removed from where they are produced. Excretion is carried out by a set of specialized excretory tissues and organs, with the help of the transport system. The organs that assist the process of excretion are as follows.

- **Lungs**—The lungs are responsible for the excretion of carbon dioxide and water vapor, which result from the process of aerobic respiration. These materials diffuse out of the blood into the alveoli and are expelled in the breathing process when exhaling.
- **Liver**—The liver's role in excretion includes the recycling of worn-out red blood cells and the production of urea, which results from the breakdown of amino acids. The chemical breakdown of amino acids is known as **deamination**.

- **Sweat glands**—The sweat glands of the skin play a role in excretion by removing water, salts, and urea from the blood and excreting them as perspiration. Heat, a by-product of metabolism, is also removed from the tissues by the sweat glands. As water in the perspiration evaporates, it carries with it body heat that would otherwise overheat the cells.

- **Kidneys**—The kidneys in the urinary system are major excretory organs in human beings. They help regulate the chemical composition of the blood and thereby the chemical composition of the body tissues. Two arteries that branch off the aorta carry blood to the kidneys for filtering. These arteries quickly branch into many capillary networks, each known as a **glomerulus**. The glomerulus is nested inside a cup-shaped structure, **Bowman's capsule**, where many soluble blood components (including water, salts, urea, and soluble nutrients) are absorbed from the blood by diffusion. Bowman's capsule is part of a larger structure, known as the **nephron**, that is the functional unit of the kidney. In a looped portion of each nephron, active transport is used to reabsorb most of the soluble nutrient molecules, certain mineral ions, and some water. These reabsorbed components are returned to the blood flowing out of the kidney by way of veins to the vena cava. The concentrated mixture of waste materials remaining in the nephron, including water, salts, and urea, is known as urine.

- **Ureters**—The two ureters, which are small tubes, conduct urine from the kidneys through the lower abdomen to the urinary bladder.

- **Urinary bladder**—The urinary bladder collects urine from the ureters and stores it for periodic excretion from the body.

- **Urethra**—The urethra is a small tube that leads from the urinary bladder to the outside of the body. Urine is released to the environment through the urethra.

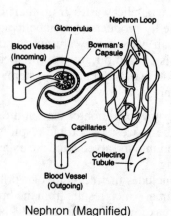

Nephron (Magnified)

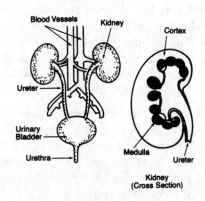

Urinary System

QUESTION SET 2.2—HUMAN PHYSIOLOGY 2
(ANSWERS EXPLAINED, P. 286)

1. When humans exhale, air passes from the trachea directly into the
   (1) bronchioles
   (2) alveoli
   (3) bronchi
   (4) pharynx

2. Which human excretory structure aids in the maintenance of normal body temperature?
   (1) sweat gland
   (2) nephron
   (3) liver
   (4) urinary bladder

3. What is the principal function of structure X represented in the diagram below?

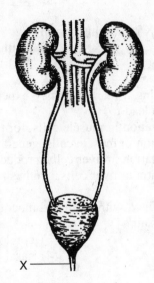

   (1) filtration of cellular wastes from the blood
   (2) transport of urine out of the body
   (3) storage of urine
   (4) secretion of hormones

4. The diagram below represents part of a capillary in a specific region of the human body.

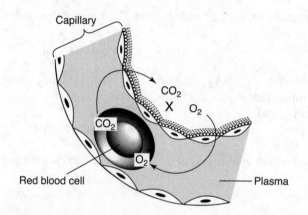

The region labeled X represents part of
(1) a glomerulus (3) a villus
(2) an alveolus (4) the liver

5. An individual running a marathon may experience periods of oxygen deprivation that can lead to
(1) anaerobic respiration in muscle cells, forming lactic acid
(2) aerobic respiration in muscle cells, generating glycogen
(3) anaerobic respiration in liver cells, producing glucose
(4) aerobic respiration in liver cells, synthesizing alcohol

6. Which structure is lined with a ciliated mucous membrane that warms, moistens, and filters air?
(1) pharynx (3) epiglottis
(2) alveolus (4) nasal cavity

7. In humans, the ureter transports urine from the
(1) blood to the kidney
(2) liver to the kidney
(3) kidney to the urinary bladder
(4) urinary bladder to outside the body

8. Which set of symptoms would most likely lead to a diagnosis of asthma?
(1) enlargement and degeneration of the alveoli
(2) constriction of the bronchial tubes and wheezing
(3) inflammation and swelling of the epiglottis
(4) constriction of the nasal cavity and watery eyes

9. Which structure shown in the diagram below contracts, causing a pressure change in the chest cavity during breathing?

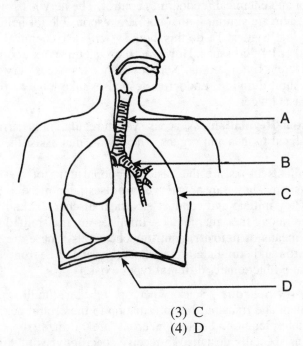

- A
- B
- C
- D

(1) A   (3) C
(2) B   (4) D

10. What is the function of the nephron?
   (1) It breaks down red blood cells to form nitrogenous waste.
   (2) It regulates the chemical composition of the blood.
   (3) It forms urea from the waste products of protein metabolism.
   (4) It absorbs digested food from the contents of the small intestine.

Regulation is the life function by which human beings control and coordinate other life functions to maintain existence. Regulation involves both nerve control and chemical (endocrine) control. The nervous system and the endocrine system are similar in that both are responsible for helping to regulate the body's activities. To do this, both systems use chemical messengers that affect other living tissues. However, the systems work at different rates and over different time periods. Nervous responses occur very rapidly and have a very short duration. Endocrine responses take longer to initiate, but their effect lasts longer.

Nervous System Regulation—Nervous system regulation is carried out by a set of specialized tissues and organs. The organs that assist the process are as follows:

- **Neurons**—Neurons are the basic functional units of the human nervous system. They transmit **nerve impulses** from place to place in the body. They initiate and conduct electrical discharges, known as nerve impulses, along their membranes. Impulses are transmitted from cell to cell by means of **neurotransmitter** chemicals. These are secreted by one neuron and stimulate an impulse in a second neuron. Three structurally and functionally different types exist.

  ➤ **Sensory neurons**—Sensory neurons receive stimuli from the environment and transmit this information to the central nervous system for interpretation. Sensory neurons are normally concentrated in organs specially designed to receive specific types of stimuli. These sensory receptor organs include the eyes, ears, nose, tongue, and skin.

  ➤ **Interneurons**—Interneurons exist primarily in central nervous system organs, although they may also be found in nerve centers in other parts of the body. The interneurons are responsible for interpreting sensory impulses brought to them by the sensory neurons. Interneurons also transmit commands to those motor neurons leading back to the body's effector organs (muscles or glands).

  ➤ **Motor neurons**—Motor neurons carry impulses from the command center in the central nervous system to effector organs (muscles or glands), where an appropriate response is initiated.

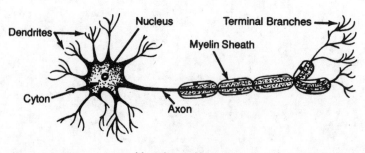

Vertebrate Neuron

86

- **Nerves**—Nerves are bundles of neurons that may contain a single type of neuron (sensory or motor nerves) or two separate types (mixed nerves). A fatty sheath protects the neurons in a nerve from coming into contact with each other and short-circuiting the impulses they carry. Nerves are specialized for conducting impulses over comparatively long distances at high rates of speed.
- **Brain**—The brain is a large organ composed of a mass of interneurons located within the cranial cavity in humans. The brain is one of the most highly specialized parts of the human body. It is responsible for regulating everything from the simplest to the most complex human activity. The brain is subdivided into three major regions, each responsible for a specific aspect of human behavior. The **cerebrum** is responsible for conscious thought, memory, sense interpretation, reasoning, and other voluntary activities. The **cerebellum** is responsible for coordinating muscular activities and helping the body maintain physical balance relative to its surroundings. The **medulla oblongata** is responsible for the regulation of automatic activities including heartbeat, breathing, blood pressure, and peristaltic activity.
- **Spinal cord**—The spinal cord is continuous with the brain. It extends downward from the base of the brain along the dorsal surface of the body. It is encased within the bony vertebral column, which protects it from mechanical damage. The major functions of the spinal cord include connecting the brain to the peripheral nerves and coordinating the reflex response. Reflexes are simple, inborn, involuntary patterns of behavior that permit immediate, unthinking responses to potentially dangerous situations.

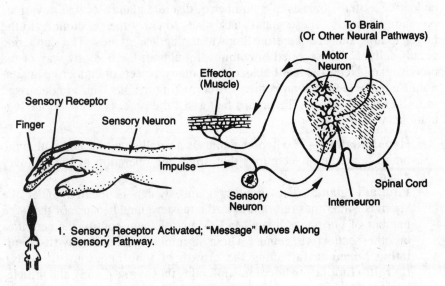

Reflex Arc

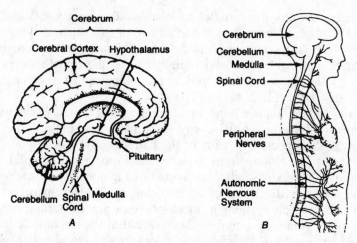

Central Nervous System

- **Peripheral nervous system**—The peripheral nervous system consists of all types of nerves that branch through the body from the central nervous system. The peripheral nervous system is organized into two divisions. The **somatic nervous system** includes the nerves that control the actions of the voluntary skeletal muscles. The **autonomic nervous system** consists of nerves regulating automatic functions such as the actions of glands and involuntary muscle.

Endocrine Regulation—Endocrine regulation is carried out by a set of specialized tissues and organs, with the help of the transport system. The endocrine system consists of a number of discrete glands located at various points in the body. These glands lack ducts to carry their secretions to the target tissues and are therefore known as ductless glands. The endocrine glands deliver their secreted **hormones** throughout the body by way of the bloodstream. **Negative feedback** controls many aspects of endocrine regulation. Negative feedback and other self-correcting mechanisms are discussed in further detail below. The organs that assist the process of endocrine regulation are as follows.

- **Hypothalamus**—The hypothalamus is a small gland located within the brain. It produces secretions that affect the operation of the pituitary gland.
- **Pituitary gland**—The pituitary gland, which is located under the brain, is sometimes referred to as the master gland because of the large number of hormones it produces. Many of them control the activities of other endocrine glands. These hormones include **growth-stimulating hormone** (affecting the growth of long bones in the body), **thyroid-stimulating hormone** (affecting the thyroid gland), and **follicle-stimulating hormone** (affecting the ovaries in females).

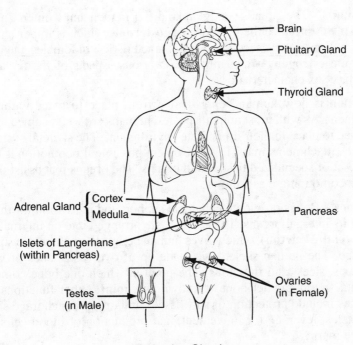

Brain
Pituitary Gland
Thyroid Gland
Adrenal Gland { Cortex
Medulla
Pancreas
Islets of Langerhans
(within Pancreas)
Testes
(in Male)
Ovaries
(in Female)

Endocrine Glands

- **Thyroid gland**—The thyroid gland is located in the neck region, surrounding the trachea. Its principal hormone is **thyroxin** (affecting the general metabolic rate of the body).
- **Parathyroid gland**—The parathyroid glands are embedded within the thyroid gland. The hormone produced by the parathyroids is **parathormone** (affecting the metabolism of calcium in the body).
- **Adrenal gland**—The adrenal glands, located on the kidneys within the abdomen, consist of two separate regions, the **cortex** (outer region) and the **medulla** (inner region). The adrenal cortex secretes steroid hormones that regulate water balance and blood pressure as well as the conversion of complex protein and fat molecules into simpler glucose. The adrenal medulla secretes the hormone **adrenaline**. It increases the rates of metabolism, heartbeat, and breathing and stimulates the conversion of complex glycogen molecules into simpler glucose.
- **Islets of Langerhans**—The islets of Langerhans, small groups of glandular tissue scattered throughout the pancreas, produce the hormones **insulin** and **glucagon**. These two hormones have opposite effects on the storage of sugar in the liver and muscles of the body. Insulin promotes the storage of excess blood sugar as glycogen, thus lowering blood sugar levels. Glucagon stimulates the conversion of glycogen back into glucose, thus raising blood sugar levels.

- **Gonads**—The gonads (sex glands) differ in men and women. In males, the **testes** produce the hormone **testosterone**, which is important in the promotion of male secondary sex characteristics. In females, the **ovaries** secrete **estrogen**, which influences various aspects of the female secondary sex characteristics.

Locomotion is how human beings move from place to place within their environment. As with most animal species, humans are well adapted to move from place to place in their immediate environment. The skeletal system and its muscle attachments make this movement possible. Locomotion is carried out by a set of specialized tissues and organs. The organs that assist the process of locomotion are as follows.

- **Bones**—Bones provide mechanical support and protection for the body and its internal organs. These bones are arranged into an internal skeleton (**endoskeleton**), which gives human beings their characteristic body shape. The human skeleton is made up of over 200 bones of varying shapes, sizes, and functions. The point at which one bone comes into contact with another bone is known as a **joint**. Other functions of the bones include protection of soft tissues and organs, anchorage sites for muscles, leverage for movement, and production of blood cells in the bone marrow.

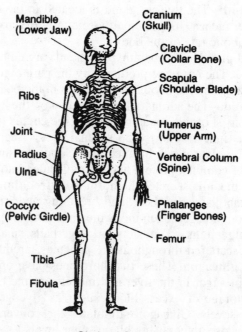

Skeletal System

- **Cartilage**—Cartilage is a flexible, fibrous connective tissue that pads the joints between bones. Cartilage is found in many of the flexible parts of the body, including the outer ear and the nose. In embryos and very young children, cartilage is abundant, making up the major portion of the skeleton. In adults, it is much reduced, being found only at bone ends and in flexible portions of the body such as the trachea.

- **Muscles**—Muscles in the human being are of three types. **Visceral** and **cardiac muscles** are involuntary muscles controlled primarily by the autonomic nervous system. **Skeletal muscles** are voluntary and are those most directly involved in human locomotion. Skeletal muscles normally work in opposing pairs, each pulling a joint in a different direction. By coordinating these muscle pairs, the body is able to move skeletal levers to accomplish movement from place to place. Muscles that extend (open) joints are known as **extensors**, while muscles that flex (close) joints are known as **flexors**.

  Muscles require energy in order to contract. During heavy exercise, oxygen supplies may not be sufficient to release energy to the muscle cells by aerobic means. At such times, the muscle cell has the capacity to use **lactic acid fermentation** to supply additional energy. The buildup of lactic acid in the muscle tissues is known as **muscle fatigue**.

- **Tendons**—Tendons are responsible for attaching muscles to the bones of the skeleton. They are composed of tough, inelastic connective tissue.

- **Ligaments**—Ligaments attach bones to other bones at the skeletal joints and are composed of tough, elastic connective tissues that are able to stretch as the joints flex.

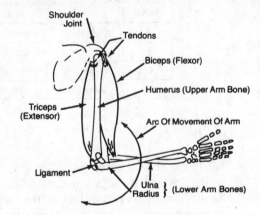

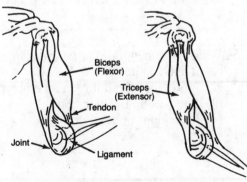

Elbow Joint

QUESTION SET 2.3—HUMAN PHYSIOLOGY 3
(ANSWERS EXPLAINED, P. 289)

1. During a race, the body temperature of a runner increases. The runner responds by perspiring, which lowers body temperature. This process is an example of
   (1) maintenance of homeostasis
   (2) an antigen-antibody reaction
   (3) an acquired characteristic
   (4) environmental factors affecting phenotype

2. Which substances are secreted at the endings of nerve cells?
   (1) antibodies   (3) neurotransmitters
   (2) antigens   (4) lipids

3. Which row in the chart below contains the words that best complete this statement?

   "The _I_ glands produce _II_, which are transported by the _III_ system."

Row	I	II	III
A	Digestive	Hormones	Circulatory
B	Endocrine	Enzymes	Lymphatic
C	Endocrine	Hormones	Circulatory
D	Digestive	Enzymes	Lymphatic

   (1) A   (3) C
   (2) B   (4) D

4. All living things carry out a variety of life functions such as coordination, excretion, digestion, circulation, and synthesis. Select *two* of the life functions listed. Define the two life functions you selected and explain how they interact to keep an organism alive. [4 points]

   _____

5. Which statement does *not* correctly describe a function of cartilage?
   (1) It anchors muscles to bones.
   (2) It provides flexibility in an embryo.
   (3) It makes up the outer ear.
   (4) It cushions bones at a joint.

6. A reflex arc is illustrated in the diagram below.

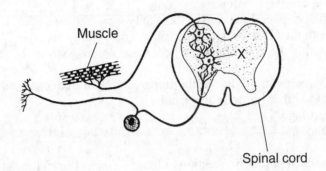

Structure X represents
(1) an effector
(2) a motor neuron
(3) an interneuron
(4) a receptor

7. One of the functions of the human endoskeleton is to
(1) transmit impulses
(2) produce blood cells
(3) produce lactic acid
(4) store nitrogenous wastes

8. A human skeleton is shown in the illustration below.

The elongation of structures A and B was stimulated by a hormone produced by the

(1) islets of Langerhans
(2) liver
(3) pituitary gland
(4) striated muscles

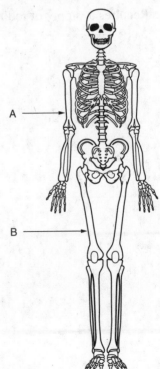

9. When a child runs to his/her mother after hearing a clap of thunder, the child is using
   (1) the central nervous system only
   (2) the peripheral nervous system only
   (3) both the central and the peripheral nervous systems
   (4) neither the central nor the peripheral nervous systems

10. The humerus, the bone in the upper arm of a human, is directly connected to other bones in the arm by
    (1) cartilage
    (2) tendons
    (3) extensors
    (4) ligaments

11–12. For each phrase in questions 11 and 12, select from the list below the answer best described by that phrase. Then record its number.

<u>*Endocrine Glands*</u>
(1) thyroid
(2) adrenal
(3) islets of Langerhans
(4) parathyroid

11. Secretes a hormone in times of emergency, accelerating metabolic activities

12. Requires a supply of iodine to synthesize its hormone

II. HOMEOSTASIS

The components of the human body, from organ systems to cell organelles, interact to maintain a balanced internal environment. To accomplish this successfully, organisms possess a diversity of control mechanisms that detect deviations and make corrective actions.

When all components of the human body are functioning at peak efficiency, each helps to establish a balanced set of conditions that supports the continuation of life. This balance, or steady state, is known as **homeostasis**. In this state, all systems work together to perform their specialized functions and to synthesize their specialized products. The resulting products are used to perform other life processes. The human body contains sensory tissues that can detect and measure the concentrations of many organic and inorganic compounds. When inappropriate levels of these materials are found in the body, a response is initiated that helps to correct the imbalance and reestablish homeostasis. Many different mechanisms, known as **feedback loops**, exist to assist in the maintenance of homeostasis. Examples of feedback loops include the following.

- As carbon dioxide builds up in the body tissues as a result of aerobic respiration, it diffuses into the blood for elimination from the body as respiratory waste. Elevated concentrations of dissolved carbon dioxide are detected by the central nervous system. Commands are sent to the breathing mechanisms that increase the breathing rate. The increased breathing rate allows dissolved carbon dioxide to exit the bloodstream at an accelerated pace and introduces atmospheric oxygen into the blood at the same time. As the central nervous system detects a reduced carbon dioxide concentration in the blood, commands are sent that slow the breathing rate to normal levels as homeostatic balance is reestablished once more.

- When human skin comes into contact with a hot surface, the body's tissues may be damaged as a result. A **reflex arc** consisting of a sensory neuron, interneuron, and motor neuron respond to this danger. The sensory neuron detects the stimulus and identifies it as heat. This information is transmitted to an interneuron in the spinal cord. A command response is then initiated that is sent along a motor neuron to muscles. The muscles are commanded to move the part of the body away from the hot surface. This reflex action is involuntary and very rapid. The quick response helps to protect the skin and underlying tissues from damage and maintain homeostatic balance.

- The mechanism that controls many aspects of endocrine regulation is **negative feedback**. It operates on the principle that the effects of a particular hormone may inhibit the further production of that hormone but may stimulate the production of another hormone. A good example of negative feedback is the insulin-glucagon feedback loop. Elevated sugar levels in the blood stimulate the production of insulin (which functions to store glucose in the liver) while inhibiting the production of glucagon. As blood sugar levels drop as a result of insulin's action, glucagon production increases. As a result, glucose is released from the liver, while insulin production is inhibited.

- Sensors in the central nervous system are specialized to monitor the temperature of the blood. When this temperature exceeds 98.6°F (38°C), by a few tenths of a degree, the central nervous system sends out a series of commands that have various effects. Breathing deepens and quickens, allowing heat to escape from the body by means of exhaled air and water vapor. Sweat glands on the face and body open. Perspiration beads begin to appear on the skin surface, where they can evaporate and cool the skin further. By initiating these responses, the body avoids overheating and damaging sensitive tissues throughout. This protection is important in maintaining overall homeostatic balance.

If there is a disruption in any human system, there may be a corresponding imbalance in homeostasis. When the balance of the body's systems is disrupted, cells, tissues, and their associated organs are affected. This may lead to failure of one or more organs or systems within the body. When such a failure occurs, the homeostatic balance of the entire organism can be thrown off. This can lead to illness or death, depending on the severity of the imbalance.

For example, untreated high blood pressure can damage the nephrons, the functional units of the human kidney. If the high blood pressure condition persists for a long period of time, the tissues of the kidneys can become so badly damaged that the kidneys cease functioning altogether. As the kidneys stop functioning to filter the blood of urea and salts, the concentration of these wastes builds up to unsafe levels in the blood. As this unfiltered blood reaches other body tissues, it poisons them and causes them to slow or stop functioning. Eventually, the body's tissues become so badly poisoned with toxic metabolic wastes that they can no longer support life activity. At this point, organs and body systems fail, and the person may die if he/she is not treated. See the section "Disease as a Failure of Homeostasis" for more information about disease as a failure of homeostasis (see page 184).

III. CELL FUNCTION AND STRUCTURE

The organs and systems of the body help to provide all the cells with their basic needs. The cells of the body are of different kinds and are grouped in ways that enhance how they function together. Each of the body systems and its associated organs mentioned provides a different functional life activity. These systems and organs work together to support life. Life really begins with the biochemical activity of the individual cells that make up a multi-celled organism such as the human being. Each living cell in the body needs food, water, and oxygen to function appropriately These cells must also be cleaned constantly to free them of toxic metabolic wastes such as urea and carbon dioxide.

To illustrate, consider how the body's systems work together to provide for the needs of an individual cell living deep inside the body of a human being. The digestive system provides food for this cell by breaking down the complex structure of ingested foods into soluble components that can be absorbed by the cell. Oxygen needed by the cell for respiration is absorbed from the atmosphere by the respiratory system. The circulatory system absorbs these soluble food and oxygen molecules from their respective points of entry into the body. It moves them near the cell so the molecules can be absorbed through the cell membrane. The cell metabolizes these molecules, producing cell components and energy needed to support the cell's life activities. The cell also produces a variety of waste materials, including carbon dioxide, salts, and urea, that must be removed from the cell and from its immediate environment. Once again, the circulatory system assists by absorbing these waste molecules and moving them to organs that specialize in the release of wastes to the outside environment. Carbon dioxide is removed from the blood in the respiratory system, while salts and urea are filtered out of the blood by the excretory system. Meanwhile, all these systems are being coordinated by the regulatory system. As a result, a homeostatic balance is struck. All systems operate in a coordinated fashion to sustain life.

Tissues

Each cell in the body is a part of the complex fabric of the organism. Cells are specialized to perform specific functions in the body. Groups of such specialized cells are known as **tissues**. They are often found clustered together within organs where they carry on their specialized activities with great efficiency. Examples of human body tissues include:

- skin tissues that provide an impervious layer surrounding and protecting interior organs;
- bone tissues that secrete bone mineral and maintain the body's endoskeleton;
- blood tissues that transport oxygen and other components as well as protect the body against disease;
- muscle tissues that move skeletal levers during locomotion and specialize in the physical movement within certain body organs;
- nerve tissues that provide a communication link among all body systems; and
- endocrine tissues that produce a variety of chemical substances that control the functioning of other body tissues.

Many other specific tissues operate within the body. Each of the tissue types listed on page 97 can be subdivided into several highly specialized tissues that carry on very narrow roles in maintaining homeostatic balance. For example, the broad category of muscle tissue contains skeletal muscle that moves skeletal levers, visceral muscle that creates internal movements such as peristalsis in the intestines, and cardiac muscle that establishes and maintains rhythmic contractions of the heart and arteries. Blood tissues include red blood cells that carry oxygen, phagocytes that engulf bacteria, lymphocytes that produce antibodies against disease, platelets that specialize to produce clots, and a variety of other cell types, each with its own special function in the body. Nerve tissues are subdivided into sensory neurons that receive environmental stimuli, motor neurons that control muscles and glands, and interneurons that process sensory input and send motor commands.

Cells have particular structures that perform specific jobs. These structures perform the actual work of the cell. Just as systems are coordinated and work together, cell parts must also be coordinated and work together. The cell is extremely complex and is composed of smaller functional parts known as cell organelles. Many of these organelles were unknown to scientists until 20th-century scientific techniques and equipment (electron microscope, ultracentrifuge) made them visible to researchers. Each type of organelle has special work to do that promotes the survival of the cell. In a sense, these tiny structures are similar to organs and organ systems. They provide specialized functions coordinated to establish and maintain homeostasis within the cell. Organelles are specialized to carry on the life functions of nutrition, respiration, circulation, excretion, regulation, reproduction, and locomotion within their particular cell.

Cell Membrane

Each cell is covered by a membrane that performs a number of important functions for the cell. These include separation from its outside environment, controlling which molecules enter and leave the cell, and recognition of chemical signals. The processes of diffusion and active transport are important in the movement of materials in and out of cells. The outer membrane of the cell, which regulates the transport of materials into and out of the cell, is known as the **plasma (cell) membrane**. The plasma membrane also serves as the interface between the cell interior and its outside environment. Recent research points to the plasma membrane's function as a site of cell-cell recognition, which has important implications for the control of disease, including cancer.

The structure of this membrane has been the subject of study for many years. Several models have been developed to describe it. The **fluid-mosaic model** (developed by Singer and Nicholson in 1972) is the currently accepted model for this structure. This model shows the principal component of the plasma membrane to be a double layer of lipid molecules with many embedded proteins, known as receptors. This structure allows the passage of small, soluble molecules such as monosaccharides, amino acids, and dissolved gases

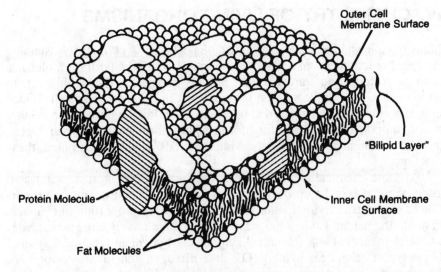

Fluid-Mosaic Model

while preventing the passage of larger molecules (for example, starches and proteins). Current research indicates that size may be only one factor affecting the passage of materials through the cell membrane. Chemical recognition factors are also important and are regulated by a "lock-and-key" mechanism in which chemical receptor sites attract, link with, and transport specific molecules across the membrane boundary to the cell interior.

Passive and Active Transport

All particles in the environment of the cell contain energy of motion (kinetic energy). As these particles collide with the plasma membrane, they may pass through by slipping between the lipid molecules making up the membrane. Because no cellular energy is expended in this absorption process, it is known as **passive transport**. **Diffusion** is a form of passive transport in which a dissolved substance passes through the plasma membrane from a region of higher relative concentration to a region of lower relative concentration of that substance. The diffusion of water molecules into or out of the cell (through the plasma membrane) is known as **osmosis**.

Some cells are capable of expending cellular energy in order to pump materials into (or out of) themselves from regions of lower relative concentration to regions of higher relative concentration. Because cellular energy must be expended to accomplish this process, it is known as **active transport**. The plasma membrane may be involved in active transport through the use of special carrier proteins embedded in the membrane's lipid layers. These carrier proteins take advantage of the specific chemical nature of desired components by attaching to them and carrying them through the membrane to the cytoplasm.

IV. CHEMISTRY OF LIVING ORGANISMS

Many organic and inorganic substances are necessary to permit life-sustaining chemical reactions to take place in the cell. Large organic food molecules such as proteins and starches must initially be broken down (digested to amino acids and simple sugars, respectively) in order to enter cells. Once nutrients enter a cell, the cell will use them as building blocks in the synthesis of compounds necessary for life. In addition, molecular oxygen, water, and other inorganic molecules provide materials needed to complete these reactions.

The same chemical elements that comprise the Earth's crust, water, and atmosphere also make up the bodies of living things. However, the proportion of elements in living matter is different from the proportion of elements in nonliving matter. Living things are made up of relatively large percentages of the elements **carbon** (chemical symbol **C**), **hydrogen** (chemical symbol **H**), **oxygen** (chemical symbol **O**), and **nitrogen** (chemical symbol **N**). Elements found in lower percentages include sulfur (chemical symbol S), phosphorus (chemical symbol P), magnesium (chemical symbol Mg), iodine (chemical symbol I), iron (chemical symbol Fe), calcium (chemical symbol Ca), sodium (chemical symbol Na), chlorine (chemical symbol Cl), and potassium (chemical symbol K).

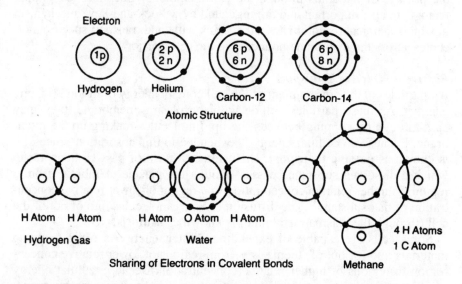

Electron

Hydrogen   Helium

Carbon-12   Carbon-14

Atomic Structure

H Atom   H Atom

Hydrogen Gas

H Atom   O Atom   H Atom

Water

Sharing of Electrons in Covalent Bonds

Methane

4 H Atoms
1 C Atom

Compounds

Compounds are substances composed of two or more different elements bound together chemically. The bonds holding these compounds together may be ionic (involving the transfer of electrons between atoms) or covalent (involving the sharing of electrons between atoms). [*Note: The chemistry of bonding is not a required understanding for Living Environment. For details on this topic, consult a chemistry textbook.*]

Two broad categories of compounds exist in nature: **inorganic** and **organic** compounds. All organic compounds contain the elements carbon and hydrogen. Inorganic compounds may contain any of Earth's elements but rarely contain carbon and hydrogen together. Living things are composed of both inorganic and organic compounds.

Inorganic Compounds

Inorganic compounds commonly found in living things include the following.

- **Water**—Living things consist of 60 to 98 percent water, which acts as a medium for transport and for chemical activities within the cell.
- **Salts**—Salts are chemicals composed of metallic and nonmetallic ions combined by ionic bonds. They are important for maintaining osmotic balance in the cell and for supplying ions necessary for many of the cell's chemical reactions.
- **Acids and bases**—These compounds are important for maintaining the proper balance of hydrogen ion concentration (acidity/alkalinity) in the cell.

Organic Compounds

Organic compounds commonly found in living things include the following.

- **Carbohydrates** are organic compounds and so contain the elements carbon and hydrogen. In addition, they contain the element oxygen. Carbohydrates generally contain hydrogen and oxygen in the ratio of 2:1. This ratio differs from that found in other organic compounds. Like many organic compounds, carbohydrates are actually formed around a skeleton of carbon atoms linked together in a chain. Some chains are quite short (six or fewer carbon atoms), while others are very long (thousands of carbon atoms).

  The simplest stable carbohydrate is the monosaccharide (**simple sugar**). Monosaccharides form around a chain of 5 or 6 carbon atoms. An example of a monosaccharide is the simple sugar **glucose**, whose chemical formula is $C_6H_{12}O_6$. Glucose is the basic chemical unit of nearly all the more complex carbohydrates.

  A more complex form of carbohydrate is the disaccharide (double sugar), composed of two chemically linked monosaccharide molecules. Disaccharides are commonly formed around a chain of 12 carbon atoms. The chemical process by which this linkage occurs is **dehydration synthesis**. The atoms making up water molecules are removed from the monosaccharide molecules and the two monosaccharides join to form a single disaccharide molecule. Examples of disaccharides are maltose and sucrose.

A complex form of carbohydrate is known as a polysaccharide (many sugar). It forms by chemically combining many monosaccharide molecules via dehydration synthesis. The chain of monosaccharide units forming a polysaccharide may be thousands of units long and may be highly branched. Starch (amylose), glycogen, and cellulose are examples of polysaccharides.

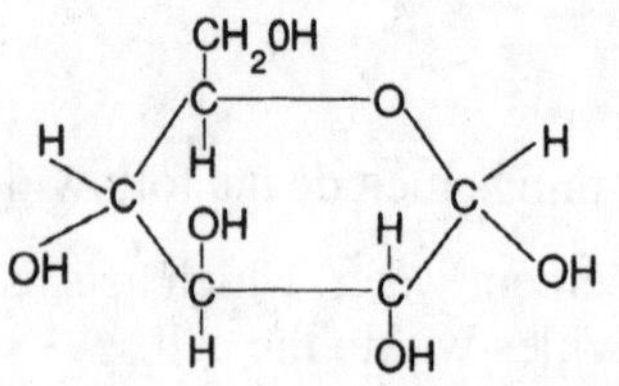

Monosaccharide
Structural Formula: Glucose

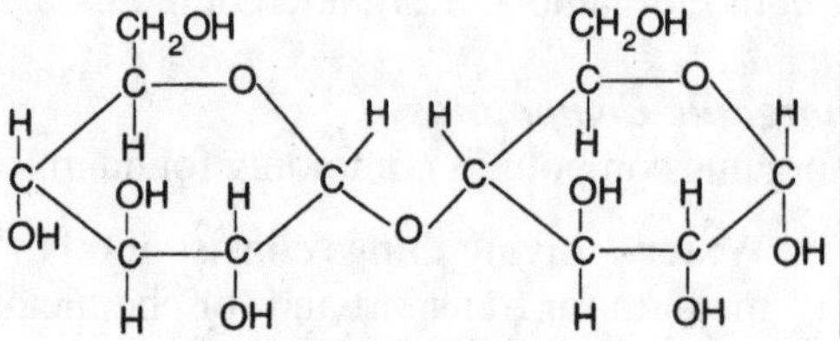

Disaccharide
Structural Formula: Maltose

- **Lipids** are composed of the elements carbon, hydrogen, and oxygen. Many of the more common lipids, known as triglycerides, are constructed of a unit of **glycerol** (a three-carbon alcohol) combined chemically via dehydration synthesis with three molecules of **fatty acid** (a hydrocarbon chain with an attached carboxyl group). Examples of lipids include animal fats, plant oils, and waxes.

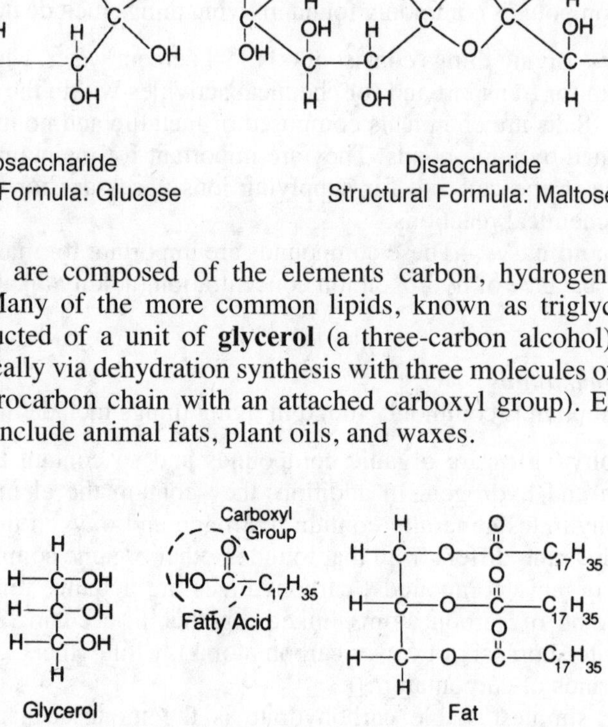

Glycerol

Fat

Lipid Structural Formula: Triglyceride

- **Proteins** contain the elements carbon, hydrogen, and oxygen, as do carbohydrates and lipids. In addition, proteins contain a substantial proportion of nitrogen. Certain proteins may also contain sulfur. Proteins are polymers made up of repeating units of other, simpler compounds known as **amino acids**. Although there are more than 20 different types of amino acids, they all share a common general structure. All amino acids contain an amino group ($-NH_2$) and a carboxyl group ($-COOH$) attached to a central carbon atom. A third attached group (radical) varies and gives each type of amino acid its unique properties.

102

Two amino acid molecules may be joined together chemically by dehydration synthesis to form a dipeptide. The term **peptide** refers to the name of the actual chemical bond between carbon and nitrogen atoms (C–N) that joins the two amino acid units together. As more and more amino acids link together by dehydration synthesis, an amino acid chain, known as a **polypeptide**, is formed. Polypeptide chains form the basis of protein molecules. Because of the almost endless variations in which amino acids may be arranged in a polypeptide, proteins are found in thousands of different forms. The extreme variability of proteins is thought to be the chemical basis for the individual variations in living things. Examples of proteins are structural proteins and enzymes. See below for additional information about the mechanisms of protein synthesis.

R = variable (radical) group of atoms

Structural Formula: Amino Acid

Structural Formula: Dipeptide Formation

- **Nucleic acids** are extremely complex polymers that function to determine the characteristics of individual cells and thereby the characteristics of the entire organism of which they are a part. Principal nucleic acids include **deoxyribonucleic acid (DNA)** and **ribonucleic acid (RNA)**. Nucleic acids are polymers composed of a simpler chemical subunit known as a **nucleotide**. Thousands of these units comprise a single DNA molecule, making DNA one of the largest and most complex of all organic compounds. DNA exists in hundreds of thousands of different forms, depending on the precise arrangement of nucleotides in the molecule. Its variability and its role in protein synthesis are the keys to genetic variation in living things.

DNA nucleotides themselves are quite complex. They are composed of three separate subunits: phosphate group, deoxyribose, and one of four nitrogenous bases, including adenine (A), thymine (T), cytosine (C), and guanine (G). DNA's function in the cell is to code for the production of specific protein types in the cell. This function is discussed in more detail in the section "Genetic Material." RNA nucleotides are similar to those comprising DNA molecules, but substitute ribose sugar for deoxyribose and uracil (U) in place of thymine.

Chemical Reaction: Hydrolysis

When taken into the body as food, complex food molecules such as starches, proteins, fats, and nucleic acids must be broken down into their simpler chemical subunits before they can be used by the living cell. These large molecules cannot pass readily through the plasma membrane, but their smaller, soluble subunits can. The act of breaking down these complex molecules is known as chemical digestion and is accomplished by the chemical process of **hydrolysis**. Hydrolysis (splitting with water) involves replacing existing chemical bonds within a complex molecule with the atoms comprising a water molecule, thereby breaking the complex molecule apart at those points and producing simpler molecules. The following word equations will illustrate this process:

$$\text{starch} + N \text{ water} \longrightarrow N \text{ glucose}$$
$$\text{maltose} + \text{water} \longrightarrow 2 \text{ glucose}$$
$$\text{lipid} + 3 \text{ water} \longrightarrow \text{glycerol} + 3 \text{ fatty acids}$$
$$\text{protein} + N \text{ water} \longrightarrow N \text{ amino acids}$$

The end products of hydrolysis are as follows:

Complex Food Molecules	Molecular End Products
Carbohydrates (starches, double sugars)	Simple sugars
Lipids (fats, oils)	Fatty acids and glycerol
Proteins	Amino acids
Nucleic acids	Nucleotides

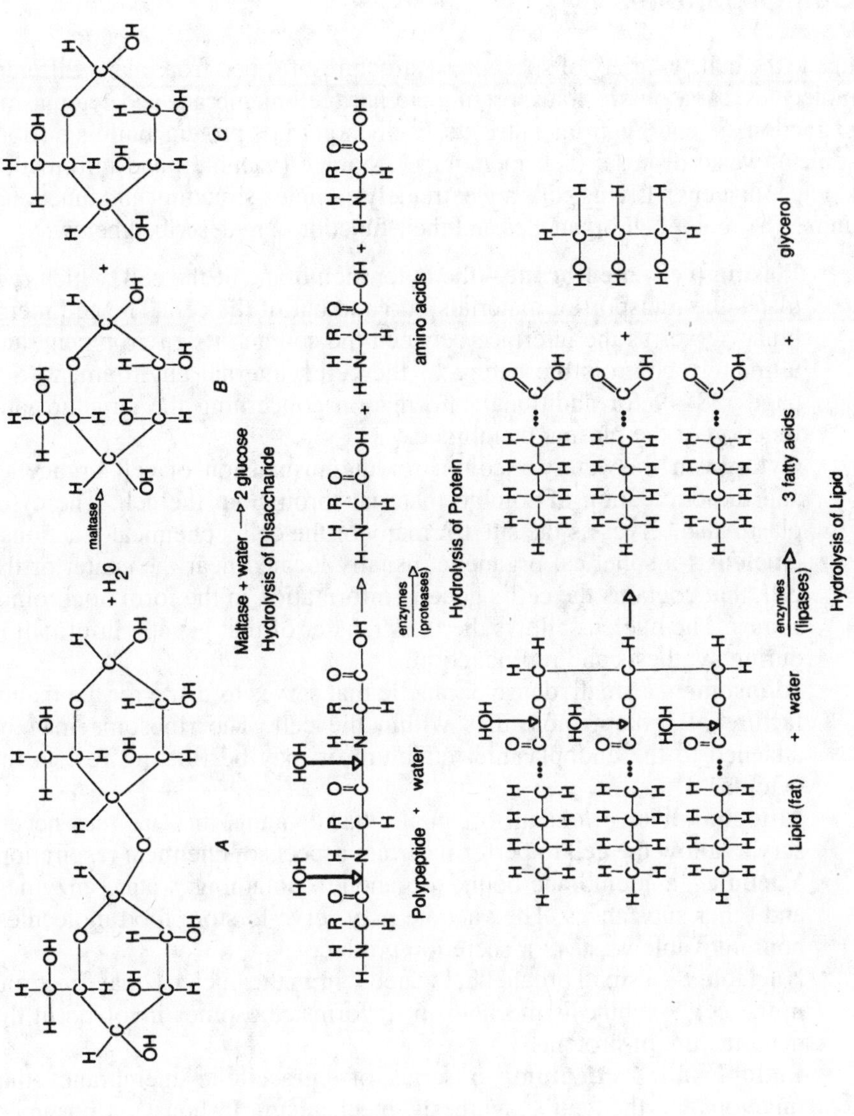

C

B

A

+ H_2O

maltase

Maltase + water ---> 2 glucose

Hydrolysis of Disaccharide

amino acids

enzymes (proteases)

Hydrolysis of Protein

Polypeptide + water

glycerol

3 fatty acids

Hydrolysis of Lipid

enzymes (lipases)

Lipid (fat) + water

Hydrolysis

Cell Organelles

Inside the cell, a variety of specialized structures, formed from many different molecules, carry out the transport of materials (cell membrane and cytoplasm), extraction of energy from nutrients (mitochondria), protein building (ribosomes), waste disposal (cell membrane), storage (vacuole), and information storage (nucleus). Living cells are extremely complex structural and functional units. The major cell organelles and their functions are described below.

- **Plasma (cell) membrane**—the outer membrane of the cell, which regulates the transport of materials into and out of the cell. The cell membrane serves as the interface between the cell and its environment, and ultimately controls the nature of the cell's internal environment. See pages 98–99 for additional information concerning the structure and function of the plasma membrane.
- **Cytoplasm**—a watery medium for the suspension of cell organelles and the circulation of soluble material throughout the cell. The cytoplasm also serves as the site for many of the cell's chemical reactions.
- **Nucleus**—a spherical organelle, usually located near the center of the cell, that contains the cell's genetic information in the form of chromosomes. The nucleus allows the free transfer of that genetic information during synthesis and reproduction.
- **Ribosome**—a small, dense organelle that serves as a site for the manufacture of protein molecules within the cell. The ribosome may be attached to the endoplasmic reticulum or may be floating free in the cytoplasm.
- **Mitochondrion**—a small organelle that contains the enzymes necessary to allow the cell to perform certain aspects of chemical respiration.
- **Vacuole**—a membrane-bound organelle containing water, enzymes, and other substances. The vacuole may serve to store food molecules, nonremovable wastes, or secretion products.
- **Nucleolus**—a small organelle, located within the nucleus, that functions in the cell's synthesis mechanism. It forms ribosomes involved in the manufacture of proteins.
- **Endoplasmic reticulum**—a series of intracellular membranes that functions in the cell's synthesis mechanism. It houses ribosomes, accepts manufactured proteins, and transports these proteins to the plasma membrane for incorporation into the membrane or for secretion to the cell exterior.
- **Golgi complex**—a series of membrane-bound organelles that functions in the cell's synthesis mechanism. It accepts manufactured proteins and transports these proteins to the plasma membrane for incorporation into the membrane or for secretion to the cell exterior.
- **Lysosome**—a specialized vacuole that aids nutrition by carrying digestive enzymes and by merging with food-containing vacuoles. The lysosome may also help to recycle aging or defective cells.

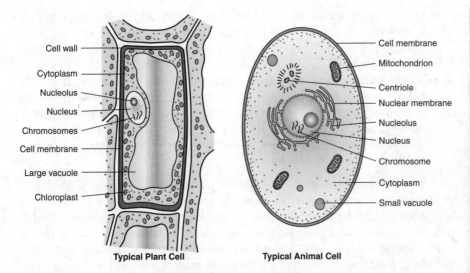

Plant Versus Animal Cells

- **Centriole**—a cylindrical structure, found primarily in animal cells, that apparently functions in the process of cell division.
- **Chloroplast**—a chlorophyll-containing structure, found primarily in plant and algae cells, in which the chemical reactions of photosynthesis occur.
- **Cell wall**—a structure, found primarily in plants, that provides mechanical support and protection for the cell.

Cellular Interaction

Receptor molecules play an important role in the interactions between cells. Two primary agents of cellular communication are hormones and chemicals produced by nerve cells. If nerve or hormone signals are blocked, cellular communication is disrupted and the organism's stability is affected.

Receptor molecules are specialized proteins embedded in the plasma membrane. These receptor proteins are able to capture specific types of molecules that come into contact with the membrane and move them to the cell interior. Once inside the cell, these captured molecules can affect the behavior of the cell if they have the capacity to do so. Two types of molecules that have this capacity are hormones and neurotransmitters.

- **Hormones** are biochemicals produced and secreted by endocrine glands. Each hormone exerts its effect on specific target tissues in the body. In order to be absorbed by the individual cells of a target tissue, these neurotransmitters must be captured by receptor molecules and moved into the cell interior. Endocrine glands are located far from the tissues that they affect, and they have no tubes to connect them to these tissues. Since no ducts carry them, hormones are distributed to other parts of the body by means of the circulatory fluid.

- **Neurotransmitters** are biochemicals produced and secreted by the neuron's terminal branches. They carry the nerve impulse from one neuron to the next. An example of such a neurotransmitter is acetylcholine. In order to be absorbed by the dendrites of a subsequent neuron, these neurotransmitters must be captured by receptor molecules and moved into the cell interior.

Because these chemicals function in controlling the body's metabolic processes, they are essential in maintaining homeostatic balance. When an endocrine gland fails to produce its hormone in appropriate quantities, the life process controlled by it does not operate properly. This can lead to organ or system failure. It may even threaten the survival of the entire organism. For example, if the pancreas malfunctions and causes insulin production to cease, then blood sugar is not converted to glycogen. This will cause the sugar concentration to build in the blood until it reaches dangerous levels. Tissues throughout the body may be affected, but some, such as kidney tissues, may be destroyed and cease to function. If the kidneys fail, then the bloodstream has no way to cleanse itself, and all the body's tissues will be damaged or destroyed.

Unicellular Life Forms

Essential Question:

- *How are unicellular life forms able to carry out life processes needed to sustain life?*

Performance Indicator 4.1.3 *The student should be able to explain how a one-celled organism is able to function despite lacking the levels or organization present in more complex organisms.*

The organellar structures present in some single-celled organisms (known as **protists** and **monera**) act in a manner similar to the tissues and systems found in multicellular organisms. This enables single-celled organisms to perform all of the life processes needed to maintain homeostasis. While these protists and monera exist in nature in the millions, a few species are well-known to scientists. These include the **paramecium**, the **ameba**, and the euglena. The paramecium is a slipper-shaped protist that inhabits freshwater environments. At nearly 400 micrometers (400 millionths of a meter), one of the largest of the paramecia is *Paramecium caudatum*. Its organelles function efficiently to allow the organism to thrive in its environment.

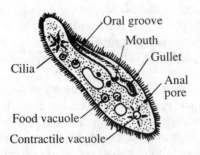

Paramecium

- Nutrition depends on the oral groove and food vacuoles. The oral grooves allow the paramecium to ingest food particles. The food vacuoles permit the digestion of these particles to a simple chemical form that can be used by the cell. Egestion is accomplished by means of the anal pore.
- Transport is accomplished by means of the plasma membrane and the cytoplasm, which permit the absorption and circulation of materials.
- Respiration takes place in the mitochondria, which contain the enzymes needed to release energy from food molecules.
- Excretion of metabolic wastes occurs directly though the plasma membrane. A specialized organelle, the contractile vacuole, expels excess water from the cell.
- Regulation is the function of the nucleus, which manufactures proteins that control the cell's metabolic activities.
- Locomotion is accomplished in paramecia by cilia that sweep in a coordinated motion to propel the cell through the watery environment.

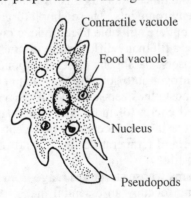

Ameba

The ameba, *Amoeba proteus*, is also a relatively large freshwater protozoan but lacks a defined shape. The ameba is also successful in its environment because it contains structural adaptations on the organella level that allow it to carry on the basic life functions efficiently.

- Nutrition is aided by the pseudopods and food vacuoles. The pseudopods allow the ameba to ingest food particles. The food vacuoles permit the digestion of these particles to a simple chemical form that can be used by the cell. Egestion is accomplished by means of reverse pinocytosis at the cell membrane.
- Transport is accomplished by means of the plasma membrane and the cytoplasm, which permit the absorption and circulation of materials.
- Respiration takes place in the mitochondrion, which contains the enzymes needed to release energy from food molecules.
- Excretion of metabolic wastes occurs directly though the plasma membrane.
- Regulation is the function of the nucleus, which manufactures proteins that control the cell's metabolic activities.
- Locomotion is accomplished in ameba by the use of pseudopods, which flow forward on a solid surface to propel the cell through its environment.
- **Algae** is a term used to refer to a large number of unicellular and colonial organisms that resemble plants in their ability to carry on photosynthesis as their nutritional process. Some algae (kelp, seaweeds) form large aggregations that mimic true multicellular plants and contain specialized structures that resemble, in appearance and function, leaves, stems, and roots. Algae are found in virtually all aquatic and marine ecosystems where light is present and are responsible for providing a significant amount of Earth's atmospheric oxygen. Like other unicellular organisms, algae are capable of performing all of the basic functions necessary to sustain life.
- **Bacteria** are simple unicellular organisms that inhabit many of Earth's least-hospitable environments. These simple organisms generally lack organized nuclei but are still able to reproduce rapidly under favorable conditions. Bacteria fill many different environmental niches, including roles as decomposers, pathogens (disease organisms), and converters of atmospheric nitrogen into soil nutrients that help plants grow. Some bacteria are chemosynthesizers capable of deriving cell energy from the chemical bond energy of inorganic compounds. A few species of bacteria even thrive in deep-ocean thermal vents under conditions of pressure, toxicity, and temperature that would prove lethal to almost any other form of life.
- **Fungi** are organisms whose structures range from unicellular to loosely multicellular. Fungi occupy an essential niche in the natural environment, along with bacteria, as nature's decomposers and recyclers. In this role, they break down the complex bodies of other living things and recycle them for reuse in the environment. Fungi gain their nutrition by sending root-like structures (mycelia) into the living or dead organic tissues of plants and/or animals and secreting digestive enzymes that

break down the complex tissues into simple, absorbable nutrient molecules. Fungi include molds, yeasts, rusts/smuts, and mushrooms.

Multicellular Life Forms

More familiar to us due to their size and proximity to human experience are the many species of multicellular organisms that inhabit Earth. Multicellular organisms may contain as few as hundreds to as many as several millions of cells. These cells are typically specialized to perform unique functions in the organisms. Despite this specialization, each cell must be able to perform the essential functions necessary to sustain life. It is not the intention of this book to provide a comprehensive survey of all multicellular species, or even all major groupings of these species. Following is a brief overview of the major species groupings (phyla and selected classes), which may be expanded upon at the discretion of the classroom teacher.

Animal-Like Organisms

Animals, in general, are multicellular, heterotrophic ("other-feeding") organisms. Because their nutritional habit involves consuming the bodies of other organisms for food, animals are known collectively as **consumers**. Animals that consume plants are known as **herbivores**, while those that consume animals are known as **carnivores**. Two major subdivisions of the animal kingdom include invertebrates and vertebrates (chordates).

Invertebrates (non-chordates) are those animal-like organisms that lack a chord or spine as part of their structure. Many invertebrate species are soft-bodied, while others develop a rigid exoskeleton to protect inner organs and to provide levers for locomotion. Invertebrates include phyla such as porifora (e.g., sponges), coelenterates (e.g., coral, jellyfish, hydra), flat and round worms (e.g., planaria, nematodes), echinoderms (e.g., sea urchins, sand dollars, starfish), annelids (e.g., earthworms), and arthropods (e.g., grasshoppers, orb spiders), among several others.

Chordates include true chordates and vertebrates. True chordates (e.g., lancelet, amphioxus) are simple multicellular animals that resemble small worms except for the significant inclusion of a simple notochord located along their dorsal aspect. The notochord provides an anchor for the attachment of simple muscles and support of body shape. The earliest direct ancestors of vertebrates are thought by scientists to have resembled these simple chordates. **Vertebrates** are chordates whose dorsal supporting structure has evolved into a jointed adaptation (vertebral column) of cartilage or bone. The vertebral column provides multiple functions for these animals, including basic body support and articulation, neural protection, and muscle attachment. Examples of vertebrates include subgroupings (classes) of animals such as fish (e.g., trout, bass, shark), amphibians (e.g., grass frog, spotted salamander, red eft), reptiles (e.g., painted turtle, garter snake, anolis lizard), birds (e.g., robin, bluebird, red-tailed hawk), and mammals (e.g., gray whale, manatee, timber wolf, white-tail deer, human being).

The adaptations for survival of complex multicellular animals (vertebrate mammals) are typified by those of the human being. These adaptations are detailed in Unit 2, Chapter 1 (pages 67–94) of this book.

Plant-Like Organisms

Plants, in general, are multicellular, autotrophic ("self-feeding") organisms. Two major subdivisions of the plant kingdom include the non-vascular plants and vascular plants. Plants function in the ecosystem as **producer** organisms, a term that relates to their ability to produce simple sugars (glucose) through the process of **photosynthesis**. The sugars are produced in sufficient quantities to serve the nutritional needs of both plants and animals. Plants also produce oxygen gas as a by-product of photosynthesis, which is used by both plants and animals in the process of respiration.

Non-vascular plants, also known as **bryophytes**, are those lacking true roots, stems, or leaves and the vascular (tubular) tissues found in these structures. Examples of bryophytes include mosses and liverworts. These organisms are found typically in environments containing poor or rocky soil, high moisture, and low light conditions. They are low-lying plants that provide ground cover for bare soils and function in these environments to break down rock into smaller fragments that form the basis of soil. This function alters the environment and paves the way for subsequent communities of plants that need deeper soils to germinate, root, and grow. In this role, bryophytes inhabit their ecological niche as pioneer organisms.

Vascular plants, also known as **tracheophytes**, are those having true roots, stems, and leaves that contain vascular (tubular) tissues. These vascular tissues (**xylem** and **phloem**) function to transport water and dissolved minerals upward from the roots through the stems to the leaves (xylem) and transport dissolved sugars downward from the leaves through the stems to the roots (phloem). Examples of tracheophytes include cone-bearing plants (e.g., white pine, blue spruce, juniper) and flowering plants (e.g., wild geranium, shadbush, sugar maple).

All vascular plants contain three types of specialized structures: roots, stems, and leaves. Each of these structures performs a set of functions that help the plant survive in its environment.

- **Roots** penetrate and hold onto the soil, absorbing from it the water and dissolved minerals needed by the plant to support its metabolic activities. Roots function to support the plant, absorb needed materials, and store complex carbohydrates (starches) for use by the plant in its nutritional activities. Roots contain the vascular tissues xylem and phloem.

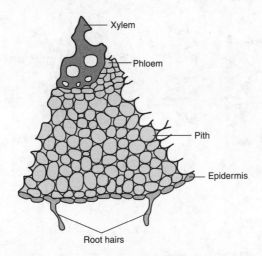

- **Stems** provide a branching structure that is attached to the roots and supports the leaves (or needles) and reproductive structures (flowers or cones) of the plant. The principal role of the stem is to provide for the transport of materials between the roots and the leaves. Stems contain the vascular tissues xylem and phloem.
- **Leaves** (or needles) are specialized to carry on the nutritional activities of the plant and are highly adapted to perform this function at maximum efficiency. Leaves of flowering plants are broad, maximizing the surface area available for the efficient absorption of light needed to energize the photosynthetic process. Leaves contain specialized tissues that further maximize the efficiency of photosynthesis. The **upper epidermis** contains the other leaf tissues and help to minimize moisture loss. The **palisade layer** contains the majority of the chloroplasts of the leaf and is the primary photosynthetic tissue of the leaf. The **spongy layer** of loosely packed cells allows for the free circulation of atmospheric

gases (carbon dioxide and oxygen) that are needed for, or produced by, the photosynthetic reactions. The **lower epidermis** helps to contain the other leaf tissues, but also contains specialized **stomates** (pores) that are bounded by specialized guard cells. **Guard cells** regulate the passage of atmospheric gases into (carbon dioxide) and out of (oxygen) the leaf through the stomates. Finally, the leaf contains **veins** that function to transport materials into (water and dissolved minerals) and out of (dissolved sugars) the leaf. Leaf veins contain the vascular tissues xylem and phloem.

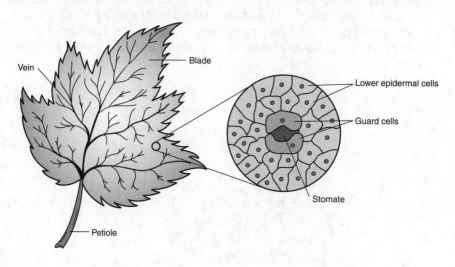

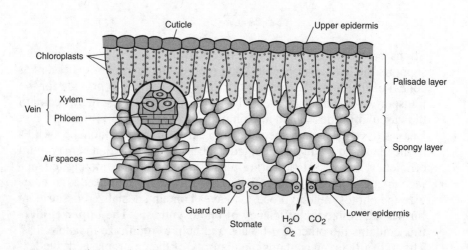

QUESTION SET 2.4—CELL ORGANELLES, BIOCHEMISTRY, UNICELLULAR LIFE FORMS
(ANSWERS EXPLAINED, P. 293)

1. Which structures could most likely be observed in cells in the low-power field of a compound light microscope?
   (1) cell walls and chloroplasts
   (2) ribosomes and endoplasmic reticula
   (3) lysosomes and genes
   (4) nucleotides and mitochondria

2. What would most likely happen if the ribosomes in a cell were not functioning?
   (1) The cell would undergo uncontrolled mitotic cell division.
   (2) The synthesis of enzymes would stop.
   (3) The cell would produce antibodies.
   (4) The rate of glucose transport in the cytoplasm would increase.

3. Which activity is illustrated in the diagram below?

   (1) a virus destroying a cell by extracellular digestion
   (2) a moss plant performing intercellular digestion
   (3) a protozoan ingesting food during heterotrophic nutrition
   (4) a lysosome egesting a food particle into the cytoplasm

4. Which activity is represented by the arrows in the diagram below?

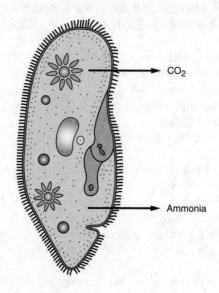

CO_2

Ammonia

(1) anaerobic respiration
(2) autotrophic nutrition
(3) deamination of amino acids
(4) excretion of metabolic wastes

5. The diagram below represents a green alga.

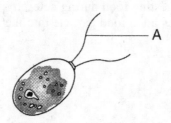

A

Which process is most closely associated with structure A?
(1) excretion   (3) locomotion
(2) transport   (4) reproduction

6. Which formula represents an organic compound?
(1) $Mg(OH)_2$ (3) $C_{12}H_{22}O_{11}$
(2) $NaCl$ (4) NH_3

7. The diagram below shows the same type of molecules in area A and area B. With the passage of time, some molecules move from area A to area B.

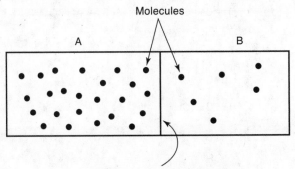

The movement is the result of the process of
(1) phagocytosis   (3) diffusion
(2) pinocytosis   (4) cyclosis

8–9. Base your answers to questions 8 and 9 on the structural formula of a molecule below and on your knowledge of biology.

8. Which statement best describes this molecule?
   (1) It has the ability to control heredity.
   (2) It has the ability to control reactions.
   (3) It has a high energy content.
   (4) It is involved in photosynthesis.

9. Which formula represents an end product derived from the chemical digestion of this molecule?

(1)　　O=C=O   (3)　H—O—H

(2)
```
          H
          |
    H—C—OH
          |
    H—C—OH
          |
    H—C—OH
          |
          H
```

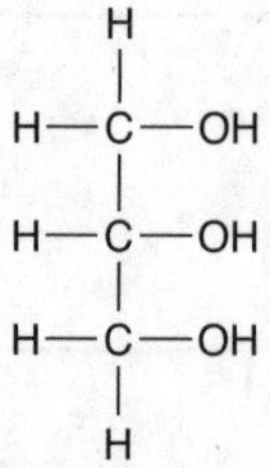

(4) 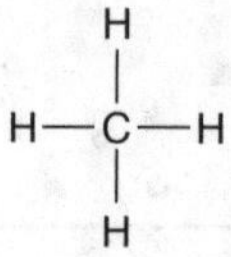
```
          H
          |
    H—C—H
          |
          H
```

10–11. Base your answers to questions 10 and 11 on the diagram below and on your knowledge of biology.

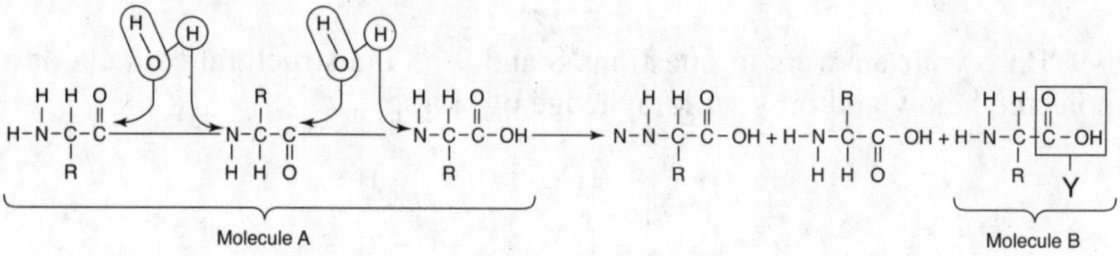

Molecule A

Molecule B

10. In molecule B, what type of group is contained in box Y?
    (1) an amino group   (3) a carboxyl group
    (2) a variable group   (4) a peptide group

11. How many peptide bonds are present in molecule *A*?
    (1) 1   (3) 3
    (2) 2   (4) 4

118

12. A student observed a wet mount of some stained plant cells in the high power field of a compound light microscope. Diagram A represents the general appearance of these cells. The student then added several drops of a liquid to the wet mount and continued the observations. Diagram B represents the general appearance of the cells a few minutes after adding the liquid.

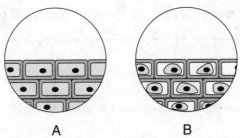

A　　　　　　　　　B

The liquid that the student added to the wet mount was most likely
(1) salt water　　　　　　　(3) pond water
(2) distilled water　　　　　(4) tap water

13. Which statement is a valid inference concerning structure X represented in the diagram below?

(1) Structure X contains guard cells that regulate glucose intake.
(2) Structure X carries out heterotrophic nutrition.
(3) Structure X produces gametes for asexual reproduction.
(4) Structure X transports materials for metabolic activities.

V. GENETIC CONTINUITY

> **KEY IDEA 2—GENETIC CONTINUITY** Organisms inherit genetic information in a variety of ways that result in continuity of structure and function between parents and offspring.

Essential Question:

- *How does genetic inheritance help ensure continuity of life?*

A significant unifying principle of biology is the concept that all living organisms possess a set of instructions, in the form of genes, that determine the characteristics of the organisms. These instructions are unique to the species. They determine the general physical and biochemical traits of the group. They are also unique to the individual, coding for the specific set of characteristics that sets one individual organism apart from all others of that species. Students should be familiar with the

- mechanisms by which genetic traits are passed from generation to generation to maintain genetic continuity of the species.
- molecular basis of genetics that is used to maintain genetic continuity mechanism by which it changes through recombination, mutation, and genetic engineering.

You should be familiar with the mechanism by which DNA replicates its structure during normal cell operations and in the process of reproduction. You must also understand the roles of DNA and RNA in the coding of cell-specific proteins and understand the role of these structural and functional proteins (enzymes, hormones, and other substances) in the operation of the cell.

Humans have long used their understanding of heredity to selectively breed new plant and animal varieties as well as to maintain hybrid varieties. Students should appreciate the economic role that these activities have played over the past centuries as well as the scientific basis of such activities. A relatively new branch of science, genetic engineering, has taken on an increasingly important and visible place in the scientific community. You should understand and appreciate this concept and make informed judgments as to the ethical considerations surrounding such research.

Genetic Material

Essential Question:

- *How does DNA control the inheritance of genetic traits in living things?*

Performance Indicator 4.2.1 *The student should be able to explain how the structure and replication of genetic material results in offspring that resemble their parents.*

The Genetic Code

Every organism requires a set of coded instructions for specifying its traits. For offspring to resemble their parents, information must be reliably transferred from one generation to the next. Heredity is the passage of these instructions from one generation to another. As scientists have continued their study of genetics, they have learned more and more of the details of the mechanisms of genetics. In one of the most significant discoveries in genetic science, **deoxyribonucleic acid (DNA)** was revealed to be the chemically active agent of the gene. As is now known, DNA replicates itself when chromosomes replicate in the early stages of cell division. DNA is passed from generation to generation during reproduction and acts as genetic factors. DNA interacts with the cell's chemical factory and produces the observable effects of the phenotype when genes are inherited by a cell or an organism. DNA regulates the production of enzymes in the cell and thereby enables the cell to perform the complex cellular chemical reactions necessary to sustain life. (See "DNA Code," page 128, for further information.)

DNA Structure

In order to understand the role of DNA in the maintenance of genetic continuity, you must first understand the currently accepted model of the DNA structure. DNA is a polymer made up of a repeating chemical unit known as the **nucleotide**. Thousands of these units are known to comprise a single DNA molecule, making DNA one of the largest of all organic compounds. DNA exists in hundreds of thousands, if not millions, of different forms, depending on the precise arrangement of nucleotides in the molecule. Its variability is the key to genetic variation in living things. DNA nucleotides themselves are quite complex, being composed of three separate subunits.

- Phosphate group—a chemical group made up of phosphorus and oxygen.
- Deoxyribose—a five-carbon sugar made up of carbon, oxygen, and hydrogen.
- Nitrogenous base—a chemical unit composed of carbon, oxygen, hydrogen, and nitrogen. Bases found in DNA are **adenine** (A), **thymine** (T), **cytosine** (C), and **guanine** (G). In chemical terms, the bases A and T are

complementary as are the bases C and G. In other words, A always links to T and C always links to G.

Phosphate

Deoxribose

Nitrogenous□
Base

Adenine ← Complements → Thymine

Guanine ← Complements → Cytosine

Complementary Nucleotides (DNA)

The Watson-Crick model, developed by James Watson and Francis Crick, is an attempt to describe the physical and chemical structure of DNA in a way that would explain its known characteristics, including its ability to replicate. Watson and Crick's model was developed using the best experimental evidence available at the time and involves the following points.

- Nucleotide units are joined end to end, forming a long chain of alternating deoxyribose and phosphate units with nitrogenous bases sticking out on one side of the chain. The specific arrangement of nitrogenous bases on the chain makes up the genetic code.

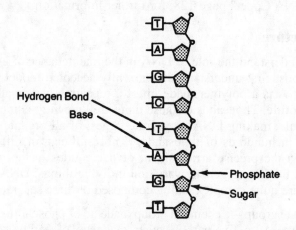

DNA: Single Strand

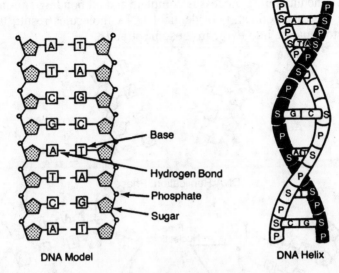

Base

Hydrogen Bond

Phosphate

Sugar

DNA Model

DNA Helix

DNA: Double Strand

- A second chain, with complementary nitrogenous bases, is aligned with the first, forming a double-stranded, ladderlike molecule. In this formation, the repeating sugar and phosphate units form the uprights of the ladder, while the pairs of linked nitrogenous bases form the rungs connecting the two strands.
- The nitrogenous bases paired to form the ladder rungs are always adenine to thymine (A-T) and cytosine to guanine (C-G). The bases are held together by weak hydrogen bonds.
- Finally, the double strands of DNA twist around each other to form a "double helix," or "twisted ladder," shape.

The processes that occur during cell division provide a reliable mechanism for transferring this genetic code from one generation to the next. Of central importance in these processes is the process of **DNA replication**. Replication is the exact self-duplication of the genetic material in the early stages of cell division. Chromosome replication is actually a function of the replication of the DNA making up the chromosome strand. DNA replication is thought to occur as follows.

- The two strands of the DNA molecule separate by unzipping between pairs of nitrogenous bases.
- Unbound (free) nucleotides floating freely in the cytoplasm are attracted to and incorporated into the unzipped portion of the DNA molecule. In the building process that follows, complementary nucleotides are attracted to each other to ensure that the new strands produced are exact duplicates of the original strands.

- When the unzipping process is complete and all bonding sites are filled with free nucleotides, two identical DNA molecules result; these are free to separate into two chromosome strands.

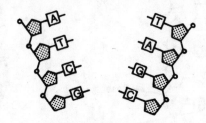

DNA Replication: Phase I

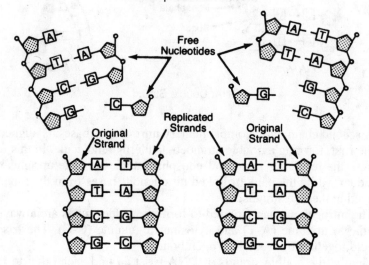

DNA Replication: Phase II

The Role of Genes in Heredity

Hereditary information is contained in genes, located in the chromosomes of each cell. An inherited trait can be determined by one or by many genes, and a single gene can influence more than one trait. A human cell contains many thousands of different genes in its nucleus. Hereditary factors known as **genes** are thought to exist as discrete portions (known as loci) of chromosomes. The term "discrete" in this case refers to the concept that genes are always located at the same point (or locus) on a chromosome. It is believed that pairs of homologous chromosomes contain linear, matching arrangements of genes exerting parallel control over the same traits. Pairs of genes that exercise such parallel control over the same traits are known as **alleles**. The control mechanism used by genes in genetic inheritance is now thought to be chemical in nature and is described in greater detail on the next page.

Genetic characteristics can be extremely complex and may require the actions of several separate genes to be expressed in the adult organism. Examples of traits that require more than a single pair of genes to be expressed are human height and human hair color. Other traits may be controlled by a single pair of genes. Some of these traits may have extreme phenotypes, such as albinism (lack of skin pigment) in humans.

Mapping of the human genome intrigued scientists for many years. In 2003, the International Human Genome Sequencing Consortium announced an essentially finished version of the human genome sequence. Among the pieces of data to be determined from this project was an exact count of the total number of separate genes found on human chromosomes. Each of these genes codes for the production of a different, unique protein. Each protein has a separate and distinct function to perform in the cell. Scientists will increasingly be able to use this information to accurately predict an individual human's predisposition to inherit certain traits, including genetic disorders, for which preventive treatments can be developed. Over time, it may even become possible to bioengineer the human genome. Many ethical questions will need to be identified, discussed, and resolved as this project matures.

Effect of the Environment

Genes are inherited, but their expression can be modified by interactions with the environment. Genetic traits are determined largely through the precise information found in the cell's gene structure. Although this information provides a basis for each individual organism's characteristics, it is not the only force at work in shaping the actual phenotype. Another major force in shaping the final phenotype is the environment in which the gene has to operate. A variety of factors in the environment can actually alter the effects of a particular gene. Some examples of this effect are as follows.

- Effect of light on chlorophyll production—Although most plants have the genetic ability to produce chlorophyll, they will do this only in the presence of light. Without light, these plants produce only a light-yellow pigment and therefore appear pale and sickly until they are exposed to sunlight. After a few days of exposure to sunlight, the chlorophyll production mechanism is enabled and green color returns.
- Effect of temperature on hair color in the Himalayan hare—In their native Arctic environment, Himalayan hares have white body hair with black hair on their extremities. However, when raised in warm climates, they are entirely white. In exploring this phenomenon, scientists shaved off some hair from the hare's back (normally white) and strapped an ice pack onto the bare skin. Under these experimental conditions, the hare's hair grew back black, indicating the role of environmental temperature on the production of hair color in this species.

Cell Division

In asexually reproducing organisms, all the genes come from a single parent. Asexually produced offspring are normally genetically identical to the parent. Asexual reproduction involves the production of a new organism of a species from a cell or cells of a single parent organism. This is made possible by a type of cell division known as **mitotic cell division**. In this type of cell division, the genetic material is duplicated, followed by the splitting of a single cell or group of cells. This results in the production of more cells with characteristics identical to those of the single parent organism.

The **Cell Theory** states that "all cells arise from preexisting cells by cell division." The type of cell division involved is known as mitotic cell division and involves two distinct stages.

- **Mitosis** is a precise duplication of the contents of the parent cell nucleus followed by an orderly separation of those contents into two new, identical nuclei. The events of mitosis are as follows.

  ➤ Replication (exact self-duplication) of each chromosome strand in the nucleus of the parent cell. This results in the doubling of each chromosome strand to form double-stranded chromosomes. Each of the two strands of a double-stranded chromosome is known as a **chromatid**. Chromatids are chemically identical to each other and carry identical genetic information. These chromatids are held together by a **centromere**.

  ➤ Disappearance of the nuclear membrane surrounding the chromosomes.

  ➤ Appearance of the **spindle apparatus**, a series of fibers that attach to the double-stranded chromosomes at the centromere during the early stages of mitosis.

  ➤ Replication of the centromere of each double-stranded chromosome. This is followed by separation of the two chromatids of the double-stranded chromosomes to form single-stranded chromosomes.

  ➤ Migration of the single-stranded chromosomes along the spindle apparatus toward opposite ends of the cell. The chromatids in each pair separate in this stage of mitosis, allowing a full set of single-stranded chromosomes to migrate to each pole.

  ➤ Reformation of the nuclear membrane around the chromosomes grouped at the ends of the cell. These two daughter nuclei are identical to each other and to the parent cell nucleus in terms of the number and type of chromosomes as well as in the particular genetic information found within these chromosomes. The significance of mitosis is the *exact duplication* of the parent nucleus.

- **Cytoplasmic division** is the separation of the two new nuclei into two new daughter cells as the cytoplasm of the parent cell divides. After the formation of the **daughter nuclei**, the cytoplasm of most (though not all) cells is divided into two roughly equal portions, each enclosing one of the daughter nuclei. The daughter cells will grow and eventually go through the cycle of mitotic cell division themselves. The mechanisms governing this cytoplasmic division are not fully understood but are known to occur differently in different types of organisms. For example, in plants and animals, mitosis occurs in much the same way as

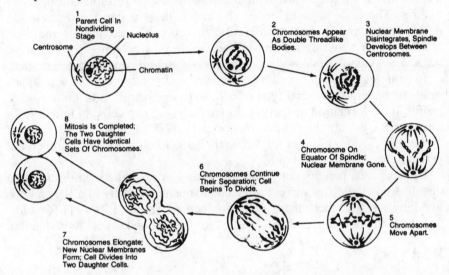

1 Parent Cell In Nondividing Stage
Centrosome
Nucleolus
Chromatin

2 Chromosomes Appear As Double Threadlike Bodies.

3 Nuclear Membrane Disintegrates, Spindle Develops Between Centrosomes.

8 Mitosis Is Completed; The Two Daughter Cells Have Identical Sets Of Chromosomes.

4 Chromosome On Equator Of Spindle; Nuclear Membrane Gone.

6 Chromosomes Continue Their Separation; Cell Begins To Divide.

5 Chromosomes Move Apart.

7 Chromosomes Elongate; New Nuclear Membranes Form; Cell Divides Into Two Daughter Cells.

described on page 126. However, in animals, the centriole is known to function in the formation of the spindle apparatus. In plants, the spindle forms without the presence of a centriole. In animals, cytoplasmic division is accomplished by the formation of a constriction that separates the two daughter cells. In plants, the division of the cytoplasm occurs after the formation of a cell plate that separates the daughter nuclei.

Mitosis

Reproduction

A common element of asexual reproduction is the fact that offspring produced by this mechanism are genetically identical to the parent organism. See the section "Mechanisms of Reproduction" (pages 159–171) for additional information on asexual reproductive patterns.

In sexually reproducing organisms, the new individual receives half of the genetic information from its mother (via the egg) and half from its father (via the sperm). Sexually produced offspring often resemble, but are not identical to, either of their parents. Sexual reproduction involves the production of a new offspring organism from the fusion of two sex cells, known as **gametes**, each of which is contributed by a different parent organism. It is made possible by a special kind of cell division known as **meiotic cell**

division. In this type of cell division, the genetic information is halved. It is restored to its full complement during fertilization.

Diploid and Haploid Explained

Each body cell of an organism contains a number of chromosomes characteristic of the species. This number is known as the **diploid** chromosome number. The term "diploid" refers to the fact that the chromosomes in the nucleus are found in pairs with similar structure. These chromosome pairs are known as homologous chromosomes; they carry genes for the same traits. The symbol **2n** is used to represent the diploid chromosome number.

In gametogenesis (formation of gametes), cells located in the sex organs (gonads) undergo a special type of cell division (meiotic cell division) that results in the formation of sperm cells (in males) or egg cells (in females). In this process, the number of chromosomes is reduced by half. This reduced number of chromosomes is known as the haploid chromosome number and is represented by the symbol **n**.

Because each parent contributes 50 percent of the genetic information to the offspring, the offspring resemble, but do not look exactly like, the parents. See the section "Mechanisms of Reproduction" (pages 159–171) for additional information concerning meiosis, fertilization, and sexual reproduction.

DNA Code

In nearly all organisms, the coded instructions for specifying the characteristics of the organism are carried in DNA, a large molecule formed from subunits arranged in a sequence with bases of four kinds (represented by A, G, C, and T). The chemical and structural properties of DNA are the basis for how the genetic information that underlies heredity is both encoded in genes (as a string of molecular bases) and replicated by means of a template. Recall that DNA is a complex organic molecule composed of thousands of repeating nucleotide molecules and that each free nucleotide carries with it one of four nitrogenous bases. The particular sequence of nitrogenous bases adenine, thymine, cytosine, and guanine (A, T, C, and G) comprising a strand of DNA provides the type of chemical code that is understood by the chemical mechanisms of the cell. The DNA code is used by these mechanisms to manufacture specific enzymes and other proteins through the process of protein synthesis. A DNA strand provides a **template** (pattern) for the formation of **messenger RNA** (mRNA). The DNA code is transcribed (read) by mRNA as the latter is synthesized in a pattern complementary to the DNA strand. The process by which the DNA code is transferred to mRNA code is known as **transcription**. In RNA molecules, uracil (U) nucleotides are substituted for DNA's thymine (T).

Each group of three nitrogenous bases, known as a triplet **codon**, provides the information necessary to code for the insertion of a single, specific amino acid into a building protein molecule. The particular sequence of triplet codons on DNA (and transcribed to mRNA) enables amino acids to be linked together in a specific sequence during protein synthesis.

Cells store and use coded information. The genetic information stored in DNA is used to direct the synthesis of the thousands of proteins that each cell requires. The cell contains many thousands of such codes in its chromosomes. Each strand of DNA in the chromosome has the potential to provide the complete chemical code for the manufacture of at least one complete protein. These proteins are highly specific. They result in the expression of some specific trait or portion of a trait in the living cell and, consequently, in the organism of which they are a part.

Protein Synthesis

The actual process of protein synthesis begins when an mRNA molecule migrates out of the nucleus, where it was formed, and attaches to a ribosome located on the surface of the endoplasmic reticulum. Recall from the previous section that mRNA carries the information, read from a genetic blueprint housed in a molecule of DNA, needed to encode the creation of a single protein strand. Once attached to the ribosome, the mRNA begins the process of protein synthesis by insertion of a "start" codon. The codon sequence of the mRNA provides the pattern upon which a new polypeptide (protein) strand will be built.

In this role, mRNA acts like a piece of manufacturing machinery in a factory, while the ribosome represents the factory within the machinery operates.

Floating free in the cytoplasm of the cell are molecules of **transfer RNA** (tRNA). Though composed of the nucleotides A, U, G, and C, tRNA differs in structure and function from mRNA. Each type of tRNA is specifically designed to link to one of 20 different types of amino acid at a chemically active site and carries an **anticodon** that is designed to link to a complimentary codon on the mRNA molecule. As protein synthesis continues to completion, tRNA molecules link in a specific order to mRNA on the ribosome. While so linked, the tRNA molecules facilitate the formation of chemical **peptide** bonds between adjacent amino acid molecules. The chemical linkage of many amino acids by means of peptide bonds results in a chain molecule known as a **polypeptide** or protein. The synthesis of an individual protein molecule ceases when the "stop" codon is reached in the mRNA sequence. The result of this process is the formation of a polypeptide chain

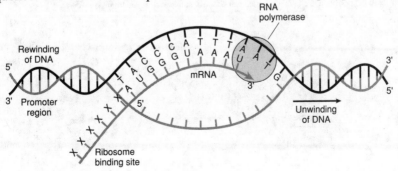

of a specific chemical composition that is released into the cytoplasm to perform a particular biochemical function. The process by which the mRNA code is decoded by tRNA and transferred to the amino acid sequence in a polypeptide molecule is known as **translation**.

To continue the factory analogy, tRNA molecules act as factory workers who carry manufacturing components (amino acids) to the mRNA machinery and determine their exact placement in the protein product being manufactured.

(1) Initiation: First, the small subunit of the ribosome binds the mRNA. Second, the first tRNA with the anticodon to the start codon binds to the mRNA. Then, the large subunit binds. The second codon is now available in the A site.

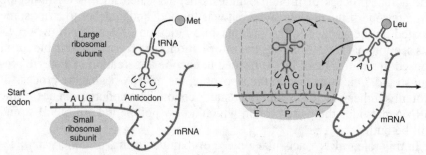

(2) Elongation: tRNAs carrying amino acids enter the A site. An enzyme in the ribosome catalyzes bond formation between the adjacent amino acids. The ribosome-mRNA complex shifts. The first tRNA exits from the E site. A new tRNA enters the A site, and the process repeats.

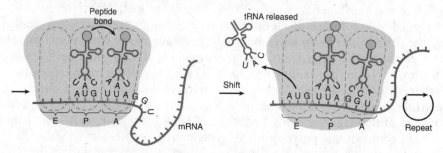

(3) Termination: When a stop codon is present in the A site, the enzyme release factor enters the ribosome and releases the polypeptide chain. After that, the ribosomal subunits separate from the mRNA.

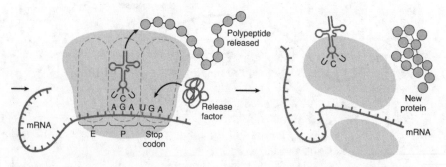

One-Gene–One-Polypeptide Hypothesis

Each gene in the cell's nucleus contains the coded information required to synthesize a single polypeptide chain. A particular gene is believed to operate throughout the life of the cell to produce its specific polypeptide and only that polypeptide. This concept is known as the **one-gene–one-polypeptide hypothesis**. The modern concept of the gene defines it as that sequence of nucleotides and codons in a molecule of DNA necessary to code for a complete polypeptide chain.

Gene Mutations

Genes are segments of DNA molecules. Any alteration of the DNA sequence is a mutation. Usually, an individual cell's altered gene will be passed on to every cell that develops from it. Gene mutations may be defined as being any changes in the nitrogenous base sequence of a molecule of DNA. When the base sequence of DNA is altered, the amino acid sequence of the polypeptide for which it codes will likewise be altered. Such an alteration may affect the operation of the resulting enzyme, preventing it from properly catalyzing its reaction and thus preventing a trait from being expressed by the cell. The majority of gene mutations are harmful because they result in the cell being impaired from performing some specific task. In most cells, the corresponding allele continues to function, and the cell continues its activities unaffected. In rare cases, a mutation may result in a lethal gene that kills the cell either by producing a substance toxic to the cell or by failing to produce a protein of vital importance to the cell.

Gene mutations are passed on to every cell that arises from the mutated cell. If the mutation occurs in **somatic** (body) tissues, its effect is limited to the tissues immediately surrounding the mutated cell. If the mutation occurs in a **primary sex cell**, it can be passed on to the offspring that result from fertilization of or by the gametes produced from that primary sex cell. In the latter case, a mutation can enter the gene pool of a population and be passed on to succeeding generations. This is now known to be a cause of variation in a species.

Amino Acids

The work of the cell is carried out by the many different types of molecules it assembles, mostly proteins. Protein molecules are long, usually folded chains made from 20 different kinds of amino acids in a specific sequence. This sequence influences the shape of the protein. The shape of the protein, in turn, determines its function. Although there are 20 different types of amino acids, they all share a common general structure. All amino acids contain an amino group ($-NH_2$) and a carboxyl group ($-COOH$) attached to a central carbon atom. A third attached group (radical) varies and gives each type of amino acid its unique properties. Two amino acid molecules may be joined together chemically by dehydration synthesis to form a dipeptide. The term "peptide" refers to the name of the actual chemical bond between carbon and nitrogen atoms that joins the two amino acid units together. As more and more amino acids link together by dehydration synthesis, an amino acid chain, known as a **polypeptide**, is formed. Polypeptide chains form the basis of protein molecules. Because of the almost endless variations in which amino acids may be arranged in a polypeptide, proteins are formed in thousands of different forms. The extreme variability of proteins is thought to be responsible for the individual variations in living things. (See page 103 for chemical structure of amino acids.)

The functions of proteins may be classified as either structural (forming a part of the cell material) or functional (having a role in the chemistry of the cell). A few common examples of proteins are insulin (a hormone), hemoglobin (an oxygen-carrying pigment in red blood cells), and enzymes (organic catalysts).

Offspring resemble their parents because they inherit similar genes that code for the production of proteins that form similar structures and perform similar functions. We have learned that DNA's principal role in the cell is to code for the production of specific polypeptides. Each enzyme or structural protein synthesized by a cell has a specific function to perform. These functions comprise the characteristics, or traits, of the cell. It follows logically, then, that the individual traits that a cell displays are a function of the particular combinations of DNA found in its cells. The particular combination of these traits gives an organism its individuality that enables us to recognize them as individual organisms. The following diagram will help to illustrate this concept.

DNA ⟶ gene ⟶ polypeptide ⟶ enzyme ⟶ reaction ⟶ reaction product ⟶ trait

Within this individuality, certain traits resemble those of the parent organisms, illustrating the genetic continuity that links one generation to the next. Species traits are similar enough for us to recognize an individual organism as part of a species group but particular enough for us to recognize family similarities and individual characteristics.

Differentiation

The many body cells in an individual can be very different from one another. This occurs even though they are all descended from a single cell and thus have essentially identical genetic instructions. This is because different parts of these instructions are used in different types of cells and are influenced by the cell's environment and past history. All cells in an individual multicellular organism arise by mitotic cell division from a single parent cell, the **zygote**. All cells in the body therefore receive the same genetic information. Despite this, the body contains many specific tissue types, many highly specialized to perform a narrow range of tasks. The process at work to produce different cells that perform different functions is **differentiation**. Differentiation is how cells become specialized into tissues to perform the various tasks needed to keep the body functioning. This will be discussed in more depth in the section "Mechanisms of Reproduction" (pages 159–171). It is mentioned here to illustrate the hypothesis that a gene will operate in the cell only when its polypeptide is required by the activities of the cell and that at other times the gene is switched off. In fact, considerable evidence has accumulated that points to just such a concept. Only the genes specific to the activities of a specific cell type are activated in that cell.

This concept maximizes the efficiency of the cell. The cell's protein synthesis mechanisms are not overloaded producing enzymes and structural proteins that have no direct function in the specialized cells in which they operate. For example, human muscle tissues must produce the proteins actin and myosin that make up the muscle cell's contractile apparatus. Therefore, the genes for the production of these proteins are switched on in muscle cells. At the same time, the genes that code for the production of hemoglobin are switched off in muscle cells since hemoglobin, which is required in red blood cells, has no direct function in muscle cells. See "Stem Cell Research," page 170.

Genetic Engineering

Essential Questions:

- *What effect will genetic engineering have on the future of genetic inheritance?*
- *What cautions should be considered in the use of genetic engineering?*

Performance Indicator 4.2.2 *The student should be able to explain how the technology of genetic engineering allows humans to alter the genetic makeup of organisms.*

For thousands of years, new varieties of cultivated plants and domestic animals have resulted from **selective breeding** for particular traits. Geneticists have applied many of the theoretical concepts of the science of genetics to the practical areas of plant and animal breeding. Many of the most productive plant and animal breeds raised for human uses have been developed through careful artificial selection of breeding individuals, crossbreeding (hybridization) between related varieties, and inbreeding of perfected strains to maintain established traits. Such traits as reproductive potential, resistance to disease, adaptability to climate, and productivity, among others, are considered desirable traits for human uses. These and other traits are purposely bred into grains (corn, wheat), flowers (roses, orchids), fruits (apples, oranges), cattle, horses, dogs, and many other species. They are maintained by careful attention to breeding patterns.

In the past, crude methods were used to mate members of a species displaying desirable traits in the hopes that some of the offspring of such a mating would display enhanced characteristics of the desired type. When desirable offspring were obtained, the hybrid animals or plants were crossbred in an attempt to produce a pure-breeding strain of the desired variety.

Genetic Manipulation

In recent years, new varieties of farm plants and animals have been engineered by manipulating their genetic instructions to produce new characteristics. Modern methods of **gene manipulation** have been developed that make isolating and inserting specific genes for particular traits within the genome of a species possible. Once inserted, perpetuating the desired trait through traditional methods of hybrid crosses is possible. **Genetic engineering** refers to the series of techniques used to transfer genes from one organism to another. It involves removing a small piece of DNA from a cell and adding it to the gene structure of another cell. The new DNA that results is known as recombinant DNA. The spliced gene in the recombinant DNA will continue to produce its polypeptide product in the new cell, thus transferring to that cell a genetic ability it lacked before.

These techniques have been used to produce new, improved varieties of certain farm plants (wheat, corn, coffee) and animals (cattle, fish). Some of these species have been bred to display enhanced characteristics such as resistance to disease. Others have been bred to produce human proteins needed for research or for medical applications.

An already successful medical technique that holds great promise for the future is the production of human hormones and other proteins in the bodies of laboratory animals or bacterial colonies. Human genes have been removed from healthy human cells and added to the gene sequences of bacteria that are then cultured in large quantities in the laboratory. The substances produced in this manner are considered to be safer than those produced by traditional methods, including those drawn from human subjects. Some examples of this process are as follows.

- Bacteria colonies have been genetically engineered to produce human insulin, interferon, hepititis B vaccine, and human growth hormone. These proteins were gathered and purified for use in treating various human disorders.
- Potato plants have been altered to produce serum albumin.
- Tobacco plants with spliced human genes produce antibodies and melanin.

Different enzymes can be used to cut, copy, and move segments of DNA. Characteristics produced by the segments of DNA may be expressed when these segments are inserted into new organisms, such as bacteria. In the mid-20th century, research with certain bacteria led to the discovery of enzymes capable of snipping DNA molecules at particular nitrogenous base sequences. Such enzymes are known as **restriction enzymes**. This knowledge was used to develop laboratory methods by which the desirable genes of one species may be snipped out of the genome of that species and inserted into the genome of a different species. Once part of the genome of the host species, the gene can function to produce the hormone or other protein coded by the inserted gene. In this way, it is possible for a human trait to be displayed by a different species, including bacteria. The DNA that is altered by this method is known as **recombinant DNA**.

Inserting, deleting, or substituting DNA segments can alter genes. An altered gene may be passed on to every cell that develops from it. Any altered sequence of DNA in a living cell can be passed along to succeeding generations of the cell by means of mitotic cell division. Like mutation, genetic engineering can result in altered sequences of nitrogenous bases on a molecule of DNA. When nitrogenous base sequences are added, deleted, or substituted in a DNA molecule, these changes are inheritable. If the genetically engineered cell is a primary sex cell, the gametes resulting from that primary sex cell may inherit the altered gene sequence, leading to the possibility that an artificial gene may be passed along from generation to generation in a host species that does not normally display the characteristic associated with the species.

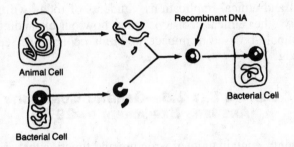

Genetic engineering involves removing a desirable gene from a cell, adding it to the gene structure of a bacterial cell, and replacing the recombinant DNA into the bacterial cell. The bacteria then have the capability to produce the protein produced by the original animal or plant cell.

Genetic Engineering

Gene Therapy

Knowledge of genetics is making possible new fields of health care. For example, finding genes that may have mutations that can cause disease will aid in the development of preventive measures to fight disease. Substances, such as hormones and enzymes, from genetically engineered organisms may reduce the cost and side effects of replacing missing body chemicals. The term **gene therapy** has become a commonly used term in biomedical research. This process involves the splicing of functional genes into cells that contain defective, nonfunctional genes for a particular trait. The clinical techniques used to accomplish gene therapy are still being worked out. However, they involve research into methods of introducing the functional genes into the person or organism affected by the defective gene. Cystic fibrosis is one human disorder successfully treated in clinical trials using this technique.

Cloning

A scientific curiosity just a few years ago, cloning has become an increasingly common method for the production of plants and animals used as food. Cloning in plants involves the growth and differentiation of undifferentiated tissues, resulting in the formation of a set of offspring with characteristics identical to those of the donor plant. In animals, cloning is accomplished by replacing the haploid nucleus of an egg cell with a diploid nucleus of a somatic (body) cell of a donor animal. Stimulated to divide, the egg produces an embryo that matures into an adult animal with characteristics identical to those of the donor animal—a clone.

This method of reproduction is controversial because it places the creation of new life in the hands of laboratory scientists who make decisions about what organisms will be selected for cloning. A potential dilemma surrounds the cloning of humans, a currently banned practice that could have widespread social and ethical implications. Students of today will need to face this and many other ethical decisions about how our expanding knowledge of genetics should be used to impact the human condition. See "Stem Cell Research," page 170.

QUESTION SET 2.5—GENETIC CONTINUITY
(ANSWERS EXPLAINED, P. 297)

1. Living things contain units of structure and function that arise from pre-existing units. This statement best describes the
   (1) cell theory
   (2) lock-and-key model of enzyme activity
   (3) concept of natural selection
   (4) heterotroph hypothesis

2. Which statement most accurately compares mitotic cell division in plant and animal cells?
   (1) It is exactly the same in plant and animal cells.
   (2) The walls of plant cells pinch in, but the membranes of animals do not.
   (3) Most plant cells use centrioles, but most animal cells do not.
   (4) In both plants and animals, the daughter cells are genetically identical to the original cell.

3. What are the normal chromosome numbers of a sperm, egg, and zygote, respectively?
   (1) haploid, haploid, and haploid
   (2) haploid, diploid, and diploid
   (3) diploid, diploid, and diploid
   (4) haploid, haploid, and diploid

4. Which statement describes the work of Gregor Mendel?
   (1) He developed some basic principles of heredity without having knowledge of chromosomes.
   (2) He explained the principle of dominance on the basis of the gene-chromosome theory.
   (3) He developed the microscope for the study of genes in the garden pea.
   (4) He used his knowledge of gene mutations to help explain the appearance of new traits in organisms.

5. Genes carried on only an X chromosome are said to be
   (1) hybrid   (3) autosomal
   (2) codominant   (4) sex-linked

6. In which situation could a mutation be passed on to the offspring of an organism?
   (1) Ultraviolet radiation causes skin cells to undergo uncontrolled mitotic division.
   (2) The DNA of a human lung cell undergoes random breakage.
   (3) A primary sex cell in a human forms a gamete that contains 24 chromosomes.
   (4) A cell in the uterine wall of a human female undergoes a chromosomal alteration.

7. The diagram below illustrates what happens to the fur coloration of a Himalayan hare after exposure to a low temperature.

Ice pack

Before   After

This change in fur coloration is most likely due to
   (1) the effect of heredity on gene expression
   (2) the arrangement of genes on homologous chromosomes
   (3) environmental influences on gene action
   (4) mutations resulting from a change in the environment

8–11. Base your answers to questions 8 through 11 on the diagram below and on your knowledge of biology. The diagram represents molecules involved in protein synthesis.

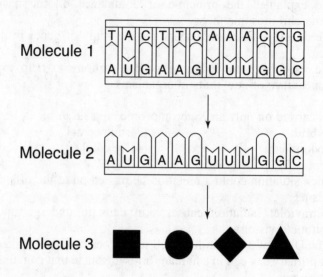

8. In plant cells, molecule 1 is found in the
   (1) centriole
   (2) nucleus
   (3) cell wall
   (4) lysosome

9. The building blocks of molecule 3 are known as
   (1) amino acids
   (2) DNA molecules
   (3) fatty acids
   (4) RNA molecules

10. Where do the chemical reactions coded for by molecule 2 take place?
   (1) in the vacuole
   (2) on the plasma membrane
   (3) in the lysosome
   (4) at ribosomes

11. Molecule 3 is formed as a result of
   (1) deamination
   (2) dehydration synthesis
   (3) enzymatic hydrolysis
   (4) oxidation

12–15. Base your answers to questions 12 through 15 on the reading passage below and on your knowledge of biology.

Female or Male, Which Will It Be?

After fertilization, *all* human embryos begin forming the basic female reproductive structures. These structures are present by the time the embryo has toes, fingers, eyes, and a heart at 35–40 days into gestation. If the egg was fertilized by a sperm containing a Y chromosome, a series of changes occurs that will produce a male.

Recent research has isolated a genetic switching mechanism that is part of the process that determines sex in humans. The Y chromosome contains a trigger factor known as the *SRY* gene, which activates the male pattern of development after 35–40 days of gestation. The *SRY* gene causes testes to develop. These, in turn, produce testosterone, which causes the development of male characteristics such as the penis, masculine muscles, and eventually, facial hair.

At this stage of development, the embryo has both male and female potential. However, the *SRY* gene sends a chemical message to another gene known as *MIS*. The *MIS* gene causes the developing female organs in the embryo to disappear. The combined action of the *SRY* and *MIS* genes results in the change of the embryo from female to male.

12. During the first five weeks after a human egg is fertilized, the embryo develops
   (1) only toes, fingers, eyes, and a heart
   (2) male reproductive structures and other organs
   (3) female reproductive structures and other organs
   (4) male or female reproductive structures depending on whether the egg was fertilized by an X-bearing or a Y-bearing sperm

13. The male pattern of development is activated by the
   (1) *SRY* gene
   (2) entire Y chromosome
   (3) entire X chromosome
   (4) *MIS* gene

14. The *MIS* gene is activated by
   (1) the X chromosome
   (2) a chemical message
   (3) the presence of male reproductive structures
   (4) the presence of female reproductive structures

15. An embryo is changed from a female to a male by the action of
    (1) two X chromosomes
    (2) all the genes on a Y chromosome
    (3) an *MIS* gene, only
    (4) *SRY* and *MIS* genes

---

16. The code of a gene is delivered to the enzyme-producing region of a cell by a
    (1) hormone
    (2) nerve impulse
    (3) messenger RNA molecule
    (4) DNA molecule

17. Which process could be used by breeders to develop tomatoes with a longer shelf life and to develop cows with increased milk production?
    (1) natural selection
    (2) sporulation
    (3) genetic engineering
    (4) chromatography

18. What would most likely result if mitosis was not accompanied by cytoplasmic division?
    (1) two cells, each with one nucleus
    (2) two cells, each without a nucleus
    (3) one cell with two identical nuclei
    (4) one cell without a nucleus

19. Breeders have developed a variety of chicken that has no feathers. Which methods were most likely used to produce this variety?
    (1) artificial selection and inbreeding
    (2) grafting and hybridization
    (3) regeneration and incubation
    (4) vegetative propagation and binary fission

20. In fruit flies with the curly wing mutation, the wings will be straight if the flies are kept at 16°C but curly if they are kept at 25°C. The most probable explanation for this is that
    (1) fruit flies with curly wings cannot survive at high temperatures
    (2) the environment influences wing phenotype in these fruit flies
    (3) high temperatures increase the rate of mutations
    (4) wing length in these fruit flies is directly proportional to temperature

21–23. Base your answers to questions 21 through 23 on the diagram below and on your knowledge of biology.

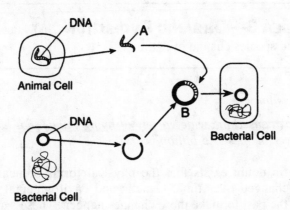

21. Structure *A* contains a
    (1) genetic code
    (2) single nucleotide only
    (3) messenger RNA molecule
    (4) small polysaccharide

22. Structure *B* represents
    (1) a ribosome
    (2) transfer RNA
    (3) recombinant DNA
    (4) a male gamete

23. The technique illustrated in the diagram is known as
    (1) cloning
    (2) genetic engineering
    (3) protein synthesis
    (4) *in vitro* fertilization

VI. ORGANIC EVOLUTION

> ## KEY IDEA 3—ORGANIC EVOLUTION Individual organisms and species change over time.

Essential Question:

- *What has caused the wide variety of living things that have inhabited Earth over the past 3.5 billion years?*

Little scientific doubt exists that the physical forms of nearly all living things have changed over time. Genetic and environmental factors have combined in the past to make these changes happen. These same forces are still at work, selecting traits that better adapt species to Earth's ever-changing environment. Organic evolution specifically refers to the mechanisms thought to govern the structural, behavioral, and functional changes in living species over geologic time. These changes may include variation within a species or the production of new species.

Theory of Natural Selection

The unifying theory underlying the modern concept of evolution is the theory of natural selection. First advanced by Charles Darwin in the 19th century, this theory is based on scientific observation coupled with an understanding of how living things respond to environmental pressures. The basic features of the **theory of natural selection** include observations of overproduction of offspring in naturally occurring species, the existence of variation among offspring, and the struggle for survival. The theory also includes logical assumptions concerning the selection of individuals whose variations provide adaptive value and the subsequent survival and reproduction of the best-adapted individuals. To Darwin's original construct has been added a wealth of new observations and a more complete understanding of the genetic basis of variation. See pages 151–152.

Many scientific observations, some only peripherally related to the science of evolution, have lent support to the concepts embodied in the theory of natural selection. The sciences of geology (study of rocks) and paleontology (study of fossils) have demonstrated the extreme age of the Earth and the existence of fossilized organisms different from, but still similar to, organisms that exist today. The sciences of anatomy (study of structure) and embryology (study of reproduction and development) have enabled scientists to classify organisms based on similarities of structure and developmental patterns. The sciences of cytology (study of cells) and biochemistry (study of chemical reactions in living things) have pointed to the basic similarity of all living things. The relative relatedness of living things shows close structural similarities.

Process of Evolution

Essential Question:

• *How do new species arise from previously existing species?*

Performance Indicator 4.3.1 *The student should be able to explain the mechanisms and patterns of evolution.*

Evolution is a natural force thought to be responsible for the dizzying array of different life-forms that have inhabited our planet. A survey of living things reveals that approximately 1.9 million different species of every description exist on Earth. Classification provides scientists with the means for sorting and grouping these organisms for easier study. The basis of biological classification is physical structure, although other criteria, such as embryonic, genetic, and biochemical similarities, are also used. Organisms that are similar in their physical traits are usually similar in other ways. Because of this fact, the characteristics of a large grouping of similar organisms can be learned by studying a few representatives of the group. When a new organism is discovered, it can be readily grouped with other, similar organisms when only a few of its characteristics are known. It is assumed that organisms sharing many traits in common are likely to share a common ancestry as well.

Scheme of Classification

The most widely accepted scheme of classification places every known organism into one of five large groupings known as kingdoms. The organisms within a particular kingdom share many broad characteristics in common, although there can be considerable diversity (difference) of form among them. The five kingdoms are as follows.

• Monera—unicellular forms having a primitive cell structure
• Protista—unicellular organisms with plantlike or animal-like characteristics
• Fungi—unicellular, colonial, or multicellular saprophytic organisms
• Plant—multicellular, photosynthetic organisms
• Animal—multicellular, heterotrophic organisms

Within each kingdom, organisms having greater similarity to each other than to the organisms in other groups are classified together into phyla. Therefore, each phylum contains groups of organisms showing characteristics distinctly different from those of other phyla. Phyla are subdivided into still narrower groupings of organisms including classes, orders, and families, and finally into groups showing high degrees of similarity known as genera.

The members of each genus are so similar that they might easily be mistaken for each other by most people. Each genus is further broken down into species. The members of a species are so similar biologically that they can share genetic information and reproduce more individuals like themselves. Fertile offspring result from reproductive activities between members of the same species.

Kingdom	Phylum	Characteristics	Examples
Monera	Bacteria Blue-green algae	Have primitive cell structure lacking a nuclear membrane.	*E. coli*
Protista	Protozoa Algae	Are predominantly unicellular organisms with plant-like and/or animal-like characteristics.	Paramecium Ameba Spirogyra
Fungi		Cells are usually organized into branched, multi-nucleated filaments that absorb digested food from their environment.	Yeast Bread mold Mushroom
Plant		Are multicellular, photosynthetic organisms.	
	Bryophytes	Lack vascular tissues; have no true roots, stems, or leaves.	Moss
	Tracheophytes	Possess vascular tissue; have true roots, stems, and leaves.	Geranium Fern Bean Trees (maples, oaks, pines, etc.) Corn
Animal		Are multicellular, heterotrophic organisms.	
	Coelenterates	Have two cell layers, hollow body cavity.	Hydra Jellyfish
	Annelids	Have segmented body walls.	Earthworm Sand worm
	Arthropods	Have jointed appendages, exoskeleton.	Grasshopper Lobster Spider
	Chordates	Possess dorsal nerve cord, internal skeleton.	Shark Frog Human being

This classification scheme is used by scientists to organize living things into easily understood groupings. To keep track of the more than 1.9 million different species that have been discovered and classified, a system of naming (nomenclature) has been developed. This system uses two names in much the same way that most people have at least two names to help others tell them apart. This naming system, known as **binomial nomenclature** (two-name naming), was first devised by Carl Linnaeus in the eighteenth century. The two names are the genus name (always capitalized) and the species name (always written in lowercase). The language used in the system of binomial nomenclature is Latin. Some examples of this system are as follows.

Genus	Species	Common Name
Homo	*sapiens*	human being
Canis	*familiaris*	dog
Canis	*lupis*	wolf
Felis	*domestica*	cat
Felis	*leo*	lion

Evolutionary Assumptions
The basic theory of biological evolution states that Earth's present-day species developed from earlier, distinctly different species. The currently accepted theories of organic evolution assume that modern life-forms have evolved from previously existing life-forms. One source of this assumption is the logical conclusions drawn from a knowledge that living things arise from other living things by reproduction. Just as present-day organisms have been produced by the generations that immediately proceeded them, so modern-day species must have developed from the ancient species of eons past. In addition to this logical conclusion, a great deal of scientific evidence supports this assumption provided by studies of geology and paleontology.

No direct evidence indicates the exact date of Earth's formation. By studying the indirect evidence drawn from the radioactive dating of Earth's rocks, however, geologists have estimated the age of Earth to be between 4.5 and 5.0 billion years. In arriving at this estimation, scientists have assumed that Earth is at least as old as the oldest rocks so far discovered. Fossils of early life-forms have been found preserved in certain of these rocks and are dated by scientists using various techniques. It is logical to conclude that the fossils are as old as the rock layers in which they are embedded.

Genetic Mechanisms of the Cell
New inheritable characteristics can result from new combinations of existing genes or from mutations of genes in reproductive cells. Mutation and the sorting and recombining of genes during meiosis and fertilization result in a great variety of possible gene combinations. We have already learned that the source of individual traits is found within the genetic mechanisms of the cell. These mechanisms include the cellular processes of mutation, meiosis, fertilization, and protein synthesis.

Mutation is important in providing the new genes that may lead to the production of new genetic traits. Although the majority of such mutations are harmful or neutral, a small percentage may provide significant adaptive advantages to a species. It is important to recognize that in sexually reproducing species, only mutations in gametes can be passed on to succeeding generations. Protein synthesis is the basic chemical process in the cell that results in production of new proteins that function in the cell and result in new variations.

Within the reproductive process, meiosis and fertilization provide the mechanism by which new combinations of both old and new traits may be tried out as new varieties within a species. These can be counted in the millions of possible individual combinations. Successful combinations are perpetuated in the species through the processes of sexual reproduction. Unsuccessful combinations are removed from the population's gene pool by the selective forces of the environment.

Gene Mutations

Mutations occur as random events. Gene mutations can also be caused by such agents as radiation and chemicals. When they occur in sex cells, the mutations can be passed on to offspring. If they occur in other cells, the mutations can be passed on to other body cells only. Gene mutations involve changes in the chemical nature of the gene. The active chemical in the gene is DNA. When this material undergoes chemical alteration, its control over cell activities and cell characteristics changes, causing alterations in the phenotype of the organism. Although these changes are likely to be small and difficult to detect, they may occasionally be great enough to be easily noticed or even to cause death. Albinism (lack of skin pigment in humans) is an example of a human trait caused by a single gene mutation whose effects are quite obvious and dramatic. The phenotype (albino) that results from the homozygous recessive allelic combination is characterized by pale white skin, yellow hair, and pink eyes.

Although most gene mutations cause changes that are neutral or harmful, occasionally their effect is beneficial. A beneficial mutation, such as one that produces a useful enzyme, causes a phenotypic change that in some way gives an organism an advantage in its environment over other organisms of the same species. An extremely beneficial gene mutation can cause major shifts in the species' genetic characteristics, a phenomenon considered by many scientists to be a major driving force in evolution.

Gene mutation can occur spontaneously by random chemical alteration of DNA during replication. In addition, a number of naturally occurring and human-produced phenomena are known to cause or accelerate gene mutation. These phenomena include:

- Radiation—Sources of radiation include cosmic rays, radon, X rays, ultraviolet radiation, radioactive radiation, and electromagnetic radiation; and
- Chemicals—Sources include benzene, formaldehyde, asbestos, dioxin, and tobacco residues.

Geologic Record

Natural selection and its evolutionary consequences provide a scientific explanation for the fossil record of ancient life-forms as well as for the molecular and structural similarities observed among the diverse species of living organisms. Found frequently within certain types of rocks are **fossils**, the preserved direct or indirect evidence of organisms that lived in the past. Fossils are most commonly discovered embedded in sedimentary rock, such as sandstone or limestone. However, the remains of organisms may also be found preserved in ice or permanently frozen soil or in naturally occurring tars, salts, or other chemical deposits. Knowing the age of the rock layers in which fossils are embedded enables scientists to determine with reasonable accuracy the age of those fossils. Fossils have been discovered that have been dated by scientific methods to be more than 3 billion years old.

In undisturbed layers (strata) of sedimentary rock (such as sandstone, limestone, slate, and shale), the lowest layers were laid down first, the middle layers next, and the topmost layers last. It follows logically, then, that fossils found embedded in the lower strata are older than those in the upper strata. In fact, deeper layers of such rock are known to contain fossils of older, simpler life-forms. Strata formed near the surface contain younger and generally more complex forms. Note that these sedimentary layers may be disturbed by seismic activity that folds, inverts, subtends, or factures them. They may also be disturbed by erosion cycles that can wash fossils out of the sediments that originally encased them.

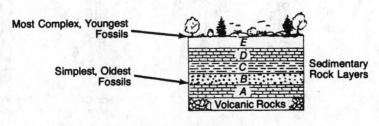

Geologic Record

Scientists have been able to identify a certain degree of continuity among the fossils in consecutive layers of fossil-bearing rock. It is possible to find fossils in upper strata that resemble those in lower strata, even though they are clearly different species. This fact lends support to the theory that genetic links exist between modern life-forms and ancient forms. It also suggests that genetic links exist among diverse modern life-forms by virtue of their common links to ancestral species. For example, all vertebrate species, from fish to mammals, share a chordate **common ancestor** that lived millions of years in the past. Modern species having similar structures share these ancestral forms in common. Even diverse species displaying few obvious similarities, such as earthworms and mollusks, are thought to share distant common ancestors. The concept of common ancestry, in which two divergent forms can trace their lineage to a single preexisting life form, is central to understanding the science of evolution.

Comparative Sciences

Added to evidence supplied through the fossil record is that stemming from the sciences of comparative cytology, anatomy, biochemistry, and genetics. The cell as a structural and functional unit is a feature that all known living things share in common. The organelles located within these cells function in much the same way in the cells of every organism. Despite the basic similarity of all cells, certain differences among cells of different species are known to exist. Organisms with a very similar cell structure are usually considered to be more closely related than organisms whose cells show many differences.

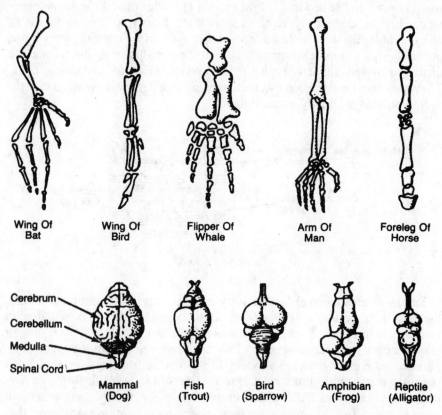

Wing Of Bat   Wing Of Bird   Flipper Of Whale   Arm Of Man   Foreleg Of Horse

Cerebrum
Cerebellum
Medulla
Spinal Cord

Mammal (Dog)   Fish (Trout)   Bird (Sparrow)   Amphibian (Frog)   Reptile (Alligator)

Homologous Forelimbs and a Comparison of Vertebrate Brains

The determination of similarities in anatomic (structural) features is perhaps the most common method of demonstrating biological relationships among organisms. This method provides the basis for biological classification, in which an organism is placed into a kingdom, phylum, class, order, family, genus, and species, based on its degree of structural similarity with other members of those groups. Similar organisms have limbs, internal organs, or other structures that are constructed similarly. Such structures, known as **homologous structures**, are believed to have originated from common ancestral forms of the same structures.

We learned that each polypeptide in the cell is coded by a unique strand of DNA. We also learned that the ability to produce such polypeptides may be passed from generation to generation through the processes of reproduction and genetic inheritance. Related organisms, therefore, having inherited their characteristics from common ancestors, may be expected to share many genes and their corresponding enzymes in common. Biochemical analysis of enzymes and other proteins shows that a great deal of similarity exists in the biochemical makeup of organisms known to be related genetically. For example, the complex protein hemoglobin is found in the blood of many vertebrate species, whereas it is less common among invertebrates. Generally, the more closely related two organisms are, the more similar is their biochemical makeup. Likewise, organisms that are not as closely related share fewer biochemical similarities.

Origin of Life

Billions of years ago, life on Earth is thought to have begun as simple, single-celled organisms. About a billion years ago, increasingly complex multi-cellular organisms began to evolve. A question that has challenged scientists and nonscientists alike is the question of the origin of life. The cell theory assumes that all cells arise from previously existing cells. However, what gave rise to the first cell? Scientists have proposed the **heterotroph hypothesis** to help explain the origin of the first primitive life-forms on the ancient Earth. This scientific hypothesis assumes that the first primitive life-forms were not able to manufacture their own food (were heterotrophic). The heterotroph hypothesis is consistent with much of the currently accepted scientific theory on the origins of the universe and with current understandings of the sciences of biology and biochemistry. However, like many hypotheses developed to explain phenomena that cannot be directly observed and measured, the heterotroph hypothesis is based on extensions of basic assumptions about Earth's origins.

The first cell is theorized to have been an aggregate of simple organic molecules that were produced in the primitive oceans from the chance combinations of inorganic molecules under extremely energy-rich conditions. Experiments conducted in the laboratory by Stanley Miller, Sidney Fox, and other scientists have confirmed that this chance aggregation is possible, even in relatively short periods of time, given the right set of conditions.

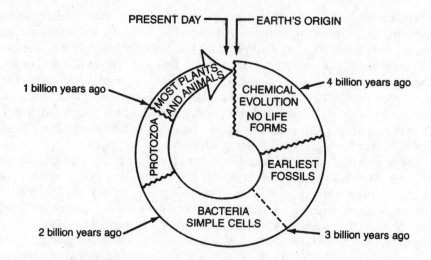

The geologic clock shows all of Earth's history from its formation 5 billion years ago to the present. Human beings appeared at "11:59," "1 minute" before the present!

Geologic Clock

It is thought that the earliest living cells obtained their energy by means of a cellular process similar to fermentation, which has the natural by-product carbon dioxide. This carbon dioxide, it is believed, built up in concentration in Earth's atmosphere as a result of widespread fermentative activity. Certain early cells, having spontaneously evolved the ability to use the newly introduced gaseous carbon dioxide to manufacture their own organic foods, became Earth's first food producers. Extensive autotrophic nutritional activity, similar to photosynthesis, added free molecular oxygen to Earth's atmosphere. This new environmental condition likely proved toxic to many of Earth's newly evolved species. A few species, having spontaneously evolved an ability to use this molecular oxygen in the respiratory process, became Earth's first aerobic organisms.

The first cell-like aggregations are assumed to have appeared about 3.5 billion years ago. Unicellular protists exist in the fossil record from about 1.8 billion years ago. The first simple multicelled organisms appear in the record approximately 1.0 billion years before the present.

Theories of Evolution

Species evolve over time. Evolution is the consequence of the interactions of (1) the potential for a species to increase its numbers, (2) the genetic variability of offspring due to mutation and recombination of genes, (3) a finite supply of the resources required for life, and (4) the ensuing selection by the environment of those offspring better able to survive and leave offspring. In the 19th century in England, a naturalist named Charles Darwin devised a theory of evolution based on variation and **natural selection**. This theory forms the basis of the modern theory of evolution. The theory of evolution by natural selection has been modified over the past 100 years to include the following points.

- *Overproduction*—Scientists observe that naturally occurring species have a tendency to produce far more offspring than can possibly survive to become reproducing adults. In fact, if all these offspring did survive, Earth's habitats and available resources would be quickly used up.
- *Competition*—Despite the tendency to overproduce, the number of individuals in natural populations tends to remain relatively constant over many generations. This suggests that within each species, there is a struggle for survival that eliminates many individuals before they reach reproductive maturity. This struggle may be termed intraspecies (within a species) competition.
- *Variation*—For centuries, scientists have been aware of the extreme variability that exists between and within species. Scientists in the time of Darwin were unaware of the mechanisms that produced this variation. This knowledge would have to wait until scientists such as Mendel, DeVries, Morgan, and others had conducted their groundbreaking work in the science of genetics. We know now that this variability results from genetic and reproductive forces. As the chemical composition of DNA changes in the cell via mutation, the resulting new traits that develop enter the gene pool and are subjected to selection pressures that either increase or decrease the frequency of the variation in the species population.

Adaptive Advantage

Each individual in a species population is engaged in a struggle for existence in its own environment. The individuals that survive (are selected) are assumed to be those best adapted to survive under the particular set of environmental conditions in question. The individuals that perish are considered to be those less well adapted for survival. When the survivors later reproduce, they tend to pass on the genes associated with their adaptive advantages. In this way, nature provides selection pressures that limit or eliminate traits that do not promote individual survival. This is the primary role of natural selection in the process of evolution.

The frequencies of these favorable genes, then, increase in the gene pool relative to the frequencies of these genes controlling less favorable traits. This shift of gene frequencies in the gene pool of a species population is thought to constitute the mechanism of evolution.

The variation of organisms within a species increases the likelihood that at least some members of the species will survive under changed environmental conditions. It is important to recognize that the adaptations in a population that are favorable under one set of environmental conditions may prove to be highly unfavorable under a different set of environmental conditions. At the same time, traits present in the gene pool of a population that have low or neutral survival value may markedly increase in value if the environment changes or if the population moves to a new environment.

Some examples of this concept are as follows.

- In the past 100 years, the environments of many insect pests (for example, houseflies, mosquitoes, roaches, weevils) changed when new chemical pesticides such as DDT were introduced into their environments. Most of the insects that came into contact with the insecticide died since they were not genetically resistant to it. A small number of the organisms were genetically resistant to the chemicals, however, and survived to reproduce offspring that were also genetically resistant. Today, such resistant strains present a problem to chemists attempting to develop other new insecticides to deal with insect infestations. In this case, the insecticide has acted as an agent of natural selection. (Note that chemical pesticides are now considered to be an environmental threat to many species other than the target insect pests, including beneficial insects and humans.)
- A similar situation has occurred in the evolution of antibiotic-resistant strains of bacteria in some hospital environments. As in the example above, newly introduced antibiotics such as penicillin represented a change in the environment of disease-causing bacteria. Nonresistant bacteria died when they came into contact with the antibiotic, but a few resistant bacteria survived. The resistant bacteria gave rise to entire strains of the bacterial species that are resistant to the antibiotic. The selecting agent in this case is the antibiotic.

Behaviors have also evolved through natural selection. The broad patterns of behavior exhibited by organisms are those that have resulted in greater reproductive success. Although structural and biochemical adaptations are those most frequently considered in discussions of genetically controlled characteristics in species, behavioral patterns are increasingly being recognized as having a genetic basis. Examples of genetically programmed behaviors include bird migration patterns, fish spawning behaviors, bird imprinting behaviors, and a host of others. The science of ethology studies the simple and complex behavior patterns of animals. Many of these behaviors promote reproductive success, whether to ensure that species members meet in favorable nesting environments or to ensure that only members of reproductively compatible species attempt mating.

Patterns of Evolution

Evolution does not result in long-term progress in some set direction. Evolutionary changes appear to be like the growth of a bush. Some branches survive from the beginning with little or no change, many die out altogether, and others branch repeatedly, sometimes giving rise to more complex organisms. The processes of mutation, allelic recombination, and natural selection are constantly at work in the production and selection of new adaptations in all species. When enough unique adaptations have been accumulated in a species population so that it becomes distinct from other populations of the same species, it may be classified as a new variety of the species. Under the right set of selection pressures, a species variety may accumulate so many variations that it becomes reproductively isolated from other varieties of the species. At this point, it has become a distinct species. The process by which new species arise from parent species is known as **speciation**. Speciation may be accelerated by factors that isolate one variety from another. These include geographic isolation and reproductive isolation.

Most scientists generally agree that the mechanisms controlling the evolutionary process are similar to those outlined above. However, considerable debate still exists concerning the time frame in which this mechanism operates. Two theories about the time frame question have been put forth.

- **Gradualism**—This theory assumes that evolutionary change is slow, gradual, and continuous. A gradualistic view of evolution is supported by fossil records of a species that display slight changes in each sedimentary layer, leading to a significant divergence between specimens found in the bottom and top layers.
- **Punctuated equilibrium**—This theory assumes that species experience long geologic periods of stability (of a million years or more) in which little or no significant change takes place. This stability is punctuated by brief periods (of a few thousand years) in which dramatic changes occur within species. During these brief periods of change, many species are thought to evolve very quickly from parent species, while other species become extinct on a massive scale. Such a view of evolution is supported by fossil evidence in which little change is noted between most sedimentary layers but sudden bursts of change (that may coincide with dramatic and cataclysmic geological events such as meteor strikes and volcanic eruptions) are evident in the fossils of a few sedimentary layers. The diagram below provides models of these two theories that may be useful in helping you understand the differences and similarities between them. The models demonstrate how an ancestral species (*X*) might have evolved into modern species (*A*, *B*, and *C*) within the context of these theories.

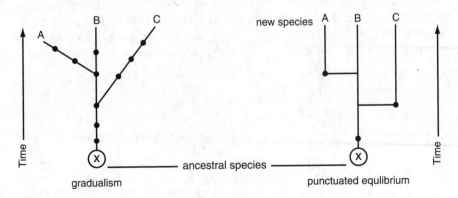

gradualism

punctuated equilibrium

ancestral species

These evolutionary events are at best incompletely understood. How they actually occurred and under what circumstances they took place can only be a matter of speculation supported by scientific evidence. What is known, however, is that modern species show a wide divergence (difference) in form and function. Modern species may be autotrophic or heterotrophic. They may be aerobic or anaerobic. They may reproduce sexually or asexually. They may differ from other species in countless ways. These varieties are thought to have come about via the evolutionary processes previously described, filling the Earth's available environments with countless species able to survive the various physical conditions they encountered.

Extinction

Extinction of a species occurs when the environment changes and the adaptive characteristics of a species are insufficient to allow its survival. Fossils indicate that many organisms that lived long ago are extinct. Extinction of species is common; most of the species that have lived on Earth no longer exist. A natural component of evolution is the **extinction** of species whose adaptations no longer suit them to the environmental conditions that they encounter.

The natural environment is always changing in subtle ways. As long as this change is slow and gradual, most species can adapt to it or migrate to nearby habitats in which the favored conditions can be reestablished. However, when environmental change is rapid or when species cannot escape gradual change by migrating to new environments, the extinction of non-adaptable species is inevitable.

Much speculation has been made about the rapid extinction of most dinosaur species within a relatively short period of time. One theory holds that an asteroid strike of sufficient magnitude to be classified as an extinction level event may have created massive environmental changes that lead to the dinosaurs' extinction. Smaller, more adaptable species are thought to have evolved rapidly during this period, including primitive bird species from ancestral dinosaur species and mammals from primitive reptilian species.

In modern times, environmental changes introduced by human activities have greatly accelerated the rate of species' extinction. Although far from normal extinction, this situation should be regarded with concern that we are altering our environment so significantly that we are endangering our own survival as well as that of many other compatible and essential species.

QUESTION SET 2.6—ORGANIC EVOLUTION
(ANSWERS EXPLAINED, P. 305)

1. The diagram below represents undisturbed rock strata in a given region. A representative fossil of an organism is illustrated in each layer.

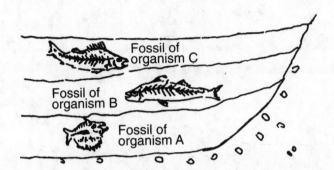

Which statement best describes a relationship among these representative organisms?
(1) Organism A was probably more structurally advanced than organism B and organism C.
(2) Organism C probably gave rise to organism A and organism B.
(3) All of these organisms probably evolved at the same time.
(4) Organism A was probably more primitive than organism B and organism C.

2. In the early stages of development, the embryos of birds and reptiles resemble each other in many ways. This resemblance suggests that they
(1) belong to the same species
(2) are adapted for life in the same habitat
(3) share a common ancestry
(4) are both animal-like protists

3. In addition to the basic ideas of Darwin, the modern theory of evolution includes a concept that
(1) variations result from mutations and gene recombination
(2) overproduction of organisms leads to extinction
(3) variations exist only in large populations
(4) competition occurs only between members of the same species

4. The diagrams below represent some structural changes that occurred over time, resulting in the development of the modern horse.

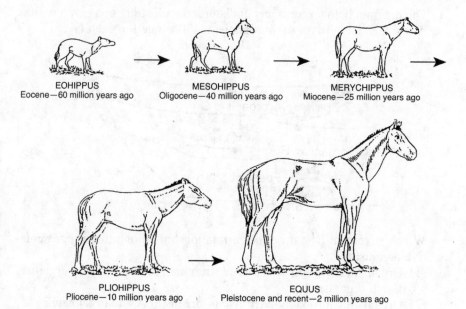

EOHIPPUS
Eocene—60 million years ago

MESOHIPPUS
Oligocene—40 million years ago

MERYCHIPPUS
Miocene—25 million years ago

PLIOHIPPUS
Pliocene—10 million years ago

EQUUS
Pleistocene and recent—2 million years ago

This sequence of structural changes best illustrates the concept of
(1) organic evolution
(2) ecological succession
(3) intermediate inheritance
(4) geographic isolation

5. Some species undergo long periods of stability interrupted by geologically brief periods of significant change. During these brief periods, new species may evolve. This pattern of evolution is part of the concept of
(1) use and disuse
(2) reproductive isolation
(3) homologous structures
(4) punctuated equilibrium

6. The diagram below illustrates the change that occurred in the frequency of phenotypes in an insect population over 10 generations.

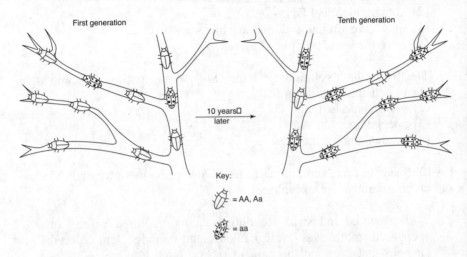

A probable explanation for this change would be that over time there was
(1) a decrease in the adaptive value of gene *a*
(2) an increase in the adaptive value of gene *a*
(3) an increase in the population of this insect
(4) a decrease in the mutation rate of gene *A*

7. According to Darwin's theory of evolution, differences between species may result from
(1) the disuse of body structures
(2) the transmission of acquired characteristics
(3) natural selection
(4) mutagenic agents

8. The concept that new varieties of organisms are still evolving is best supported by the
(1) increasing need for new antibiotics
(2) increasing number of individuals in the human population
(3) decreasing number of new fossils discovered in undisturbed rock layers
(4) decreasing activity of photosynthetic organisms due to warming of the atmosphere

9. Two nucleotide sequences found in two different species are almost exactly the same. This suggests that these species
   (1) are evolving into the same species
   (2) contain identical DNA
   (3) may have similar evolutionary histories
   (4) have the same number of mutations

10. The theory that evolutionary change is slow and continuous is known as
    (1) punctuated equilibrium
    (2) geographic isolation
    (3) speciation
    (4) gradualism

11–12. Base your answers to questions 11 and 12 on the information below and on your knowledge of biology.

> Before the Industrial Revolution, a light-colored variety of peppered moth was well camouflaged among light-colored lichens that grew on the bark of trees around London. A dark-colored variety of the peppered moth probably existed but was rarely observed because it was so easily seen by birds and eaten. When industry was introduced in London, soot killed the pollution-sensitive lichens, exposing dark tree bark. As a result, the dark-colored variety of the moth became the better camouflaged of the two moth varieties.

11. In this situation, what is the relationship between the birds and the moths?
    (1) producer-consumer
    (2) predator-prey
    (3) parasite-host
    (4) autotroph-heterotroph

12. Identify one way in which humans influenced the change in the populations of the peppered moth.

---

13. When Charles Darwin was developing his theory of evolution, he considered variations in a population important. However, he could not explain how the variations occurred. Name *two* natural processes that can result in variation in a population and explain how these processes actually cause variation. [2]

VII. CONTINUITY OF LIFE BY MEANS OF REPRODUCTION

> **KEY IDEA 4—REPRODUCTIVE CONTINUITY** The continuity of life is sustained through reproduction and development.

Essential Question:

- *What mechanisms are at work to create and perpetuate life on Earth?*

Species produce more of their own kind through the process of reproduction. In a sense, species are able to transcend the life spans of individual members of the species by passing along the genetic information contained in their reproductive cells. In asexual reproduction, the parent organism passes an exact copy of its genetic makeup to the offspring. In sexual reproduction, the father and mother each contribute 50 percent of the genetic information needed to produce the new individual. For this reason, the offspring of sexual reproduction resemble, but are not exactly like, the parents. A **zygote** (fertilized egg) contains all the genetic information needed to produce a complete organism. Development of the embryo is a highly regulated process that involves mitosis and differentiation. It is sensitive to the environmental conditions in which it is occurring. Reproductive technologies such as cloning, genetic engineering, and stem cell research have medical, agricultural, and ecological applications.

Mechanisms of Reproduction

Essential Question:

- *How do living things produce more living things of the same kind?*

Performance Indicator 4.4.1 *The student should be able to explain how organisms, including humans, reproduce their own kind.*

Reproduction and development are necessary for the continuation of any species. Cells have finite life expectancies. After formation by mitosis or fertilization, each new cell undergoes a period of growth followed by mitotic cell division. Growing evidence points to the fact that such cells have a genetically programmed limit to the number of times they can reproduce before they age and die. When cells specialize, their reproductive potential becomes significantly limited and aging becomes a more influential factor in their life expectancy. For all cells and all organisms, death is a certainty.

Therefore, the only way that a species can be continued is through the reproductive process. The living material and genetic information passed from
one generation to the next literally perpetuate the species.

Asexual Reproduction

Some organisms reproduce asexually with all the genetic information coming
from one parent. Other organisms reproduce sexually with half the genetic
information typically contributed by each parent. Cloning is the production
of identical genetic copies. All types of asexual reproduction have in common
the production of new organisms from a single parent organism. However,
the process is carried on differently by different living things. Some types of
asexual reproduction are as follows.

Binary fission is accomplished when a single cell undergoes mitosis followed
by equal cytoplasmic division. This forms two daughter cells each having
roughly the same size and shape and containing identical genetic information.
Binary fission is carried on by many species of unicellular organisms, including the paramecium and the ameba. Bacteria are also known to reproduce in
this manner.

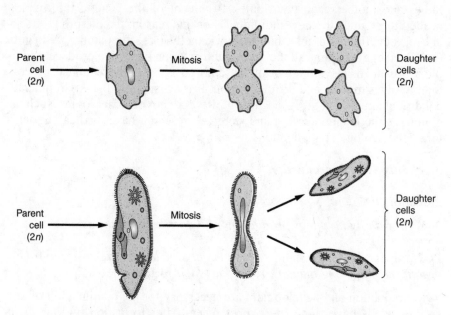

Binary Fission in Ameba (above) and Paramecium (below)

Budding is accomplished when mitosis is followed by unequal cytoplasmic
division. This results in daughter cells of unequal size that contain identical
genetic information. The larger of the two cells may divide rapidly several
more times, producing a chain or colony of daughter cells. Budding of this
sort occurs commonly in yeast. A second form of budding occurs in certain

simple multicelled animals, including the hydra. Undifferentiated cells begin to divide rapidly, forming a new, smaller organism from the tissues of the parent. This bud, genetically identical to the parent, then separates from the parent and begins to undergo rapid growth as an independent organism.

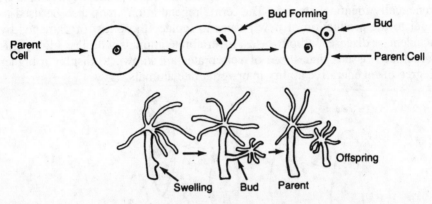

Budding in Yeast (above) and Hydra (below)

Sporulation is the formation of specialized reproductive cells, known as spores, within the parent organism. Each spore contains a nucleus surrounded by cytoplasm. When the spores are released from the parent plant and land in an environment containing conditions favorable to their growth, they begin to undergo mitotic division. The spores of most species require moisture and warmth to germinate. The mitotic divisions result in the formation of a new multicellular organism genetically identical to the original parent organism. This type of sporulation is carried on by fungi, mosses, and ferns.

Regeneration involves the production of one or more new organisms from the severed parts of a single parent organism. The pieces of cut-up planaria worm may frequently grow back lost tissues to produce as many as four or five new, identical planaria worms. Starfish cut into pieces that contain part of the central disk area may undergo regeneration that produces several new, genetically identical starfish. The term "regeneration" may also be used to refer to the production of new tissues to replace those lost or damaged by accident or disease. In both cases, invertebrate animals with less highly differentiated tissues than those of vertebrates are known to display a higher degree of regenerative ability than vertebrate animals.

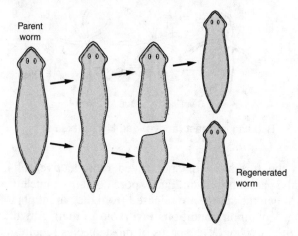

Parent worm

Regenerated worm

Vegetative propagation is a general term referring to any of several forms of asexual reproduction carried on by multicellular plants. An aspect common to all types of vegetative propagation is the production of new plant organisms from the leaves, stems, or roots (vegetative parts) of the parent plant rather than from the flower (reproductive part). As in other forms of asexual reproduction, the new organisms produced by vegetative propagation are genetically identical to the parent. Examples of vegetative propagation include cuttings, bulbs, tubers, runners, and grafting.

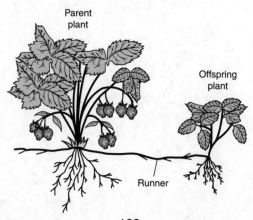

Parent plant

Offspring plant

Runner

Cloning is a term that refers to the production of a group of genetically identical offspring from the cells of a single parent organism that would normally reproduce by sexual means. Cloning has been attempted with varying success with a number of different organisms, both plant and animal. Although the process remains experimental in animals, it has proven to be quite successful in plants. It is used extensively in the production of certain commercial crops. Its main advantage is that organisms with desirable combinations of traits, which would otherwise be changed in sexual reproduction, can be reproduced rapidly, with no alteration of their phenotypic combinations. Each of the genetically identical offspring is known as a clone.

The diagram below represents the process used in 1996 to clone the first mammal, a sheep named Dolly.

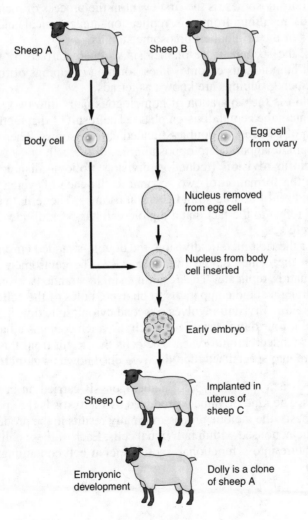

Sheep A

Sheep B

Body cell

Egg cell from ovary

Nucleus removed from egg cell

Nucleus from body cell inserted

Early embryo

Sheep C

Implanted in uterus of sheep C

Embryonic development

Dolly is a clone of sheep A

Sexual Reproduction

The processes of meiosis and fertilization are key to sexual reproduction in a wide variety of organisms. The process of meiosis results in the production of eggs and sperm that each contain half of the genetic information. During fertilization, gametes unite to form a zygote, which contains the complete genetic information for the offspring. Sexual reproduction involves the production and fusion of sex cells (**gametes**) of two parent organisms. This process ensures variations in a species. Key to this process is the production of **haploid** (n) gametes from the diploid ($2n$) cells of the parent, using the process of **meiotic cell division**. Meiotic cell division, also known as meiosis, occurs in distinct stages.

- Replication of the cell's single-stranded chromosomes, forming double-stranded chromosomes, is the first event in the process of meiosis. The chromatids resulting from this replication are chemical and genetic duplicates of the original chromosome strands.
- **Synapsis**, the second phase of meiosis, is characterized by the close pairing of homologous chromosomes, forming groupings of four **chromatids**. Such groupings are known as tetrads.
- **Disjunction** is the separation of homologous pairs into two groups as the cell enters the actual division phase. The result of disjunction is the formation of two sets of double-stranded chromosomes. These migrate along a spindle apparatus to opposite poles of the cell.
- **First meiotic division** (reduction division) follows disjunction and results in the formation of two separate cells, each of which contains the haploid set of double-stranded chromosomes. The term "**reduction division**" refers to the fact that a diploid cell has been divided into two haploid cells.
- Soon after the first meiotic division, the double-stranded chromosomes line up in the center of the two new cells. The centromeres joining the chromatids replicate. The resulting single-stranded chromosomes migrate along a spindle apparatus to opposite poles of the cell.
- **Second meiotic division** involves a second cytoplasmic division and the formation of four **daughter nuclei**, each of which contains a haploid set of single-stranded chromosomes. The cells that result from this process will mature into specialized reproductive cells known as **gametes**.

Spermatogenesis is a specific type of gametogenesis carried on in the male gonad, or testis. The kind of gamete produced by the **testis** is the sperm cell. In spermatogenesis, the meiotic process normally results in the production of four haploid nuclei housed within individual cells. Each of these cells has the potential to mature into a functional, motile **sperm cell** containing its own haploid nucleus.

Oogenesis is a specific type of gametogenesis carried on in the female gonad, or ovary. The kind of gamete produced by the **ovary** is the **egg cell**. In oogenesis, the meiotic process normally results in the production of four haploid nuclei housed within individual cells. Only one of these cells has the potential to mature into a functional egg cell containing a haploid nucleus. The other three cells, known as **polar bodies**, degenerate and are eventually reabsorbed by the body.

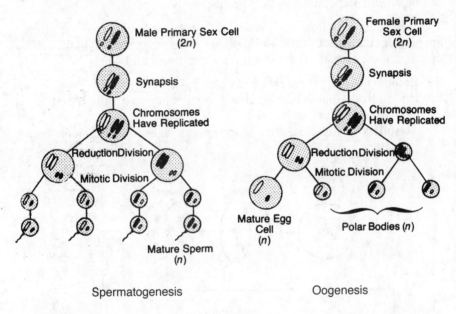

During **fertilization**, fusion of a haploid sperm cell and a haploid egg cell results in the formation of a diploid **zygote**, or fertilized egg. The restoration of the diploid condition allows the zygote to undergo mitotic divisions that eventually lead to the production of a new organism. Because the genes of the father and mother are combined in the offspring, the offspring resembles both parents but is not identical to either. In reference to meiotic cell division, this union of male and female genes is known as **recombination**.

Early Development

The zygote may divide by mitosis and differentiate to form the specialized cells, tissues, and organs of multicellular organisms. After fertilization, the zygote begins to undergo a series of rapid mitotic divisions. This process, known as **cleavage**, increases the number of cells in the growing cell mass. Little cell growth accompanies cleavage, resulting in a reduction of cell size with each division. A zygote undergoing cleavage is an **embryo**. As cleavage continues, distinct stages of development may be noted that differ with different species:

- **Blastula**—a hollow ball of cells during which the basic body plan becomes established.
- **Gastrula**—an indented blastula in which cells begin to move into the positions they will have in the adult offspring. Basic cell differentiation is thought to occur as this phase is reached.
- **Embryo**—a stage of development in which the new living organism takes on the final shape and differentiation it will have when it hatches or is born, depending on the species involved.

Differentiation is how embryonic cells become specialized into tissues that perform the various tasks throughout the body. As shown on the following chart, these tissues arise from three embryonic cell layers:

- **Ectoderm**—outer epidermis, nervous system
- **Endoderm**—digestive tract lining, respiratory tract lining, portions of the liver and pancreas
- **Mesoderm**—muscles, circulatory system, skeleton, excretory system, gonads

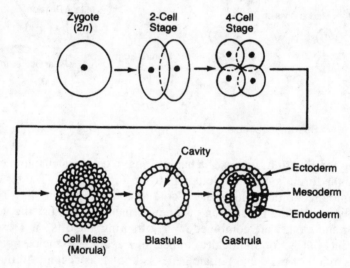

Cleavage and Differentiation

Human Reproduction

Human reproduction and development are influenced by factors such as gene expression, hormones, and the environment. The reproductive cycle in both males and females is regulated by hormones such as testosterone, estrogen, and progesterone. Development of a human embryo is governed by many factors. Principal among these factors is the precise genetic information provided to the embryo by the parents. The physical and functional traits inherited by the embryo will comprise a combination of the traits of the father and the mother.

Evidence supports the probability that basic behavioral traits are inherited in a similar fashion.

Hormones produced by the embryo as it develops, as well as those of the mother, will also help to shape the embryo. The environment of the embryo is also of extreme importance in determining the characteristics of the child. Growing evidence suggests that the health, diet, and use of drugs, tobacco, and alcohol by the mother during pregnancy can have a powerful negative effect on the developing embryo and subsequently on the child after birth.

The reproductive process in adult humans, like most physiological activities, is governed by hormones. **Testosterone** in males is important in the production of male secondary sex characteristics, including the production of healthy sperm cells. In females, the hormones **estrogen** and **progesterone**, among others, are essential in the production of female secondary sex characteristics, including the production and maturation of healthy egg cells.

Human Female   The structures and functions of the human female reproductive system, as in almost all other mammals, are designed to produce gametes in ovaries, allow for internal fertilization, support the internal development of the embryo and **fetus** (late-stage embryo) in the uterus, provide essential materials through the placenta, and provide nutrition through milk for the newborn. In human females, ova (egg cells), the female gametes, are produced by meiosis in the **ovary**. Two ovaries are located in the lower abdomen. After it has been formed, the egg cell is stored in the follicle of the ovary until it matures and is released. Once released, the egg cell travels along a short tube, the oviduct (fallopian tube), to the uterus.

The **menstrual cycle**, which begins at puberty, is a hormone-controlled process in the human female that is responsible for the monthly release of mature eggs. The average duration of the menstrual cycle is 28 days. The four stages of the menstrual cycle are as follows.

- **Follicle stage**—As the cycle begins, a single egg matures within the ovarian follicle under the influence of follicle stimulating hormone (FSH) produced by the pituitary gland. The ovary begins to produce estrogen, which stimulates the lining of the uterus to thicken and vascularize. This portion of the cycle takes approximately 14 days to complete.
- **Ovulation stage**—This portion of the cycle involves the release of the mature egg from the follicle into the oviduct. On average, this event occurs about day 14 of the cycle.
- **Corpus luteum stage**—After ovulation, the cells that made up the ovarian follicle begin to change under the influence of lutenizing hormone (LH), which is produced by the pituitary gland. The resulting structure, known as the corpus luteum, secretes the hormone progesterone, which helps to ready the uterine lining for the possible implantation of a fertilized egg. This portion of the cycle continues for about 8 to 10 days.

- **Menstruation stage**—If no fertilized egg is received in the uterus within a few days after ovulation, the uterine lining begins to break down. The disintegrated tissue and blood are expelled from the body via the vaginal canal during menstruation. This aspect of the cycle normally occurs over a few days.

The lower end of the female reproductive tract consists of the **vagina**, which functions to receive sperm cells during intercourse and later as the birth canal. **Internal fertilization** normally occurs in the **oviduct**, where sperm cells and the egg cell meet. The **uterus** acts as the internal development chamber after implantation of a fertilized egg in its lining. The lower end of the uterus is bounded by the muscular **cervix**.

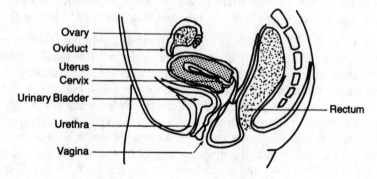

Human Reproductive System: Female

Prenatal development includes all the processes of embryonic development that occur before birth. Prenatal development begins as soon as the egg is fertilized. The resulting zygote begins to undergo cleavage. Within 10 days after fertilization, the ball of cells resulting from cleavage, now known as an embryo, implants itself in the lining of the uterus. Once embedded in the uterine lining, the embryo undergoes the process of gastrulation. At the same time, the cells of the embryo begin to differentiate into the specialized tissues and organs of the adult organism.

The tissues of the embryo and the mother grow together to form the **placenta**, the connection that allows nutrients and oxygen to pass from mother to embryo during the 9-month gestation period. Wastes such as carbon dioxide and urea pass by diffusion from the embryo's blood to that of the mother for excretion through the mother's excretory processes. The fetus (late-stage embryo) is connected to the placenta by the **umbilical cord**. The fetus is enclosed in an **amnion** sac filled with amniotic fluid.

Birth occurs after a period of development, known as the **gestation** period, of approximately 9 months. Strong contractions of the muscular uterus force the baby headfirst through the cervix and vagina to the outside of the body, where the baby begins to breathe, eat, and excrete wastes on his/ her own. Milk produced by the mammary glands provides food specialized for newborns.

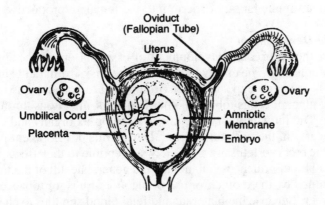

Placental Development

Human Male   The structures and functions of the human male reproductive system, as in other mammals, are designed to produce gametes in **testes** and make possible the delivery of these gametes for fertilization. Two testes are located in an internal pouch, the **scrotum**, which extends from the wall of the lower abdomen. Temperatures in the scrotum are generally 1°C to 2°C cooler than normal body temperature. This reduced temperature is optimum for sperm production. When first formed, the sperm are inactive and are stored in the testis until activated and released. Once released, the sperm travel along a series of tubes, the **vas deferens**, to the exterior of the body. Along this route, a number of glands add various fluids to the sperm cells to activate them and increase their volume. This mixture of sperm cells and fluid is known as **semen**.

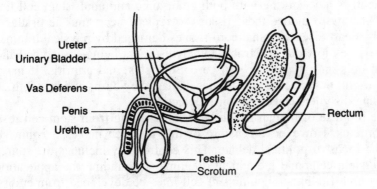

Human Reproductive System: Male

The **urethra** is a tube that carries the activated sperm along the last portion of its journey in the body. The urethra extends from the urinary bladder, at its connection with the vas deferens, to the exterior of the body through the penis. In human beings, the **penis** is the structure that permits internal fertilization through direct implantation of sperm into the female reproductive tract.

Embryonic Development

In humans, the embryonic development of essential organs occurs in early stages of pregnancy. The embryo may encounter risks from faults in its genes and from its mother's exposure to environmental factors such as inadequate diet, use of alcohol/drugs/tobacco, other toxins, or infections throughout her pregnancy. The human embryo begins to develop its organs and organ systems very early in the gestation period. During this development period, the tissues of the embryo/fetus are extremely susceptible to the effects of outside influences. These influences can include the general health of the mother, the diet maintained by the mother, or use of alcohol, drugs, or tobacco products by the mother. Because the maternal and fetal blood supplies exchange soluble materials through the tissues of the placenta, any substances that enter the mother's blood can easily diffuse across the placental membranes to affect the fetal tissues. Substances easily tolerated by adult tissues can cause severe damage to the developing fetal tissues. A direct connection has been drawn between these environmental influences and learning disabilities, attention deficits, health problems, and physical deformities experienced by infants and young children.

Stem Cell Research

In recent years, much scientific attention has been paid to the area of **stem cell research**. Stem cells are undifferentiated cells that reside in the tissues of the body and give rise to new, specialized cells. Scientists are studying stem cells to determine how their characteristics can be used to treat a variety of conditions in the body, ranging from cancer and other diseases, to damaged tissues and organs, to the effects of aging.

Scientists are working with both embryonic and adult stem cell lines to find out how to induce these tissues to replace heart muscle tissues damaged by heart disease, spinal cord tissues damaged by physical trauma, and brain tissues lost due to stroke. Other disorders being researched include Parkinson's and Alzheimer's diseases, diabetes, and arthritis. It may also prove useful in replacing skin tissues damaged by burns and in stabilizing or reversing disorders caused by birth defects.

The promise of human health benefits resulting from stem cell research is enormous. However, the nature of this research makes it controversial and the focus of political debate. This controversy includes the sources of human stem cells and the ethical dilemma surrounding the applications of the research findings. Should stem cell lines be developed from embryonic tissues, from adult tissues only, or not at all? Who will be able to receive the therapies resulting from this research? Who will make such decisions? These are ethical questions that will be answered by today's generation of young people.

QUESTION SET 2.7—CONTINUITY OF LIFE BY MEANS OF REPRODUCTION (ANSWERS EXPLAINED, P. 310)

1. Which diagram represents binary fission?

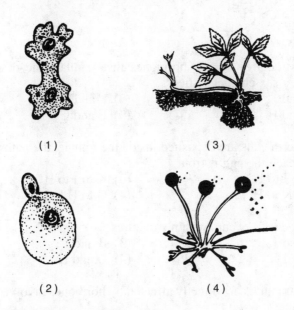

(1)   (3)

(2)   (4)

2. The process of meiotic cell division in a human male usually forms, from each primary sex cell,
   (1) one diploid cell, only
   (2) four diploid cells
   (3) one haploid cell, only
   (4) four haploid cells

3. Which sequence represents the correct order of events in the development of sexually reproducing animals?
   (1) fertilization → cleavage → differentiation → growth
   (2) cleavage → fertilization → growth → differentiation
   (3) growth → cleavage → fertilization → differentiation
   (4) fertilization → differentiation → cleavage → growth

4–7. Base your answers to questions 4 though 7 on the diagrams below and on your knowledge of biology.

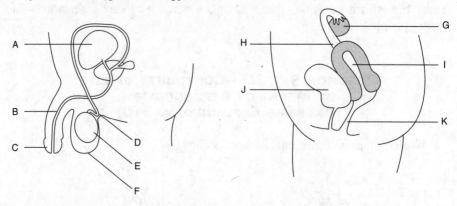

4. Which structures secrete hormones that regulate the development of secondary sex characteristics?
   (1) A and J
   (2) D and H
   (3) F and I
   (4) E and G

5. After sperm cells are deposited inside the female, the pathway they follow to reach the egg is from
   (1) H to I to K
   (2) J to K to H
   (3) K to I to H
   (4) G to H to I

6. Gametogenesis occurs within structures
   (1) A and J
   (2) E and G
   (3) B and I
   (4) D and H

7. Which structures are directly affected by hormones involved in the menstrual cycle?
   (1) C and E
   (2) A and D
   (3) G and I
   (4) I and J

8–10. Base your answers to questions 8 through 10 on the information in the chart below and on your knowledge of biology.

STAGES OF THE MENSTRUAL CYCLE

Stage	Event
A	Periodic shedding of the thickened uterine lining
B	Release of the egg
C	Production of progesterone by tissue in a follicle
D	Maturation of the egg and secretion of estrogen

8. Which structure does the egg released in stage *B* normally enter first?
   (1) cervix   (3) uterus
   (2) vagina   (4) oviduct

9. Which stage is represented by letter *A*?
   (1) ovulation   (3) follicle
   (2) menstruation   (4) corpus luteum

10. Which sequence best represents the order of stages in the menstrual cycle?
    (1) $D \rightarrow B \rightarrow C \rightarrow A$ (3) $C \rightarrow A \rightarrow B \rightarrow D$
    (2) $A \rightarrow B \rightarrow D \rightarrow C$ (4) $A \rightarrow B \rightarrow C \rightarrow D$

---

11. In human males, sperm cells are suspended in a fluid medium. The main advantage gained from this adaptation is that the fluid
    (1) removes polar bodies from the surface of the sperm
    (2) activates the egg nucleus so that it begins to divide
    (3) acts as a transport medium for sperm
    (4) provides currents that propel the egg down the oviduct

12. Substances can diffuse from the mother's blood into the fetal blood through the structure known as the
    (1) amnion   (3) yolk sac
    (2) fallopian tube   (4) placenta

13–15. Base your answers to questions 13 through 15 on the diagram below and on your knowledge of biology.

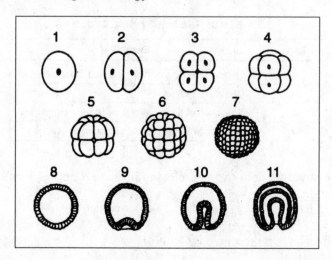

13. Which diagram shows the first appearance of the distinct layer of cells that will form the muscular, skeletal, and circulatory systems?
    (1) 11   (3) 6
    (2) 8   (4) 4

14. If stages 1 through 4 represent developmental stages of a human, where in the human female would these stages normally occur?
    (1) ovary   (3) oviduct
    (2) vagina   (4) uterus

15. Which events must occur immediately before the sequence represented in the diagrams can take place?
    (1) gametogenesis and fertilization
    (2) menstruation and menopause
    (3) prenatal development and gestation
    (4) placental formation and metamorphosis

16–18. Base your answers to questions 16 through 18 on the information and data tables below and on your knowledge of biology. Use one or more complete sentences to answer each question.

      Drinking alcohol during pregnancy can cause the class of birth defects known as fetal alcohol syndrome (FAS). Scientists do not yet understand the process by which alcohol damages the fetus. Evidence, however, shows that the more a pregnant woman drinks, the greater the chances that the child will be affected and the birth defects will be serious. Some evidence indicates that even low levels of alcohol consumption can cause intellectual and behavioral problems.

INFANT CHARACTERISTICS

Characteristics (Average)	Alcohol Use During Pregnancy	
	Drinker	Nondrinker
Weeks of development before birth	36.9	38.7
Birth weight (g)	2,555	3,094
Birth length (cm)	46.8	50.1
Head circumference (cm)	32.1	34.5

PHYSICAL ABNORMALITIES DETECTED IN INFANTS AT BIRTH

Physical Abnormalities	Alcohol Use During Pregnancy	
	Drinker (Percentage of 40 Infants)	Nondrinker (Percentage of 80 Infants)
Low birth weight	73	12
Small brain	33	0
Flattened nasal bridge	8	0
Abnormal facial features	15	0
Spinal defects	8	0
Heart defects	8	0

16. Do the data in the tables justify scientists' conclusions that alcohol causes physical abnormalities at birth by interfering with the normal development of the fetus? Defend your position with supporting data.

17. What additional data would be needed to better support the scientists' conclusions?

18. Explain why alcohol consumption by the mother is especially harmful during the early stages of pregnancy.

---

19. Write one or more paragraphs that compare the two methods of reproduction, asexual and sexual. Your answers must include: [4]

   - *one* similarity between the two methods [1]
   - *one* difference between the two methods [1]
   - *one* example of an organism that reproduces by asexual reproduction [1]
   - *one* example of an organism that reproduces by sexual reproduction [1]

VIII. DYNAMIC EQUILIBRIUM IN LIVING THINGS

> ### KEY IDEA 5—DYNAMIC EQUILIBRIUM AND
> ### HOMEOSTASIS Organisms maintain a dynamic equilibrium
> that sustains life.

Essential Question:

- *How do living things create a balanced steady state that maintains the living condition?*

Energy and materials must be constantly provided to living systems to keep them operating. Within the cell, biochemical processes occur within a very narrow range of conditions. These conditions are maintained by the very biochemical processes that require their stability. This maintenance is accomplished by the responsiveness of living systems to the environmental changes that occur around the cell. These responses can range from simple biochemical reactions to complex behaviors and result in a **dynamic equilibrium** known as **homeostasis**. When the feedback mechanisms that create and maintain homeostasis fail, the living condition is threatened, and disease or death can result.

Biochemical Processes and Homeostasis

Essential Question:

- *What biochemical reactions are common in living cells and promote the living condition?*

Performance Indicator 4.5.1 *The student should be able to explain the basic biochemical processes in living organisms and their importance in maintaining dynamic equilibrium.*

Photosynthesis
The energy for life comes primarily from the sun. Photosynthesis provides a vital connection between the Sun and the energy needs of living systems. The Sun, actually a medium-sized star, is the ultimate source of all energy on Earth. It produces radiant energy in many forms, some of which is detectable by humans as light. Visible sunlight appears white but actually contains the wavelengths needed to produce all colors. In **photosynthesis**, the energy of sunlight is trapped and converted into the chemical bond energy of organic

compounds such as sugar. Once trapped in the chemical bonds of these compounds, the energy of sunlight is made available to all living things in a more stable form. Green plants provide a vital link between the Sun's energy and the life processes of all living things.

Plant cells and some one-celled organisms carry on photosynthesis. An organelle found in the cells of green plants and algae is the **chloroplast**. The chloroplast absorbs sunlight and converts the energy to chemical bond energy. **Chlorophyll**, a green pigment found in the chloroplast, absorbs light energy. In the presence of the proper enzymes, the atoms of carbon dioxide (CO_2) and water (H_2O) are rearranged to form the more complex organic molecule glucose ($C_6H_{12}O_6$). A by-product of this process is molecular oxygen (O_2).

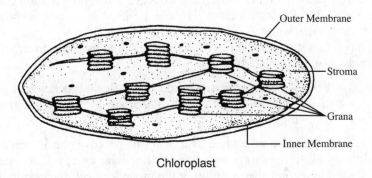

Chloroplast

A more detailed study of the photosynthetic process reveals that it actually consists of two separate processes. Each is characterized by its own set of chemical reactions.

Photochemical (light) reactions Stacked layers of chlorophyll-containing membranes within the chloroplast, known as **grana**, contain the enzymes that catalyze this process. The light energy absorbed by the chlorophyll and other pigments on these grana is used to split water molecules into their component elements, hydrogen and oxygen. This process is known as **photolysis** (splitting with light) and involves the production of an energy-carrying molecule known as adenosine triphosphate (ATP). Atoms of oxygen recombine to form molecular (atmospheric) oxygen, which is released as a gas. The hydrogen released is transferred to the next phase of the process by a hydrogen carrier compound.

Carbon-fixation (dark) reactions A second set of photosynthetic reactions combines the released hydrogen atoms with the atoms making up carbon dioxide. An intermediate product of this reaction is a carbohydrate-like, three-carbon compound known as **phosphoglyceraldehyde (PGAL)**. This compound may be used by the cell to synthesize several other compounds such as glucose. The name carbon-fixation reactions is derived from the fact that carbon atoms are fixed in place in a stable form as a result of this process. This set of reactions occurs in the **stroma** of the chloroplast. The stroma lies between the grana.

In all organisms, organic compounds can be used to assemble other molecules such as proteins, DNA, starch, and fats. The chemical energy stored in bonds can be used as a source of energy for life processes. The compound glucose, which results from photosynthesis, may be used in many ways by the cell. Some of these include the following.

- Direct use as a source of energy in cellular respiration, which is carried on by all living things, plants as well as animals.
- Conversion, for purposes of storage, into more complex forms of carbohydrates, such as starch, by the process of dehydration synthesis. Such storage carbohydrates must be reconverted to simpler molecules by intracellular digestion before they can be used in metabolic processes or transported to other sites in the organism.
- Conversion into other types of metabolic compounds, such as proteins, lipids, and nucleic acids.

Photosynthesis is perhaps the single most significant biochemical process carried on by living things. Significant results of the photosynthetic reactions include the following.

- Carbohydrates manufactured as a result of photosynthesis are used by virtually all living things as a source of cellular energy. This energy is released in a controlled fashion in the process of cellular respiration.
- Oxygen gas released as a by-product of the photosynthetic reactions is the principal source of atmospheric oxygen required by most living things for cellular respiration. Reductions in photosynthetic rate can have disastrous effects on such life-forms by reducing the concentration of oxygen in the air or water environment.

Respiration

In all organisms, the energy stored in organic molecules may be released during cellular respiration. This energy is temporarily stored in ATP molecules. The chemical bonds between atoms of food molecules provide the energy used by all living organisms to sustain life. Respiration is the biochemical mechanism by which this energy is released for use in the cell. Once released, it becomes available to the cell for use in all other life functions. Respiration takes place in cell organelles known as the **mitochondria**.

Cellular respiration is a biochemical process that includes the reactions used by cells to release energy from organic molecules such as glucose. The energy released as a result of these reactions is temporarily stored in the bonds of molecules of **adenosine triphosphate (ATP)**. This energy may be released for use in cell processes when ATP is converted to adenosine diphosphate (ADP) and phosphate (P). Like other reactions that occur in the cell, the reactions of respiration are controlled by enzymes. The following equation illustrates the reactions involving the formation of ADP and the release of cell energy:

$$H_2O + ATP \xrightarrow{\text{ATPase}} ADP + P + energy$$

The two forms of cellular respiration carried on by living things are known as **aerobic** (with oxygen) and **anaerobic** (without oxygen).

1. **Anaerobic respiration** is a form of cellular respiration carried on by cells in the absence of molecular oxygen. This process operates by means of chemical reactions catalyzed by enzymes located in the cytoplasm of the cell. Organisms typically employing this form of cellular respiration include certain bacteria and fungi (yeasts). Human beings have learned to take advantage of this process by using these organisms in the manufacture of foods such as cheese, buttermilk, and yogurt. These organisms also play an important role in the baking, wine-making, and brewing industries. Two principal types of anaerobic respiration occur.

   • **Alcoholic fermentation** is so called because one of the major by-products produced in this process is ethyl alcohol:

$$glucose \xrightarrow{\text{enzymes}} 2 \text{ ethyl alcohol} + 2\ CO_2 + 2\ ATP$$

   • **Lactic acid fermentation**—which produces lactic acid as one of its by-products:

$$glucose \xrightarrow{\text{enzymes}} 2 \text{ lactic acid} + 2\ ATP$$

   Both types of anaerobic respiration are considered to be relatively inefficient. The end products of both (ethyl alcohol and lactic acid) contain considerable amounts of energy that are not effectively released for use by the cell.

2. **Aerobic respiration** is a form of cellular respiration carried on by certain cells in the presence of molecular oxygen. This process operates by means of chemical reactions catalyzed by enzymes located primarily in the mitochondria of the cell. As in anaerobic respiration, the principal product of aerobic respiration is the energy released from organic molecules for use by the cell in other processes. Unlike anaerobic respiration, however, aerobic respiration uses molecular oxygen to release substantially more energy per glucose molecule metabolized. For this reason, aerobic respiration is a relatively more efficient form of respiration than anaerobic respiration. In general, aerobic respiration may be illustrated as follows.

$$\text{glucose} + 6 \text{ oxygen} \xrightarrow{\text{enzymes}} 6 \text{ water} + 6 \text{ carbon dioxide} + 36 \text{ ATP}$$

Like anaerobic respiration, aerobic respiration is characterized by a series of enzyme-catalyzed reactions that progressively convert glucose to end products and energy. These reactions are conceptually divided into two phases, known as the anaerobic phase and the aerobic phase.

- **Anaerobic phase**—Glucose molecules are converted to two molecules of **pyruvic acid**. The process is initiated through release of the energy stored in two molecules of ATP. Four molecules of ATP are formed in this conversion, resulting in a net gain of two molecules of ATP. A summary equation of this phase is as follows:

$$\text{glucose} + 2 \text{ ATP} \xrightarrow{\text{enzymes}} 2 \text{ pyruvic acid} + 4 \text{ ATP}$$

- **Aerobic phase**—The cell uses molecular oxygen to break down pyruvic acid further to form energy and the waste end products water and carbon dioxide. In this part of the process, an additional 34 molecules of ATP are formed. A summary equation of this phase is as follows:

$$2 \text{ pyruvic acid} + 6 \text{ oxygen} \xrightarrow{\text{enzymes}} 6 \text{ water} + 6 \text{ carbon dioxide} + 34 \text{ ATP}$$

The net gain of ATP molecules in aerobic respiration is 36, making it 18 times more efficient than anaerobic respiration. The significance of aerobic respiration is that usable energy is stored in the chemical bonds of the ATP molecules.

The energy from ATP is used by the organism to obtain, transform, and transport materials and to eliminate wastes. Cellular operations, primarily biochemical reactions, are energy-consuming activities. The energy in ATP is released to initiate these reactions if they are exothermic (energy-releasing) or to provide sustained energy if they are endothermic (energy-consuming). Active transport, in which materials are moved into or out of the cell against the concentration gradient, is a good example of an energy-consuming process. By using this process, the cell can obtain and concentrate needed materials, move them around the cell interior, and expel potentially toxic wastes to the exterior environment.

Enzyme Action
Biochemical processes, both breakdown and synthesis, are made possible by a large set of biological catalysts called enzymes. Enzymes can affect the rates of chemical change. The rate at which enzymes work can be influenced by internal environmental factors such as pH and temperature.

The controlled nature of cellular chemical reactions is made possible through the actions of cellular enzymes. Enzymes regulate the rate at which the cell's chemical reactions occur. In this role, enzymes function as organic catalysts. Each chemical reaction in the cell requires its own, specific type of enzyme in order to operate.

The rate at which enzymes catalyze their reactions changes as conditions inside the cell change. Such conditions as the temperature of the cell, the relative concentrations of enzyme and substrate within the cell, and the pH of the cell can alter the rapidity with which enzymes work.

Temperature The cell and its enzymes are very sensitive to temperature conditions. Extreme cold can slow enzyme action nearly to a halt. As the temperature of the cell rises, its enzymes begin to operate more and more rapidly. An optimum (best or most efficient) temperature allows the most rapid reaction rate. In humans this optimum temperature is 98.6°F (38°C). Extreme heat—at or above 104°F (40°C)—can halt enzyme action by deforming the molecular shape of the enzyme. This distortion is known as **denaturation** and is a permanent condition that makes the enzyme incapable of further catalytic action. The effect of temperature on a typical human enzyme is shown below.

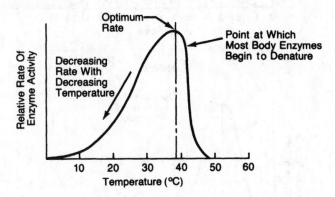

Enzyme Activity Versus Temperature

Concentration of enzyme and substrate In order to operate, enzyme molecules require the presence of substrate. In a system in which the concentration of enzyme is constant, increasing the substrate concentration from zero will result in a steady rise in reaction rate. This rise continues until the concentration roughly equals that of the enzyme. At this point, the enzyme and substrate are in equilibrium and the rate of increase slows and levels off. Increases in substrate concentration beyond this equilibrium point have little or no effect on the rate of enzyme action.

Cell acidity (pH) The cell environment may be characterized by its level of acidity or alkalinity. To measure this characteristic accurately, scientists have devised methods of measuring the relative concentration of hydrogen ion (H^+), which is given off by acids in solution with water. The pH scale has been devised to indicate the relative hydrogen-ion concentration. This scale runs from 0 to 14, 0 being extremely acidic and 14 being extremely alkaline. A pH of 7 indicates a neutral condition (neither acidic nor alkaline). A pH less than 7 indicates an acidic condition; a pH greater than 7, an alkaline condition.

Most human enzymes seem to work best at or near a pH of 7 (neutral). Despite this fact, there are many enzymes for which the optimum pH is highly acidic (stomach enzymes) or highly alkaline (intestinal enzymes). In general, however, as the cell's acidity increases above 7 or decreases below 7, most enzymes show a decrease in activity.

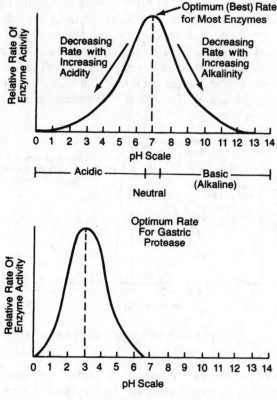

Enzyme Rate Versus pH

Enzyme Structure Enzymes and other molecules, such as hormones, receptor molecules, and antibodies, have specific shapes that influence both how they function and how they interact with other molecules. **Enzymes** are protein molecules. They are both extremely complex and extremely variable. Certain enzymes are also known to contain nonprotein components known as coenzymes; vitamins frequently function as coenzymes. Enzymes are usually named for the particular chemical (substrate) whose reactions they catalyze; enzyme names end in the suffix *-ase*.

Enzyme molecules are thought to contain specific areas responsible for linking to the substrate molecules. These reacting areas are known as **active sites**. They can be thought of as pockets or slots on the enzyme molecule into which the substrate molecules fit during the reaction catalyzed by the enzyme.

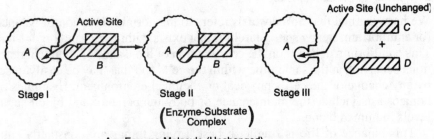

A = Enzyme Molecule (Unchanged).
B = Substrate Molecule.
C & D = Products of Hydrolysis of Molecule B.

Catalytic Action of an Enzyme

The catalytic action of enzymes may be depicted in many ways. However, the following model, known as the **lock-and-key model**, is widely accepted as being a reasonable summary of this action.

- To begin the process, the enzyme molecule must link with the substrate in a temporary, close physical association known as the **enzyme-substrate complex**.
- This association is thought to occur at the enzyme's **active site**. The chemical fit between this active site and the substrate must be exact in order for the desired reaction to occur, in much the same way that a key must fit a lock in order for the lock to be opened (hence the name "lock-and-key" model).
- During the existence of the enzyme-substrate complex, the reaction involving the substrate takes place. The substrate is chemically altered in this reaction, and products are formed. The enzyme molecule, however, remains unchanged by the reaction.
- When the reaction is completed, the enzyme and product(s) separate. The enzyme molecule, being unchanged, is free to form additional complexes and to catalyze additional reactions of this substrate.
- A single enzyme molecule may catalyze millions of reactions in this manner during its lifetime. Eventually, however, the structure of the enzyme molecule begins to deteriorate. The molecule must then be replaced.

Disease as a Failure of Homeostasis

Essential Questions:

- *How does the body defend itself against disease?*
- *What happens when a living thing gets sick?*

Performance Indicator 4.5.2   *The student should be able to explain disease as a failure of homeostasis.*

We have learned that **homeostasis** refers to the normal condition of balance (or equilibrium, steady state, stability) that exists within cells and organisms. This equilibrium depends on the coordination of thousands of chemical reactions occurring at the same time within the cells and that may be easily upset by any change of the cell's physical or chemical environment. Examples of homeostasis include the maintenance of body temperature and blood sugar levels in human beings.

This balance of life is fragile at best. Living things are constantly subjected to changes in their environments that threaten to interfere with homeostatic balance. Living systems have been developed that provide feedback mechanisms to help the regulatory system monitor the interior and exterior environment and send commands to the body to adjust to changes it detects. If these feedback mechanisms fail to provide the appropriate responses, then cell functions could become less efficient than normal, which in turn could lead to illness or death.

Organisms That Interfere with Cell Activity

Viruses, bacteria, fungi, and other parasites may infect plants and animals and interfere with normal life functions. In addition to internal disruptions in the cell that can trigger an imbalance, other organisms and materials can also disrupt the cell's operation. Many human diseases are caused by viruses, bacteria, fungi, and parasites of various types. These organisms interfere with cell activity in a number of ways.

Viruses   enter the body by inhalation of airborne viruses, consumption of contaminated water or foods, or through direct contact with infected tissues. They act by commandeering the cell's protein synthesis mechanism, redirecting it to produce many copies of the virus that go on to infect other cells. The host cells are destroyed in the process, impacting the operation of the affected tissues and organs. Examples of viral diseases in humans include the common cold, measles, hepatitis, polio, smallpox, rabies, viral pneumonia, yellow fever, herpes, and AIDS.

Bacteria   invade healthy tissues through breaks in the skin, through consumption of contaminated food or water, or by inhalation of airborne bacteria. They begin to reproduce rapidly within the host. Bacterial infections are dangerous because they produce toxins that can kill healthy host tissues. Some bacteria actually enter host cells and consume them from within. Examples of bacterial diseases in humans include Lyme disease, giardiasis, whooping cough, tetanus, typhoid fever, diphtheria, bacterial pneumonia, bubonic plague, tuberculosis, gonorrhea, chlamydia, and syphilis.

Fungi feed on healthy tissues of the body by sending rootlike tendrils into the tissues and digesting the tissues extracellularly. Common examples of fungal infections in humans include athlete's foot and ringworm.

Parasites of various kinds can invade the body through the skin via insect bites or through consumption of contaminated water or foods. Some parasites bore directly through the exposed skin to reach internal tissues. Examples of parasitic diseases in humans include hookworm, tapeworm, trichinosis, malaria, typhus, and African sleeping sickness.

Plants and other animals can be similarly infected by disease-causing agents that are specific to the tissues of those host organisms. Scientists are constantly at work developing new treatments for the diseases caused by these agents, both to protect human health and to safeguard the health of agricultural products such as grains and livestock.

The Immune System

The immune system protects against antigens associated with pathogenic organisms or foreign substances and against some cancer cells. A special function of the blood is to defend the body against disease, to provide **immunity**. In addition to the other functions associated with the transport system, the blood provides the immune response to help it react to foreign invaders. Such invaders may include the viruses, bacteria, fungi, and parasites discussed above. They may also include the chemical antigens and toxins produced by those organisms. Special cells in the body can also be important in defending the body against its own cells, when those cells become abnormal through cancer.

Some white blood cells engulf invaders. Others produce antibodies that attack the invaders or mark them for killing. Some specialized white blood cells will remain, able to fight off subsequent invaders of the same kind. White blood cells exist in several different forms with several different functions. Two such types are described here, both important in the control of disease in humans.

- **Phagocytes** engulf and destroy bacteria that enter the bloodstream. These phagocytic white blood cells gather in large numbers at sites of bacterial infection in the body.
- **Lymphocytes** produce **antibodies** specifically designed to fight off particular types of foreign proteins, known as antigens, that enter the bloodstream by various routes.

Once these cells react to a foreign invader, the immune system develops a chemical memory of the invader's proteins (antigens). When this happens, an **active immunity** develops, which is a long-lasting protective condition created when the body recovers from an infectious disease. The lymphocytes carry their chemical memory of that antigen and churn out specific antibodies at the first sign of its return to the bloodstream.

Vaccination uses weakened microbes (or parts of them) to stimulate the immune system to react. This reaction prepares the body to fight subsequent invasions by the same microbes. Many years ago, it was discovered that inoculating people with solutions containing dead or weakened disease microbes (or fragments of these microbes) would impart the same level of immunity to the inoculated person as does recovery from the disease but without many of the risks associated with contracting the full-blown disease. This inoculation of dead or weakened microbes is known as vaccination. Active immunity acquired through vaccination usually results in the same long-lasting immunity as does the body's own immune reaction.

Some viral diseases, such as AIDS, damage the immune system, leaving the body unable to deal with multiple infectious agents and cancerous cells. AIDS is an acronym for acquired immune deficiency syndrome and is caused by the HIV virus. As its name implies, AIDS leads to a reduced efficiency of the immune system, leaving its victims unable to ward off disease-causing agents that enter the body. Certain types of cancer can also gain a foothold under these conditions. As the body's ability to fight off infectious diseases diminishes, an AIDS victim can contract multiple diseases, leaving him/her in a progressively weakened condition. If untreated, this situation can gradually lead to death. There is currently no cure for AIDS, although treatments have been developed that slow the disease's progress.

Some **allergic reactions** are caused by the body's immune responses to usually harmless environmental substances. Under certain circumstances the immune system may attack some of the body's own cells or transplanted organs. Allergies are the result of the body's reaction to the chemical composition of such materials as pollen, dust, animal dander, insect saliva, foods, drugs, molds, and many other substances. Many of these materials are harmless allergens. Sensitive people produce antibodies or antitoxins to these allergens as though they were the antigens of disease-causing organisms. An additional bodily response to the antigens is the production of histamines, which may cause irritation and swelling of the mucous membranes, a symptom typical of allergic reactions that may be more life-threatening than the allergens themselves.

Organ and tissue transplants, including blood transfusions, can be safely accomplished only if the antigen types of the donor and recipient are the same or very similar. If a close match does not exist between the recipient's and the donor's antigens, the recipient's body will react to the transplanted tissues as though they were foreign, disease-causing organisms and produce antibodies. The result of such a reaction is the rejection of the transplanted tissue or organ.

Other Disorders

Disease may also be caused by inheritance, toxic substances, poor nutrition, organ malfunction, and some personal behaviors. Some effects show up right away; others may not show up for many years. There are other causes of diseases and disorders. Some examples are as follows.

- **Inheritance:**
  - ➤ sickle cell anemia—abnormal red blood cells are crescent-shaped and hemoglobin is defective
  - ➤ Tay-Sachs disease—fatty deposits build up around nerves, leading to mental retardation
  - ➤ hemophilia—blood does not clot due to its inability to produce clotting factors
  - ➤ arthritis—cartilage breaks down and skeletal joints deteriorate and/or fuse together
  - ➤ phenylketonuria (PKU)—mental retardation caused by inability to metabolize phenylalanine

- **Toxic substances:**
  - ➤ ulcers—erosion of the stomach/intestine from irritation by alcohol or other substances
  - ➤ asthma—narrowing of the bronchial tubes due to swelling caused by allergic reaction
  - ➤ emphysema—a general deterioration of the lung structure that results from exposure to air pollutants or tobacco

- **Poor nutrition:**
  - ➤ goiter—enlargement of the thyroid caused when that gland fails to produce sufficient quantities of thyroxin due to a lack of iodine in the diet
  - ➤ gallstones—deposits of hardened cholesterol that lodge in the gallbladder and cause pain
  - ➤ anemia—a reduced ability of the blood to carry oxygen that results from an iron-deficient diet

- **Organ malfunction:**
  - ➤ constipation—the large intestine absorbs too much water from feces, leading to difficult elimination
  - ➤ diarrhea—large intestine fails to absorb sufficient water from feces, leading to watery elimination
  - ➤ coronary thrombosis—blockage of the coronary artery, which feeds the heart muscle
  - ➤ angina—chest pain caused by gradual deterioration of coronary circulation

➤ gout—painful inflammation of the skeletal joints caused by the buildup of uric acid

➤ stroke—condition caused when a portion of the brain fails due to inadequate or blocked blood supply

➤ diabetes—inability to maintain proper sugar balance in the blood due to failure of the pancreas to produce sufficient quantities of insulin

- **Personal behavior:**

  ➤ high blood pressure—caused by fatty buildup in arteries, is aggravated by stress, smoking, and diet, among other activities

  ➤ tendinitis—inflammation of the tendon-bone junction caused by repeated physical stress on the affected part

Cancer

Gene mutations in a cell can result in uncontrolled cell division, known as **cancer**. Exposure of cells to certain chemicals and radiation increases mutations and thus increases the chance of cancer. Normal mitotic cell division results in the production of new cells for growth and for the repair of damaged or worn out body tissues. It is controlled within the cell itself and occurs countless times in every living cell without flaw. However, in some cells under certain conditions, the mitotic process appears to break down and begins to occur so rapidly that insufficient time is available for normal replication and chromosome separation. This rapid, abnormal cell division is known as cancer. In a very short time, cancer may produce a very large number of such abnormal cells, which begin to crowd out the normal tissues. This results in damage to these tissues and often in the death of the host organism.

Research to Control and Eliminate Disease

Biological research generates knowledge used to design ways of diagnosing, preventing, treating, controlling, or curing diseases of plants and animals. A large and increasingly important branch of biology deals with the study and control of disease in humans and other organisms. Some scientists study the bacteria and viruses that cause disease in an attempt to determine how these organisms work within the body to disrupt systems. Other scientists are hard at work decoding the genomes of these organisms to learn how their genetic makeup affects other living things. The human genome has also been mapped and decoded, enabling scientists to trace the root causes of genetically based diseases and to design gene therapy techniques to treat them. Branches of science deal with finding effective treatments for disease in the form of drugs and antibiotics. These and other research activities are providing medical and agricultural science with a sophisticated and effective arsenal to help us to combat diseases in living things.

Body Systems and Homeostasis

Essential Question:

- *How do the activities that sustain the life of a cell contribute to the survival of a multicellular organism?*

Performance Indicator 4.5.3   *The student should be able to relate processes at the system level to the cellular level in order to explain dynamic equilibrium in multicelled organisms.*

Dynamic equilibrium   results from detection of and response to stimuli. Organisms detect and respond to change in a variety of ways both at the cellular level and at the organismal level. A **stimulus** is any change that occurs in the environment of an organism that elicits a response in that organism. Such stimuli may be external or internal to the body of the organism. Examples of stimuli include light, sound, and chemical stimuli. A **response** is a reaction that an organism makes to a specific stimulus. Responses may include physical movements or glandular secretions. By reacting to its environment, an organism is able to adjust to changing conditions with the result that its systems remain in a **dynamic equilibrium** (energy-charged and adapting steady state). This equilibrium is an important prerequisite to homeostasis.

A unique feature of living matter is its sensitivity to its environment and its ability to adjust to changes both inside and outside of itself. As is the case with all life functions, this sensitivity has its basis at the cellular level. Independent one-celled organisms display sensitivity appropriate to their lifestyle and environment. This phenomenon is known as cytoplasmic sensitivity. It allows the cell to receive environmental stimuli and perform simple actions, such as avoidance behaviors, in response to them. Some one-celled organisms have specialized sensory areas that can detect environmental stimuli, such as light or vibration. Unicellular organisms use simple structures such as cilia or flagella to effect their responses to these stimuli.

In multicelled organisms, a variety of specialized adaptations assist the organism to detect stimuli and to effect responses to those stimuli. In simple organisms such as the hydra, these adaptations are limited to specialized cells that sense stimuli and communicate to other cells (nerve net) and other cells that produce crude movement (contractile fibers) or release stinging barbs (nematocysts). In more complex organisms such as the human, specialized sensing organs take in a wide range of environmental stimuli and send complex information, via nerves, to the central-processing centers of the brain and spinal cord. After analysis, this information provides the basis for complex responses including movements and secretions. The effectors of these responses (muscles and glands) receive their commands from the brain. Constant feedback allows the system to adjust its reactions based on the effects of the initial response.

In addition to sophisticated nervous responses, the bodies of complex multicelled organisms use a system of endocrine glands to control the metabolic activities of specific tissues, organs, and systems. By controlling the body's metabolism, the balance needed to sustain life is maintained.

Feedback mechanisms have evolved that maintain homeostasis. Examples include the changes in heart rate or respiratory rate in response to increased activity in muscle cells, the maintenance of blood sugar levels by insulin from the pancreas, and the changes in openings in the leaves of plants by guard cells to regulate water loss and gas exchange.

The term **feedback mechanism** refers to any process through which the body is able to adjust its actions in response to internal stimuli. The examples given represent just a few ways that the body monitors its own internal environment and adjusts to the changes it detects. Examples include:

- As the body becomes physically active, the muscle cells carry on respiration at an elevated rate. This produces additional energy but also elevated levels of carbon dioxide. As this carbon dioxide enters the bloodstream, its concentration increases and blood chemistry changes as a result. This change in the body's internal environment is detected by nerve centers in the brain and spinal cord. These nerve centers initiate commands to the breathing apparatus and heart that cause these organs to increase their rates of activity. The body's elevated heart rate causes the blood to circulate more rapidly around the body, enabling it to carry more oxygen to the muscle cells and carry away more carbon dioxide from these tissues. Breathing rate increases to accelerate the rate of gas exchange in the lungs. This response reduces carbon dioxide levels and increases oxygen levels in the blood. When carbon dioxide levels drop to a sufficiently low level, this change is also detected by the nerve centers. They then issue commands to the breathing apparatus and heart muscle to slow their rates of activity, returning the body to its resting levels. This is a clear example of a feedback loop.
- In a like manner, the body monitors the concentration of sugar (glucose) in the blood. When blood sugar levels become elevated, as they would after a person consumes and digests a meal containing carbohydrates, the pancreas is stimulated to release the hormone insulin. Insulin causes the liver to increase the rate at which blood sugar is converted to glycogen (animal starch). A second hormone, glucagon, is released from the pancreas when blood sugar concentrations drop too low, as they would long after a meal has been consumed. Glucagon stimulates the conversion of glycogen to glucose, which then enters the bloodstream to increase the blood sugar concentration. This is a clear example of a feedback loop.
- Stomates are tiny openings on the leaves of green plants that allow gas exchange between the environment and the photosynthetic cells of the

leaf interior. These openings are regulated by pairs of **guard cells** that flank each stomate. Guard cells contain chloroplasts. When photosynthetic activity is high, the concentration of sugar builds up in the guard cells. This change causes the guard cells to absorb additional water in an effort to maintain a constant sugar concentration in the cytoplasm. The absorption of this water causes the guard cells to become turgid, opening the stomate and promoting rapid gas exchange. When photosynthetic activity is low, the guard cells lose sugar and water, causing them to become limp. In this condition, the guard cells are no longer able to keep the stomates open, so gas exchange is minimized.

Both single-celled organisms and multicelled organisms sense and respond appropriately to their environments. The responses of a multicelled organism are a function of the responses of the many individual cells that make it up.

QUESTION SET 2.8—DYNAMIC EQUILIBRIUM IN LIVING THINGS
(ANSWERS EXPLAINED, P. 316)

1. What is a major distinction between living and nonliving matter?
   (1) Living matter is unable to diffuse materials.
   (2) Living matter is able to control chemical activities with organic catalysts.
   (3) Living matter is able to create energy.
   (4) Living matter is unable to use energy for metabolic activities.

2–3. Base your answers to questions 2 and 3 on the diagram below and on your knowledge of biology. The diagram represents some processes occurring in the leaf of a plant.

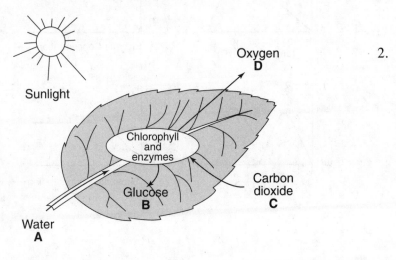

191

Which equation illustrates a process of nutrition carried out within the leaf?

(1) $B + D \rightarrow A + C$ (3) $B + C \rightarrow A + D$

(2) $A + C \rightarrow A + B + D$ (4) $A + B + D \rightarrow B + C$

3. Which letters indicate substances needed by the leaf to carry out the process of aerobic cellular respiration?

(1) A and C   (3) B and C

(2) C and D   (4) B and D

---

4. All producers and consumers use the chemical process of respiration to synthesize

(1) $C_6H_{12}O_6$ (3) alcohol

(2) ATP   (4) oxygen

5. Brine shrimp live in shallow coastal waters or near the surface of the ocean where light penetrates. Under laboratory conditions, brine shrimp are attracted to areas with the greatest light intensity and avoid areas of low light intensity. The movement of the brine shrimp to bright light is an example of

(1) negative feedback   (3) a stimulus

(2) a response   (4) active transport

6–7. Base your answers to questions 6 and 7 on the graphs below and on your knowledge of biology. The graphs represent human enzyme activity.

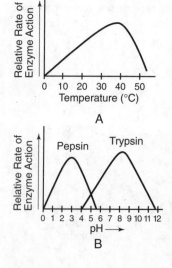

A

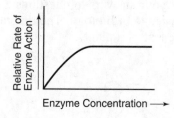

C

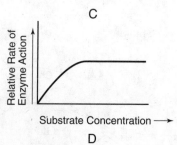

D

6. Human enzymes would most likely begin to denature at a
   (1) temperature of 40°C   (3) pH of 3
   (2) temperature of 23°C   (4) pH of 2

7. Certain enzymes work best within an acidic or a basic environment. This concept is illustrated in graph
   (1) A   (3) C
   (2) B   (4) D

---

8. The equation below summarizes some of the reactions involved in a specific biochemical process.

   hydrogen + carbon dioxide → carbohydrate + water

   The source of the hydrogen in the equation is most likely
   (1) $C_6H_{12}O_6$ (3) PGAL
   (2) H_2O (4) ATP

9. In which structure of the cell shown below do photolysis and carbon-fixation reactions occur?

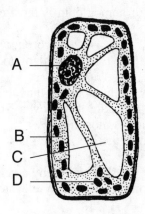

   (1) A   (3) C
   (2) B   (4) D

10–11. For each statement in questions 10 and 11, select the immune response, chosen from the list below, that is most closely associated with that statement. Then record the name of the immune response in the space provided.

Immune Response

Active immunity
Passive immunity
Allergies
Tissue rejection

10. A vaccine containing a weakened disease-causing organism is injected into the body.

_____

11. Chemicals known as histamines are released as a result of antibody production.

_____

12. An insufficient amount of hemoglobin is most closely associated with the disorder known as
    (1) angina   (3) coronary thrombosis
    (2) anemia   (4) high blood pressure

13. In a certain community, a number of humans have an abnormally enlarged structure under the skin of the lower front side of their necks. The cause of this condition is most likely
    (1) an excess of calcium in the diet, which has caused a muscle deformity
    (2) deposits of fat under the skin caused by a vegetable diet
    (3) inherited neck deformities caused by elevated environmental temperatures
    (4) a lack of iodine in the diet, which has caused the development of a goiter

14. What does photosynthesis produce?
    (1) starch, which is metabolized into less complex molecules by dehydration synthesis
    (2) protein, which is metabolized into less complex molecules by dehydration synthesis
    (3) glycerol, which is metabolized into more complex molecules by dehydration synthesis
    (4) glucose, which is metabolized into more complex molecules by dehydration synthesis

194

15. Which statement best describes the enzymes represented in the graphs below?

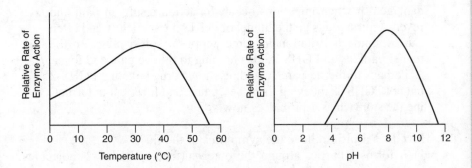

Temperature (°C)   pH

(1) This enzyme works best at a temperature of 35°C and a pH of 8.
(2) This enzyme works best at a temperature of 50°C and a pH of 12.
(3) Temperature and pH have no effect on the action of this enzyme.
(4) This enzyme works best at a temperature above 50°C and a pH above 12.

16–19. Base your answers to questions 16 through 19 on the reading passage below and on your knowledge of biology.

Take Two and Call Me in the Morning

Hippocrates observed that pain could be relieved by chewing the bark of a willow tree. We now know that this bark contains salicylic acid, which is similar to acetylsalicylic acid, the active ingredient in aspirin. Over 2,300 years after this observation by Hippocrates, scientists have learned how aspirin works.

When people get the flu or strain their backs, the body responds by making prostaglandins (PG), a group of hormone-like substances. The presence of certain prostaglandins may result in fever, headaches, and inflammation. Scientists have determined that aspirin interferes with prostaglandin H2 synthase (PGHS-2), an enzyme that the body uses to make pain-causing prostaglandins. In 1994, the structure of this enzyme was found to be a crystal with a tube running up the middle of it. Raw materials move through this tunnel to reach the core of the enzyme, where they are transformed into prostaglandin molecules. Research has shown that aspirin blocks this tunnel. Part of the aspirin molecule attaches to a particular place inside the tunnel, preventing the raw materials from passing through the tunnel. This blockage interferes with the production of prostaglandins, thus helping to prevent or reduce fever, headaches, and inflammation.

The body makes two forms of the enzyme. PGHS-1 is found throughout the body and has a variety of uses, including protecting the stomach. PGHS-2 usually comes into play when tissue is damaged or when infections occur. Its action results in pain and fever. Aspirin plugs up the tunnel of PGHS-1 completely and often causes stomach irritation in some people. Aspirin plugs up the tunnel partially in PGHS-2, thus helping to relieve pain and fever.

Perhaps further research could result in a drug targeting PGHS-2 but not PGHS-1, relieving the aches, pains, and fever but not irritating the stomach as aspirin does now.

16. How does aspirin relieve the symptoms of the flu?
    (1) It forms a barrier around the outer surface of PGHS-2 molecules, separating them from the prostaglandins.
    (2) It dissolves the crystal of the enzyme, preventing it from producing prostaglandins.
    (3) It is an acid that dissolves the prostaglandins that cause the symptoms.
    (4) It reduces the amount of raw material reaching the active site of the enzyme that produces prostaglandins.

17. Why does aspirin irritate the stomach of some people who take it?
    (1) It interferes with the activity of an enzyme that helps to protect the stomach.
    (2) It is the only acid in the stomach and eats away at the stomach lining.
    (3) It stimulates prostaglandin production in the stomach.
    (4) It is obtained from willow bark, which cannot be digested in the stomach.

18. By using one or more complete sentences, describe the molecular structure of prostaglandin H2 synthase.

19. By using one or more complete sentences, explain why chewing the bark of a willow tree could help relieve the symptoms of headache and fever.

---

20. Which human disorder is characterized by a group of abnormal body cells that suddenly begin to undergo cell division at a very rapid rate?
    (1) albinism   (3) hemophilia
    (2) cancer   (4) color blindness

21. An allergic reaction characterized by the constriction of the bronchial tubes is known as
    (1) coronary thrombosis
    (2) arthritis
    (3) asthma
    (4) emphysema

22. In some regions of the world, children suffer from a protein deficiency known as kwashiorkor. This deficiency occurs when a child's diet is changed from high-protein breast milk to watery cereal. Even though the child is receiving calories, the child becomes sick and less active, and growth ceases. These symptoms are probably due to
    (1) too many nucleic acids in the diet
    (2) an overconsumption of complete protein foods
    (3) not enough carbohydrates in the diet
    (4) a lack of essential amino acids in the diet

23. A photograph of a slide of human blood taken from a healthy individual is shown below.

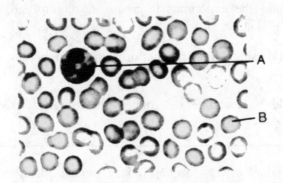

Which statement best describes the change that would be observed if the slide contained blood from an individual with anemia?
    (1) Cell type A would be fewer in number and larger in size.
    (2) Cell type B would be fewer in number and lighter in appearance.
    (3) Cell type B would be larger in size and greater in number.
    (4) Cell type A would be larger in size and darker in appearance.

24–26. Base your answers to questions 24 through 26 on the information below and on your knowledge of biology. Use one or more complete sentences to answer each question. [3]

> When a drug manufacturer develops a new drug to treat some form of disease, the drug should be tested to ensure that it does what it is supposed to do. Usually, the drug is tested on animals. If these tests are successful, it is then tested on humans.
>
> A drug called Lowervil was developed by a drug company to lower blood pressure. Lowervil has been tested successfully on animals, and the drug company is now ready to test it on humans. The drug company claims that one dose of Lowervil per day will decrease blood pressure in individuals experiencing high blood pressure.
>
> A researcher has been hired to determine whether or not Lowervil lowers blood pressure. Answer the following questions related to the experimental testing of the new drug Lowervil.

24. How should the experimental group and control group be treated differently? [1]

25. Why would using a large number of people be important in this experiment? [1]

26. How could the researcher determine if the drug is effective in reducing blood pressure? [1]

27–30. Base your answers to questions 27 though 30 on the information below and on your knowledge of biology.

Organ Transplants of the Future

While most people take good health for granted, thousands of others desperately need to replace a failing organ with one that is healthy. Most healthy organs come from people who agreed to donate them upon their death, although it is possible to remove some tissue and organs (such as kidneys and bone marrow) from living donors. Unfortunately, organs for transplant are in short supply. As of 2020, almost 114,000 Americans are currently on the waiting list for an organ transplant.

Although increasingly common, transplants are risky procedures. During the operation, veins and arteries must be blocked to prevent blood loss. This deprives parts of the body of oxygen and nutrients and may result in permanent damage. In addition, the body may recognize the transplanted organ as foreign and mount an immune response in which specialized white blood cells (T cells) attack the transplanted organ.

Drugs called immunosuppressants are given to transplant patients to prevent their immune system from rejecting the transplanted organ. However, these drugs weaken the ability of the body to fight disease and leave the patient less able to fight infection.

Scientists are exploring new technology for producing transplanted tissues and organs. Unspecialized cells called stem cells are removed from the patient and then grown in a laboratory. Treating stem cells with the appropriate chemicals causes them to differentiate into various specialized tissues. In the future, scientists hope to develop chemical treatments that will cause stem cells to grow into complete organs needed for transplants. Transplants produced by this process would not be foreign material and, therefore, would not be rejected by the immune system of the patient.

27. Explain why a transplant might be dangerous to the health of a patient.

28. State one reason that transplant patients might take an immunosuppressant drug.

29. State one specific disadvantage of taking an immunosuppressant drug.

30. Explain why doctors would consider using tissues or organs grown from stem cells.

IX. INTERDEPENDENCE OF LIVING THINGS

> ## KEY IDEA 6—INTERDEPENDENCE OF LIFE
> Plants and animals depend on each other and their physical environment.

Essential Questions:

- *How do species of different kinds depend on each other and contribute to each other's survival?*
- *How do conditions in the nonliving environment contribute to the survival of plant/animal communities?*

Living things in the environment interact with each other, developing an interdependence crucial to the survival of these populations and to the overall health of the environment. The energy and materials needed to drive this living system derive from the nonliving, abiotic factors present in the environment. Living systems serve to cycle materials and provide a mechanism to absorb and transfer energy among organisms.

Populations of organisms interact with each other in many ways in the natural environment, including competition, symbiosis, and nutritional relationships. Each species population in a natural community has a role to play in the environment, further enhancing the interdependence of species.

The Ecosystem

Diversity and Ecosystem Stability

Essential Question:

- *How do the varied species that make up an ecological community affect environmental stability?*

Performance Indicator 1.1 *The student should be able to explain how diversity of populations within ecosystems relates to the stability of ecosystems.*

When studying ecological interactions, the basic unit of study is the **ecosystem**. The ecosystem is the lowest level of ecological organization in which all environmental factors, both living and nonliving, are represented and interact freely. Earth contains many different ecosystems that provide habitats for millions of different species of plants, animals, and simpler life-forms. The diversity of life directly results from the vast variety of different ecosystems that exist on Earth.

Organisms can be categorized by the functions they serve in an ecosystem. Each species in an ecosystem has a role for which it is best suited. The environmental role of an organism is known as its **niche**. In general, no two species inhabit the same niche in an ecosystem. When two species attempt to inhabit the same niche, **interspecies competition** between them will normally result in the elimination of one of the two from the ecosystem. This fact allows different species to coexist successfully and helps maintain the stability of the ecosystem. Producers, consumers, and decomposers exist within food webs in these ecosystems, carrying out either autotrophic nutrition or heterotrophic nutrition. Each species population in an ecosystem has a different role to play in the transfer of energy and the cycling of elements in the environment.

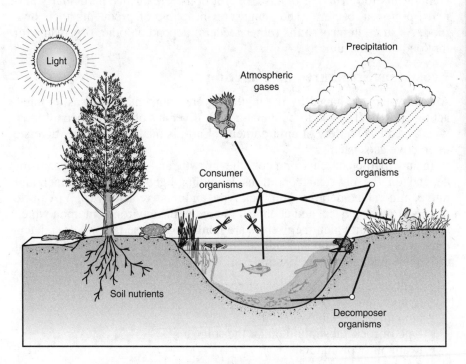

Balanced Ecosystem

Producers are the green plants and algae (the autotrophs) in the community responsible for trapping the Sun's radiant energy and using it to manufacture organic compounds that are used for their own consumption, as well as that of animals. **Consumers** include **herbivores**, **carnivores**, **omnivores**, and **decomposers** (the heterotrophs). Herbivores are **primary**

consumers because they are the first consumers to tap the energy trapped by the producers. Carnivores are **secondary consumers** because they do not tap plant energy directly but obtain it through their consumption of primary-consumer organisms. Omnivores may be either primary consumers or secondary consumers, depending on whether they consume plant matter or animal matter, respectively. **Decomposers** include saprophytic fungi and bacteria responsible for breaking down the complex structure of the bodies of living things into simpler forms that can be used by other living things. In a sense, decomposers are responsible for operating a recycling system that reuses the chemical substances of life over and over. This action is essential to the continued functioning of the ecosystem.

Organisms capable of producing their own food are known as **autotrophs** (self-feeders). Green plants and algae, the principal autotrophic organisms, manufacture organic molecules that serve as food for all the other organisms in the environment. This food is the main source of energy and structural components for all living things. Heterotrophic organisms, particularly animals, protozoa, bacteria, and fungi, are incapable of producing their own food. As such, **heterotrophs** (other feeders) depend on other organisms for the food they consume.

Ecosystem Structure and Function

An ecosystem is shaped by the nonliving environment as well as its interacting species. The world contains a high diversity of physical conditions, which creates a variety of environments. Earth is made up of many diverse, interactive ecosystems.

In any particular environment, the growth and survival of organisms depend on the physical conditions including light intensity, temperature range, mineral availability, soil/rock type, and relative acidity (pH). **Abiotic** factors are the physical and chemical factors in the environment upon which life depends, but which themselves are nonliving. These factors often determine what type of plant and animal community can become established and thrive in a particular area. Examples of abiotic factors include:

- light intensity available for photosynthesis
- temperature range
- amount of available moisture
- type of rock substratum under the soil
- availability of minerals
- availability of atmospheric gases
- relative acidity (pH) of the system

The abiotic conditions determine to a large extent what plant species can exist in a particular environment. In turn, the animal populations that exist in an environment are determined largely by the plant community. For example, a dry, sunlit field with a clay-loam soil in a temperate zone might support a wide diversity of wildflower species, such as goldenrod, aster, and wild daisy, that would thrive under these conditions. On the other hand, species such as mosses and ferns would have a difficult time existing in such an environment because of the need of these species for moist, shady conditions such as those found in a forest environment.

The living species that are successful in any given ecosystem are those best adapted to survive in that set of environmental conditions. The living, or **biotic**, factors abide by natural biological laws that govern all life-forms. For example, all known life-forms are composed of cells that serve as the organisms' functional unit. All living things depend on their ability to carry out basic life functions. Living things depend on the presence and activities of other living things to provide them with nutrients and other essential components. Relationships between organisms may be negative, neutral, or positive. Some organisms may interact with one another in several ways. They may be in a producer/consumer, predator/prey, or parasite/host relationship; or one organism may cause disease in, scavenge, or decompose another. Biotic factors include all the living components of the environment that affect the ecological community, either directly or indirectly, and help to limit the species that inhabit an area. Examples of biotic factors include:

- the population levels of an individual species
- the particular set of food requirements of a species
- the interactions that a species has with other species
- the wastes produced by the members of a species

Nutritional relationships between species involve the transfer of nutrient materials between one organism and another within the environment. Organisms may be classified as follows in terms of the type of nutritional relationships they have with other organisms.

Symbiotic relationships between species involve the ways that different types of organisms can live together in close physical association. Symbiosis is the term used to describe such relationships. Types of symbiosis include parasitism, commensalism, and mutualism.

Biomes
The Earth's natural environments are organized into broad climatic zones known as **biomes**. Each biome is categorized by a particular range of abiotic conditions, such as temperature, rainfall, solar radiation, and altitude. Each biome is also known by the type of dominant, or **climax**, vegetation that establishes itself in the region as well as by the typical, or index, animal species it supports. The biome that makes up the majority of New York State is the temperate deciduous forest biome. It has moderate temperature and rainfall, abundant solar radiation, climax vegetation made up of a mixture of broadleaf hardwood trees such as oaks, maples, and beeches, and index animals such as the deer, squirrel, and hawk. Other terrestrial biomes found on Earth include the tundra, taiga, grassland, desert, and tropical rain forest biomes. The chart that follows summarizes the characteristics of these terrestrial biomes.

CHARACTERISTICS OF TERRESTRIAL BIOMES

Biome	Abiotic Features	Climax Vegetation	Index Animals
Tundra	Permanently frozen subsoil	Mosses, lichens	Caribou, owl
Taiga	Long, severe winters, thawing subsoil	Pine, spruce, cedar	Moose, bear, eagle
Temperate deciduous forest	Moderate precipitation, cold winters, warm summers	Oak, hickory, maple, beech trees	Squirrel, fox, deer, hawk
Grassland	Considerable variation in rainfall and temperature, strong prevailing winds	Grasses, tumbleweed	Prairie dog, antelope, bison
Desert	Sparse rainfall, extreme variation in temperature	Cactus, mesquite	Roadrunner, snake, lizard
Tropical rain forest	Abundant rainfall, constant warmth	Broad-leaved plants, palm trees	Monkey, parrot, jaguar

Interaction of Populations and the Environment

In all environments, organisms compete for vital resources. The linked and changing interactions of populations and the environment compose the total ecosystem. A vital characteristic of all natural ecosystems is the interaction of species populations found within them. A **population** is defined as all the members of a species inhabiting a particular place at a particular time. In an ecosystem, some species take on the role of producers, others the role of consumers. Among the consumers, some species are predators, others prey. Still others may be scavengers or decomposers that break down complex organic matter into simpler components usable by other living things. Some species produce oxygen, while others consume it. The wastes left by one type of living thing may be vitally important to the life functions of another. The complexity of these interactions is extreme and the dynamic balance to which they contribute is delicate. Specific types of interactions based on nutritional relationships are material cycles, food chains, and food webs.

Competition

When different species living in the same environment (habitat) use the same limited resources, **competition** occurs. These resources may include requirements such as food, space, light, water, oxygen, and minerals. The more similar the requirements of the competing species, the more intense the competition is likely to be. Of course, when such resources are abundant, different species may share them without creating significant competition. Only when the resources become scarce does the competition become intense. Such competition might be termed **interspecies competition** (competition *between* species) since it involves members of different species. This type of competition should not be confused with the concept of **intraspecies competition** (competition *within* a species).

As an example, consider the variable growth patterns of goldenrod plants growing in different but neighboring environments. Goldenrod is a species of wildflower that commonly inhabits sunny fields. It is a hardy plant that competes well against other field plants, often growing to heights of five or six feet and crowding out other light-loving species. Goldenrod seeds carried into a nearby forest may germinate and produce young plants, but these plants grow to be only a few inches tall. Their stems are thin and weak, their leaves are small and sparse, and their flowers, if they form at all, are underdeveloped. How can one species of plant, inhabiting the same general area, have such different growth patterns? The key is in the different amounts of light available in the forest and the field. As long as light is abundant, as it is in the field, the goldenrod is a good competitor. When light becomes scarce, as it is in the forest, goldenrod does not compete well, and will eventually be eliminated from that environment altogether.

The interdependence of organisms in an established environment often results in approximate stability over hundreds and thousands of years. For example, as one population increases, it is held in check by one or more environmental factors or another species. Over a period of many years, a steady cycle of interspecies competition will result in a relatively stable environment in which populations coexist under sets of environmental conditions that will support their long-term survival. In this sense, a successful plant species is one whose adults can grow and thrive and whose seeds can germinate and survive to adulthood. A plant species that dominates as an adult population in an environment, but whose seedlings cannot grow to adulthood in that environment, will not continue to be a dominant species in the area for more than a few generations.

Cyclic Change

Ecosystems, like many other complex systems, tend to show cyclic changes around a state of approximate equilibrium. Ecosystems are dynamic and complex. The many biotic and abiotic factors that characterize a given environment interact in complex ways that create a relatively stable system. This stable state is said to be at **equilibrium**. These factors balance each other. When one factor changes, all the remaining factors are affected. Periodic changes in these factors can set off repeating cycles of change that become part of the dynamic state of the environment. Such periodic changes can include food availability, seasonal temperature change, cyclic rainfall patterns, and population reductions caused by disease, among many others.

Perhaps the easiest cyclic change to understand is that caused by seasonal change. As spring arrives in most of New York State, natural ecosystems emerge from a dormant state to an actively growing one. Seeds lying in the moist earth germinate, producing new annual and perennial plants. The buds of mature trees and shrubs swell with new growth and open to produce new leaves and flowers. Ice, snow, and ground frost give way to warming temperatures. The abundance of liquid water, warm temperatures, and new plant food provide an inviting environment that supports herbivores. The presence of herbivores attracts populations of carnivores. The environment continues to warm and stabilize throughout the summer months, bringing still more change. As these periodic fluctuations occur, they affect the ability of other species populations in the area to thrive or survive, including migratory birds and other species. Swings in population numbers represent another aspect of the changes in the environment. Such changes do not always promote the survival of individuals; death is a normal part of the cycle of change.

Another example of cyclic change is the predator-prey relationship. Each environment has a carrying capacity for its animal populations. The carrying capacity of herbivores, such as rabbits, is the abundance and type of plant matter (such as clover) available in the environment. If left unchecked, the rabbit population would increase as a function of the amount of such food until all clover plants were consumed. Obviously, the destruction of the food supply would cause the destruction of the rabbit population as well. In a balanced environment, predators such as foxes exist that feed on the populations of herbivores, reducing their number. This promotes the growth of clover as the numbers of rabbits decrease. As more rabbits are consumed, the well-fed fox population increases. As the rabbit population declines due to hunting by the foxes, the fox population soon also declines. As the fox population declines, hunting pressure on the rabbits is diminished, so there is a resurgence in the rabbit population. Clover populations are subsequently thinned by the resurgent rabbit population. This cycle repeats itself as the populations of clover, rabbits, and foxes respond to increases or decreases in the other populations. Such cyclic and interrelated changes ultimately lead to relatively stable populations of all three species over time.

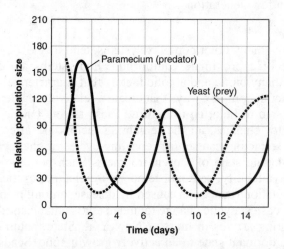

This graph illustrates how populations of predators and their prey vary in a balanced environment. Note that increases in prey population usually precede increases in predator populations, indicating that high numbers of prey directly stimulate the growth of predator populations.

Ironically, the result of these small periodic changes is normally the perpetuation of a relative stability in the environment. Each of the small changes contributes to a dynamic tension to which other environmental factors respond. The cumulative result of all these changes is ecosystem stability.

Direct and Indirect Interactions

Every population is linked, directly or indirectly, with many others in an ecosystem. Disruptions in the numbers and types of species and environmental changes can upset ecosystem stability. The presence or absence of other species populations in an environment can affect the stability of an individual species. Species populations interact in many ways, sometimes directly and sometimes indirectly. Since most organisms depend on other organisms for food, the absence of a species that is used as a food source by other species will affect the ability of those other species to survive. For example, the Karner Blue butterfly, a species unique to New York pine barrens, depends on the growth of a plant, known as lupine, for food and breeding. Lupine, in turn, can tolerate only a narrow range of soil conditions. If these conditions change, the population of lupine may be adversely affected. If the availability of lupine decreases, the survival of the Karner Blue is also impacted adversely.

Some direct relationships are **nutritional**. A nutritional relationship is where one organism uses the bodies of other organisms for food. In nature, all organisms fit into food chains that involve patterns of consumption. For example, grass is a common plant that is used by many herbivorous species for food. An example of a grass-eating herbivore is a grasshopper. Grasshoppers may serve as a food source for a carnivore, such as a frog. The frog, in turn, may become a food source for a raccoon. (See page 210). An outside force that affects any of the populations in a food chain will affect all other populations in the food chain. A blight that kills the grass will cause the grasshoppers to die of starvation. The lack of grasshoppers will drive frogs to new environments to find food. Without the frogs, the raccoons will have to seek out alternative food sources, a step that puts them into direct competition with other top predators for limited food resources. Food chains that intersect are known as food webs. (See page 210.)

Other direct relationships are **symbiotic**. Symbiosis refers to interacting species populations where some aspect of their lives is shared. Common symbiotic relationships include parasitism, mutualism, and commensalism.

In addition to these direct relationships, many interrelationships in an ecosystem are indirect. Sometimes, many steps intervene in the connection between two organisms, but a connection nearly always exists. When no direct link is obvious, it is easy to dismiss such interactions as being of little consequence to human survival. However, the potential impact of species destruction on human survival should not be discounted. There is every reason to assume that when the environment becomes inhospitable to one species, it becomes inhospitable to all similarly situated species, including humans.

Limiting Factors

Limiting factors are those conditions in an ecosystem that impose caps on the sizes of the populations that inhabit it. Limiting factors generally fall into two categories: materials and energy.

Essential Question:

- *What forces of nature manage the balance of species populations?*

Performance Indicator 4.6.1   *The student should be able to explain factors that limit the growth of individuals and populations.*

Materials as Limiting Factors

The chemical components that make up living matter are limited. In order for living things to survive over millions of years of Earth's history, these chemicals must be recycled between the living and the nonliving world. The atoms and molecules on the Earth cycle among the living and nonliving components of the biosphere. For example, carbon dioxide and water molecules used in photosynthesis to form energy-rich organic compounds are returned to the environment when the energy in these compounds is eventually released by cells. Continual input of energy from sunlight keeps the process going.

Material cycles function in nature to make chemical substances available to living things for their continued growth and reproduction. Material cycles consist of sequential chemical reactions that allow for such periodic recycling. Some examples of material cycles are as follows.

Carbon-hydrogen-oxygen cycle—In previous sections, we studied the processes of respiration and photosynthesis and discovered that they are quite similar in terms of the chemical substances involved. In photosynthesis, carbon dioxide and water combine with the aid of solar energy, producing glucose and oxygen gas. In respiration, glucose and oxygen combine to produce carbon dioxide and water, releasing cellular energy. One process produces waste products that serve as raw materials for the other. In this way, environments that contain balanced communities of plants and animals should be self-sustaining in terms of the supply of the elements carbon, hydrogen and oxygen.

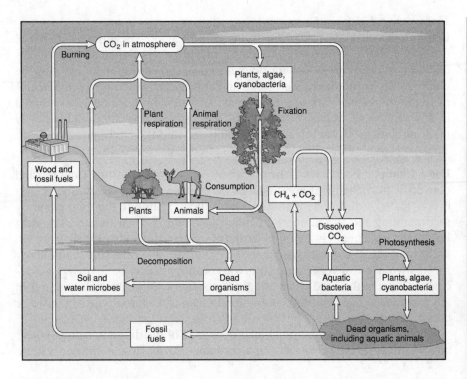

Nitrogen cycle—Nitrogen is an elemental component of the class of compounds known as proteins. The nitrogen cycle makes nitrogen available to organisms for use in protein synthesis. As in the carbon cycle, living organisms are an important part of the nitrogen cycle. Decomposers and other soil bacteria are essential in converting nitrogenous wastes of animals and plants into a form of nitrogen usable by plants as fertilizer. Animals of various types round out the nitrogen cycle as consumer organisms in the food web.

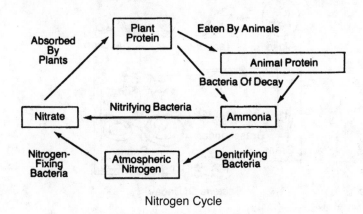

Nitrogen Cycle

Energy as a Limiting Factor

Energy is a necessary part of the living environment. The ultimate source of energy on Earth is the energy of sunlight. Unlike the chemical components of living things, energy cannot be recycled in the environment, so must be continuously resupplied. The chemical elements that make up the molecules of living things pass through **food webs** and are combined and recombined in different ways. At each link in a food web, some energy is stored in newly made structures. However, much is dissipated into the environment as heat. This concept may be illustrated with an **energy pyramid**. (See page 211.)

Food Chains Food chains begin when green plants absorb sunlight and convert it into chemical bond energy of glucose in photosynthesis. A herbivore that consumes the plant represents the next link in the food chain by incorporating the stored energy of the plant into its own tissues. The next link in the food chain is a carnivore that consumes the body of the herbivore, thereby taking in the energy-containing molecules and releasing their energy for its own uses. Several more carnivorous animals may be involved as successive links in the food chain as animals consume other animals and in turn are consumed. The final link in any food chain is a saprophytic organism or bacteria of decay.

This food chain may be illustrated by means of the diagram below, where arrows illustrate the direction of material and energy flow in the ecosystem:

grass ⟶ grasshopper ⟶ frog ⟶ raccoon ⟶ bacteria of decay

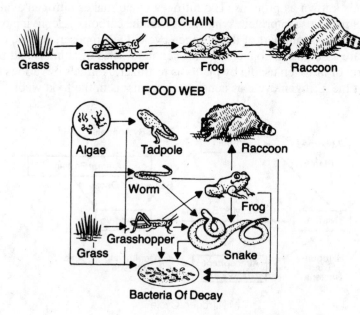

FOOD CHAIN

Grass Grasshopper Frog Raccoon

FOOD WEB

Algae Tadpole Raccoon

Worm

Frog

Grass Grasshopper Snake

Bacteria Of Decay

Food Chain and Food Web

Food Webs   To illustrate more realistically the complex nature of nutritional relationships in a natural community, the food web is used. The food web concept recognizes that many plant species are present in any ecological community, all producing energy-rich organic compounds for consumption by many different species of herbivore. At the same time, multiple combinations of carnivorous and omnivorous species interact to consume the herbivorous species. Saprophytes of many different varieties are responsible for consuming the decaying bodies of plants and animals alike. In fact, different species may compete strongly for the same type of food available in the ecosystem. When the names of all the species present in a community are written onto a sheet of paper and lines are drawn between the species for which a nutritional relationship exists, a pattern resembling a web emerges.

The solar energy available to the producer organisms in a food chain is considerable. At each step of the food chain, however, some of this energy is lost. Some is used in the life processes, some is radiated as heat, some is lost in excretory waste. In any food chain, the producer level contains the greatest amount of energy. Primary consumers contain only about 10 percent of the energy found in the producers. In turn, secondary consumers contain only about 10 percent of the energy housed in the bodies of the primary consumers and only about 1 percent of the energy in the producers. Since each feeding level contains less energy than the level below it, this phenomenon can be illustrated as a pyramid known as the **pyramid of energy**, also known as a **food pyramid**. Eventually, all energy received by the ecosystem will be lost and will radiate as heat into the atmosphere. A constant resupply of energy from the Sun is necessary to sustain the ecosystem.

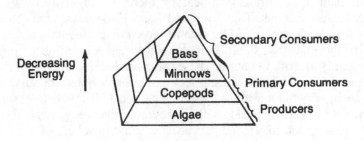

Energy (Food) Pyramid

Carrying Capacity

The number of organisms any habitat can support is limited by the available energy, water, oxygen, and minerals and by the ability of ecosystem to recycle the residue of dead organisms through the activities of bacteria and fungi. The **carrying capacity** of an environment is the number of organisms of different species that can be supported by that environment. Various factors, including available sunlight, moisture, minerals, oxygen, organic food, and other abiotic factors, are instrumental in determining the number of each species that can be supported by that environment. For example, in a grassland environment, the number of buffalo that can be supported is determined by the available food, water, and other components required by the buffalo. If these requirements are limited by natural disaster or human intervention, then the number of buffalo that can be supported may be severely limited.

Also critical to the carrying capacity of the environment is the rate of turnover of the material components of the ecosystem. This turnover is accomplished by organisms of decay—fungi and bacteria. These organisms use their specific characteristics to convert the complex polymers characteristic of most plant and animal life to simpler components that can be absorbed by plants and other organisms capable of synthesizing organic molecules. These activities are directly related to the material cycles discussed previously.

Diversity of Species and Habitat

Preservation of Species Diversity and Habitat

Essential Question:

* *Why is it important to protect endangered and threatened species and habitats?*

Performance Indicator 4.6.2 *The student should be able to explain the importance of preserving diversity of species and habitats.*

As a result of evolutionary processes, there is a diversity of organisms and roles in ecosystems. This diversity of species increases the chance that at least some will survive in the face of large environmental changes. **Biodiversity** increases the stability of the ecosystem. Evolutionary processes are constantly at work in nature. These forces create new varieties, test them against environmental pressures, select those best adapted for survival, and ensure their perpetuation in the species gene pool. This process ensures that many varieties of each species will be present in the ecosystem at any given time, increasing the likelihood that at least some members of each species will survive in the event of a drastic change in the environment.

Environmental Niche

Each species of organism has a role to play in the environment. This role is the organism's **niche** in the environment. Although the niches of some organisms may, on the surface, seem insignificant to human survival, scientists have come to understand that all populations in a natural community depend on all other populations to provide a balanced set of conditions conducive to the survival of all members of the community. Removal of any one species degrades the quality of the ecosystem because its elimination removes the role that species plays and makes it that much more difficult for the environment to recover from drastic change. The elimination of any species reduces biodiversity and makes survival of other species in that environment problematic. See page 229 for a more complete discussion.

Putting Ecological Principles to Work

Scientists have learned much about the interactions among diverse species in the ecosystem and the effects of removing species from a balanced system. Today's system of identifying and protecting endangered and threatened species is a manifestation of our growing understanding of the vital roles that all species, from snails to whales, play in our global environment. We are beginning to understand that an environment hostile to any natural species is potentially hostile to our own species, as well.

Biodiversity also ensures the availability of a rich variety of genetic material that may lead to future agricultural or medical discoveries with significant value to humankind. As diversity is lost, potential sources of these materials may be lost with it. Medical science is developing techniques to analyze and exploit the genetic makeup of animals and plants to treat diseases in humans and other species. An essential aspect of this research is the discovery of new species. Each new discovery is studied in order to determine whether it might contain genetic information leading to the production of antibiotics and other medicinal biochemicals. These newly discovered species can then be cultured in the laboratory in order to maximize the production of the desired chemical. The specific genes responsible for the manufacture of these biochemicals may also be isolated and spliced into the genome of common bacteria in order to reduce the time and expense of this production.

Ecosystem Formation

Ecosystem Change

Essential Question:

- *How do Earth's environments affect, and how are they affected by, the living community?*

Performance Indicator 4.6.3 *The student should be able to explain how the living and nonliving environments change over time and respond to disturbances.*

Ecosystem Stability

An individual species' reproductive rate may be limited by the microclimate created by the dominant community. It may also be limited by the microclimate created by such factors as moisture, temperature, light, latitude, and altitude. For example, the seedlings of hardwood trees such as oak and hickory grow well in environments characterized by low altitudes, sandy soils, and high light intensities. As these species mature, they create a microenvironment that is so deeply shaded that their own seedlings can no longer thrive. Instead, the seedlings of beech and maple, which tolerate low-light conditions, can successfully grow until they replace the populations of oak and hickory that once dominated. However, these same beech and maple seedlings do not compete well at higher altitudes and can easily be crowded out by tree species, such as spruce and hemlock, that are more tolerant of the conditions found at high altitudes. At very high altitudes, tree species in general do not thrive and are replaced by alpine wildflowers and other extreme altitude plant populations.

A stable ecosystem can be altered, either rapidly or slowly, through the activities of organisms (including humans) or through climatic changes or natural disasters. The altered ecosystem can usually recover through gradual changes back to a point of long-term stability. A climax community will normally remain intact as long as conditions in the environment do not change appreciably. If, however, the environment changes drastically because of some catastrophic event, then the climax community may die out or be swept away, making way for a new succession of communities. If the alteration is temporary, the same climax community may result. If the change is permanent, then a new group of climax organisms may be favored.

Ecological Succession

The term **ecological succession** refers to a process in which an established ecological community is gradually replaced by another. It, in turn, is replaced by other communities until a stable, self-perpetuating community is formed.

The beginning stages of an ecological succession frequently occur in barren, almost lifeless environments that may have been swept clean by glaciation, erosion, fire, volcanic eruption, or some other destructive event. The first living things to invade such an area and establish themselves in it are known as **pioneer organisms**. A typical pioneer organism populating bare rock is lichen, a fungus-alga symbiotic association that can tolerate this harsh, dry, soilless environment. A pioneer organism such as lichen, although small, can significantly alter the environment it has invaded. The lichen produces a mild acid that acts on the rock surface and erodes it into grains of sand. At the same time, the material comprising the lichen adds organic matter to the sand grains, producing a crude form of soil. This soil gradually fills the rock crevices, providing favorable areas for seed germination and thereby paving the way for later stages of ecological succession, including mosses and grasses.

The germinating spores of mosses and seeds of grasses and other herbs give rise to the second succession stage. As this stage dominates over several generations, its members contribute organic matter to the soil. The root system protects the thin soil layer from erosion. Weathering and root pressure continue to fragment the rock substratum, adding still more substance to the thickening soil layer. Eventually, the soil layer is able to support the growth of shrubs and other woody plant species. When soil depth and quality are sufficient, varieties of trees and the animal species associated with them begin to invade and then dominate the area.

In this way, each successive community of plants and animals modifies the environment to make it less favorable for its own offspring but more favorable for the establishment of the next succession stage. In such a pattern, plant species tend to dominate since they provide the basis of the food chain for the area. The animal species that can live in the area are normally those that depend on the plant community. A succession stage is normally named for the dominant plant types in the environment since these species exert the most influence on the environmental conditions of the area.

Eventually in each succession, a community becomes established that is self-perpetuating and relatively stable. Such a community is known as a **climax community**. A climax community remains the dominant community for an indefinite period, maintaining a relatively stable set of environmental conditions for both plant and animal species. The predominant climax community in New York State is the oak-hickory forest.

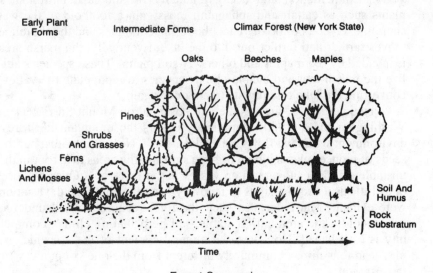

Forest Succession

Examples of this phenomenon can be readily observed in nature in a variety of settings. Keen observation and careful identification and inventorying of plant species in an area will provide evidence that succession is an ongoing biological activity in many ecosystems of the world. A few examples include the following.

- An abandoned farm field undergoes rapid succession, first playing host to wild grasses and wildflowers, then to invading shrubs and light-loving tree species such as aspen and white pine. After a relatively short

period, the former field resembles a dense, young forest and the seedlings of climax stage trees are sprouting and competing with the early
tree species. Often, such a succession pattern requires only a few
decades to revert to the low-elevation oak-hickory climax forest characteristic of much of New York State.

- A burned mountain forest tract quickly sprouts new pitch pine and
  other fire-resistant species. The seed cones of pitch pine are specially
  adapted to protect the seeds of this species during a fire but to release
  them rapidly after the ground has cooled. The new pine seedlings act
  to stabilize the soil, reducing erosion. Seed of other tree species, such
  as birch and aspen, from nearby unburned forest are transported to the
  burn area and quickly germinate to help stabilize the soil still further.
  Eventually, the seeds of slow-growing hemlock, beech, and maple trees
  take root and reestablish this high-elevation climax community typical
  of the Catskill and Adirondack Mountains of New York State.

- A stream valley flooded by a landslide that dams the stream quickly
  takes on the characteristics of a shallow lake. Over a period of many
  years, sedimentation and decay gradually fill the lake until marsh
  plants such as cattail and sphagnum moss gain a foothold in the wet
  shoreline. Over a period of years, the filling continues and the shoreline
  moves farther and farther out into the lake. Eventually, the marsh area
  is invaded by red maples and other wet-soil plants. These species stabilize the marsh mat and enable climax species to repopulate the valley,
  converting the lake to a forest stream once again.

- A volcanic eruption, such as the one observed on Mount St. Helens in
  Washington State (1980), completely covers the mountainside forest
  environment with lava flows and volcanic ash. No soil remains, but the
  seeds of pioneer plants blown in from surrounding areas take root in the
  unstable ash deposits and in crevasses in the lava rock. Over a period
  of as little as a decade, signs of recovery from the complete devastation
  were evident. Many decades later, the forest ecosystem will undergo
  succession in the area until a stable, self-perpetuating climax community is established. However, the mineral content of the volcanic ash
  may lead to a climax community different from the one swept away by
  the eruption.

1. Which statement best describes some organisms in the food web shown below?

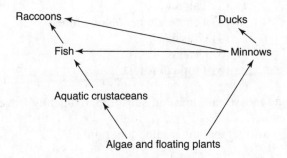

   (1) Minnows and fish are primary consumers.
   (2) Algae and floating plants are decomposers.
   (3) Aquatic crustaceans are omnivorous.
   (4) Raccoons, fish, and ducks are secondary consumers.

2. The graph below shows the changes in two populations of herbivores in a grassy field.

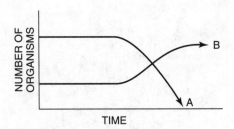

   A possible reason for these changes is that
   (1) all of the plant populations in this habitat decreased
   (2) population B competed more successfully for food than population A did
   (3) population A produced more offspring than population B did
   (4) population A consumed the members of population B

3–5. For each description in questions 3 through 5, select the biome, chosen from the list below, that best fits that description. A biome may be used more than once or not at all.

Biomes

(1) Tundra
(2) Taiga
(3) Temperate deciduous forest
(4) Grassland
(5) Desert
(6) Tropical forest

3. Lichens and mosses are present; subsoil permanently frozen.

4. Constant, warm temperature; abundant rainfall.

5. Wide variation in daily temperature; little rainfall.

---

6. The diagram below shows a milkweed plant and some of the insects that live on it or visit it.

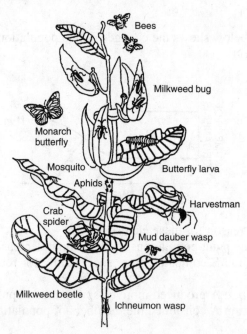

Which term best describes the group of organisms in the diagram?
(1) biosphere   (3) habitat
(2) community   (4) biome

7. Hawks and owls living in the same area compete for the same type of mouse for food. Which situation would lead to the greatest problem in the food supply?
   (1) an increase in the owl population
   (2) an increase in the mouse population
   (3) a decrease in the hawk population
   (4) a decrease in the owl population

8. Which group represents a population?
   (1) all the vertebrates living in New York State
   (2) all the *Homo sapiens* living in New York State
   (3) all the plant and animal species found in New York State
   (4) all the flowering plants found in New York State

9. A student measured some abiotic factors present in an aquarium in a biology laboratory. Which data did the student most likely record?
   (1) the weight and color of each type of scavenger
   (2) the number of each type of green plant and each type of snail
   (3) the size and number of each species of fish
   (4) the temperature and oxygen content of the water

10. Energy stored in organic molecules is passed from producers to consumers. This statement best describes an event in
    (1) the process of photosynthesis
    (2) natural selection
    (3) a food chain
    (4) ecological succession

11. Which material cycle relies *least* on the processes of photosynthesis, transpiration, evaporation, respiration, and condensation?
    (1) oxygen cycle   (3) water cycle
    (2) nitrogen cycle   (4) carbon cycle

12–14. For each symbiotic relationship in questions 12 through 14, select the type of symbiosis, chosen from the list below, that best identifies that relationship. A number may be used more than once or not at all.

Types of Symbiosis
   (1) Commensalism
   (2) Mutualism
   (3) Parasitism

12. A tapeworm lives in the digestive tract of a human.

13. Nitrogen-fixing bacteria live in the nodules on the roots of legumes.

14. A flea sucks blood from the skin of a dog.

---

15. By starting on bare rock, what is the usual ecological succession of organisms?
    (1) grasses → shrubs → lichens → trees
    (2) lichens → shrubs → grasses → trees
    (3) grasses → lichens → shrubs → trees
    (4) lichens → grasses → shrubs → trees

16. An ecosystem is represented in the diagram below.

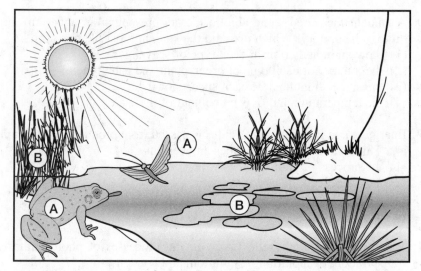

This ecosystem will be self-sustaining if
    (1) the organisms labeled A outnumber the organisms labeled B
    (2) the organisms labeled A are equal in number to the organisms labeled B
    (3) the type of organisms represented by B are eliminated
    (4) materials cycle between the organisms labeled A and the organisms labeled B

17. A certain plant requires moisture, oxygen, carbon dioxide, light, and minerals in order to survive. This statement shows that a living organism depends on
    (1) biotic factors
    (2) abiotic factors
    (3) symbiotic relationships
    (4) carnivore-herbivore relationships

18. Events that take place in a biome are shown in the diagram below.

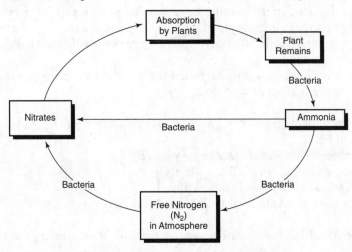

Which information is represented in the diagram?
(1) Respiration and photosynthesis are interrelated.
(2) Transpiration and condensation are related to the water cycle.
(3) Decomposers release a material that is acted on by other organisms.
(4) Predators and their prey are involved in many interactions.

19. The diagram below represents an energy/biomass pyramid.

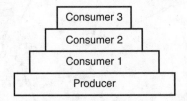

Which statement concerning the energy in this pyramid is correct?
(1) The producer organisms contain the least amount of stored energy.
(2) Stored energy decreases from consumer 2 to consumer 3.
(3) Consumer 3 contains the greatest amount of stored energy.
(4) Stored energy increases from the producer to consumer 1.

20. Which statement concerning the climax stage of an ecological succession is correct?
(1) It changes rapidly.
(2) It persists until the environment changes.
(3) It is the first community to inhabit an area.
(4) It consists entirely of plants.

21. In a pond, which change would most likely lead to terrestrial succession?
    (1) a decrease in the number of suspended particles in the pond water
    (2) an increase in current velocity of the pond water
    (3) a decrease in the number of diverse organisms in the shallow water of the pond
    (4) an increase in sediment, fallen leaves, and tree limbs accumulating on the bottom of the pond

22. The most likely result of a group of squirrels relying on limited resources would be
    (1) an increase in the number of squirrels
    (2) competition between the squirrels
    (3) increased habitats for the squirrels
    (4) a greater diversity of food for the squirrels

23. Nutritional relationships between organisms are shown in the diagram below.

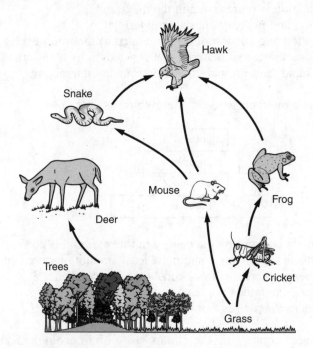

Which organisms are primary consumers?
    (1) mouse, snake, and hawk
    (2) snake, hawk, and frog
    (3) cricket, frog, and deer
    (4) mouse, deer, and cricket

24. Information relating to an ecosystem is contained in the diagram shown below.

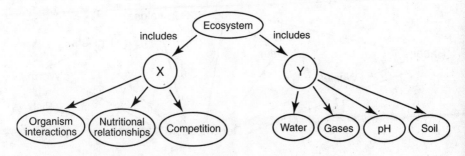

Which information belongs in areas X and Y?
(1) X—biotic factors; Y—abiotic factors
(2) X—ecological relationships; Y—biotic relationships
(3) X—abiotic factors; Y—interacting populations
(4) X—energy flow; Y—biotic factors

25. The American dogwood, a flowering tree of New York State's woodlands, has been attacked by a fungal disease specific to this tree species. Many dogwoods have died because fungicides have not proven effective in fighting the spread of this disease. Which term best describes the relationship between the dogwood trees and the fungus?
(1) commensalism
(2) mutualism
(3) parasitism
(4) saprophytism

26. The food web below shows some of the relationships that exist between organisms in a field and pond ecosystem.

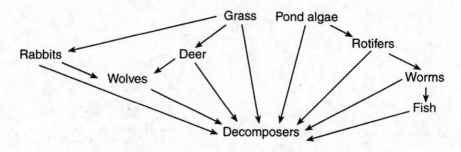

a Write one or more paragraphs describing some of the relationships in this food web. In your answer, be sure to:
   • identify a carnivore from the food web
   • describe the complete path of energy from the Sun to that carnivore
   • explain why decomposers are necessary in this food web

b A significant decrease in the wolf population occurs. After a period of one year, what change in the grass population would most likely be observed?

c A farmer sprayed pesticides onto a field next to the pond. By using one or more complete sentences, explain why several years later the fish population would contain higher pesticide levels than any other pond organisms would contain.

X. HUMAN IMPACT ON ECOSYSTEMS

> **KEY IDEA 7—HUMAN IMPACT ON THE ENVIRON-
> MENT** Human decisions and activities have had a profound
> impact on the physical and living environment.

Interdependence of Human and Natural Systems

Essential Questions:

- *What is the proper role of the human species in the ecological community?*
- *What responsibility does the human species have to protect the natural ecosystem?*
- *How do humans positively and negatively affect the ecosystem?*

Performance Indicator 4.7.1 *The student should be able to describe the range of interrelationships of humans with the living and nonliving environment.*

The human species is just one of Earth's 1.9 million living life forms. It has arisen by the same mechanisms as other species. It has the same physical requirements as other, similar species. It is dependent for its survival on its successful interactions with Earth's plant and animal species. Ultimately, it is subject to the same limitations to growth as any other species.

However, humans are also unlike any other species because of their niche as thinking, planning, and technological beings. Due to their ability to use technology, the human population is growing virtually unchecked by the natural factors that limit other species populations. Human technologies have had significant impacts on the natural world by producing materials that pollute the air, water, and soil. Its activities are increasingly displacing or destroying natural habitats and their ecological communities, reducing biodiversity and endangering the survival of many, if not all, of Earth's living things.

As a species, it is essential that we understand the necessity of preserving the natural environment and its living species as a means of ensuring our own survival. While some progress has been made in correcting certain environmental problems, much remains to be done. Education and environmental awareness on a global level is essential. Governments, industries, and the general public must come to grips with the long range impact of human activity that destroys the very fabric of biological life on Earth.

Technological Oversight and Human Population Growth

Essential Question:

- *How has human technology and the growth of the human population affected the quality of life for other species?*

Performance Indicator 4.7.2 *The student should be able to explain the impact of technological development and growth in the human population of the living and nonliving environment.*

Human Population Growth

Human population growth, unlike that of all other species, has risen at a rapid rate over the past two centuries. Human inventiveness has allowed our species to avoid or extend the limiting factors that keep natural species populations in check. For example, the development of medical technology has reduced the incidence of disease in many parts of the world, modern agriculture has reduced the likelihood of starvation in developed nations, construction technologies have allowed humans to avoid exposure to the elements and to extend their range into regions that would be otherwise uninhabitable by our species. This rapid increase in human population and the technology that supports that growth has put extreme pressures on the natural world, threatening the survival of natural species and habitats.

In many less developed areas of the world, the human population has grown faster than the food supply, causing widespread famine and threatening to eliminate large portions of the populations. Poverty and hunger in these areas has led to the resurgence of diseases that have been all but eradicated in the world's developed nations. These same conditions have forced people to abandon their homes in search of food and to be exposed to the elements, further reducing their ability to survive. Such occurrences are becoming more frequent and severe. It remains to be seen when similar collapses will occur in developed nations. Apparently, the human species is rapidly approaching a point where it will be unable to sustain continued growth.

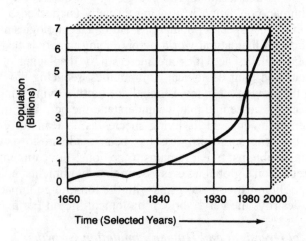

Human Population Growth

Technological Oversight

Humans have used their ability to alter the environment to support their continued survival on Earth. Technologies have been developed to assist humans to this end. Inadvertent or purposeful misuse of technology in regard to the natural world is known as **technological oversight** and is illustrated in the following paragraphs.

In creating their technologies, humans have used natural materials at an alarming rate with little concern for conservation or reuse. Many of these resources are **nonrenewable**, meaning that their supply on Earth is finite and, once used up, cannot be replaced. Nonrenewable resources that have been mined or pumped from the ground to serve human needs include metals (e.g., gold, silver, copper, iron), salts (e.g., limestone, granite, seasalt), petroleum (e.g., oil, coal, natural gas), and fresh water. **Renewable** resources are those resources that can, with proper care and attention to the future, be replaced. Timber (used to make wood, paper, cardboard) and other plant and animal resources are considered to be renewable, though positive action by humans is needed to expedite that renewal.

Environmental Pollution and Destruction

Technological and industrial processes have led to the production of chemicals and by-products that are harmful (toxic) to living things. Such toxic chemicals have contributed to global **environmental pollution**. Examples of pollutants include acidic ions, pesticides, hydrocarbons, growth hormones, pharmaceuticals, and radioactive particles. When released into the environment, these pollutants contaminate water, air, and/or soil, and make survival of natural species difficult or impossible. To the extent that humans are exposed to these pollutants, we are also subject to their toxic effects. In New York State in the early 20th century, the agricultural use of the pesticide DDT negatively impacted many native bird species by interfering with their reproductive cycle. Today, industrial by-products such as dioxin pose similar dangers to many species, including humans. The combining of acidic ions with atmospheric water has resulted in acid precipitation (acid rain), which has acidified lakes and disrupted natural ecosystems downwind of industrial sites.

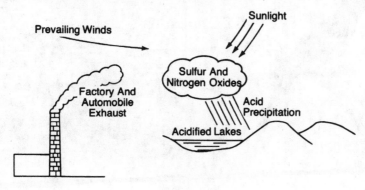

Acid Precipitation

Habitat Destruction

In order to support the rapid growth in their numbers, humans have used their technologies to remove trees and other plant life from wide areas of natural habitat for the purpose of creating agricultural lands, industrial zones, roadways, or residential and commercial developments. Humans have drained, filled, or redirected wetlands, ponds, streams, and other natural water habitats for these same purposes. Collectively known as **habitat destruction**, these activities have had the effect of displacing plant and animal species important to the maintenance of a balanced ecosystem. Virtually all present-day developed lands in New York State have come about through human technologies that have resulted in habitat destruction.

Human agriculture has produced a variety of negative effects due to inattention to its impact on the natural environment. Failure to use **cover crops** (crops that protect soil from erosion) between planting cycles has exposed bare soil to erosion, resulting in loss of topsoil and sedimentation of water sources. **Overcropping** is the failure to allow soil to recover nutrients and organic matter (humus) content between plantings. **Overgrazing** is the practice of allowing large numbers of domestic animals to graze an area too small to support them.

Non-native Animals and Plants

Humans' desire to possess non-native exotic animals and plants or their products has been enabled by modern transportation, communication, and economic technologies. This has resulted in the **exploitation** of exotic species of primates such as the spider monkey, big cats such as the ocelot, birds such as the Columbian parrot, and others as pets. In addition, rhinoceroses have been killed for their horns, elephants for their tusks, tigers for their reproductive organs, and uncounted mammal species (e.g., bison, fox, beaver) for their pelts. Each species loss has impacted its ecosystem in a negative way. Each exploited species is, to one degree or another, in danger of extinction because of this exploitation.

Transportation technology has accelerated the movement of non-native species of plants and animals from foreign countries to our own. Such organisms are known as **invasive species**, and enter our environment either through intentional introduction or inadvertently on the bottoms of ships. Because they often have no natural enemies in their adopted environment, these newly introduced organisms have disrupted local ecosystems and in some cases have totally eliminated native species. Examples of invasive species in New York include the Japanese beetle, gypsy moth, zebra mussel, Asian clam, European starling, Eurasian milfoil, purple loosestrife, chestnut blight, and the virus that causes Dutch elm disease.

Diminishing Biodiversity

In an activity known as **direct harvesting**, humans have removed plants from the ecosystem for their economic value without regard for the effect of this removal on the natural ecosystem. This removal has negatively impacted the ecosystem by decreasing **biodiversity** (variety of life) in these habitats. In the 19th century in New York State, forest trees were directly harvested from the Adirondack and Catskill mountains to provide lumber for building purposes. New York's mountains and mountain valleys were denuded in the process, allowing fragile soils to be washed away. This activity resulted in the destruction of entire mountain habitats. Internationally, direct harvesting activities for exotic hardwoods in tropical rainforests has resulted in similar habitat destruction and loss of biodiversity.

When this concept is discussed in relation to animal species, the term **overhunting** is often used to describe it. During the 17th and 18th centuries in New York State, beavers were trapped from the Hudson, Mohawk, Genesee, and Susquehanna river watersheds in large numbers for the purpose of obtaining their pelts for export. This activity nearly drove the beaver to extinction in these areas; the beaver population only began to recover in the mid-late 20th century. In the 19th century, bounties imposed on the hunting of gray wolves and mountain lions effectively removed these species from New York. Internationally, species such as the passenger pigeon, the great auk, and the dodo were hunted to extinction in this same period. Today, the overhunting of several species of whale in the open ocean has threatened to drive them into extinction. Whenever a species is eliminated from its native habitat, the species' environmental niche is vacated and ecosystem balance is disrupted.

For hundreds if not thousands of years, the community of plants and animals on the Kaibab plateau in northern Arizona remained in balance. Each species filled its appropriate niche and interacted as predator, prey, producer, or decomposer. Misguided conservation policies introduced in the early 1900s placed a bounty on mountain lions and other natural predators of the Kaibab deer. As hunters reduced the numbers of predatory species on the plateau, the deer population, with fewer natural enemies, exploded. The thousands of additional deer that resulted quickly outstripped the food supply of this fragile environment. As edible plants were consumed and root systems destroyed by the foraging deer, the deer population declined severely due to starvation.

Warning Signs of Environmental Contamination
Many warning signs exist that indicate that a dangerous disequilibrium is being created in our ecosystems. Some researchers suspect that toxic chemical by-products, heavy metals, pesticides, hormones, pharmaceuticals, and other contaminants are responsible for an increased frequency in deformation of, illness in, and death of natural populations worldwide. For example, dramatic declines and deformations in frog populations have been observed in aquatic ecosystems around the world. The culprits suspected by some scientists are hormone-disrupting chemicals in agricultural runoff. Other examples include:

- reproductive cycle disruption in birds as a result of high concentrations of DDT and other pesticides;
- deaths and population declines of birds and fish linked to contamination from oil spills;
- deaths and disorientation of marine mammals (porpoises, whales, and manatees), possibly due to coastal pollutant runoff;
- high concentrations of polychlorinated biphenyls (PCBs) in fish taken from contaminated waters.

In our own species, adverse health effects linked to chemical exposure are increasingly prevalent among children. Infancy and early childhood are periods of vulnerability to environmental pollution. Because they expend more energy and require more food, water, and oxygen per unit of mass than adults, young children are likely to take in and store more toxic chemicals per pound than do adults. In recent years, the incidence of birth defects and diseases linked to environmental pollution has risen among children living in environmentally-contaminated areas of the world. Childhood cancer has become the second leading cause of childhood deaths. Other health problems being experienced today by children living in contaminated communities include chronic lung disease, childhood asthma, attention deficit disorder, and too-early or delayed sexual maturation. Chemical pollutants such as PCBs, dioxin, lead, pesticides, and solvents are increasingly suspected as contributors to these issues among children.

Energy Consumption

Industrialization has brought an increased demand for and use of energy resources. This usage has had both positive and negative effects on humans and ecosystems. The world's industrialized nations use tremendous quantities of energy to fuel their economies. Countries such as the United States and England have long been the major consumers of energy. China, formerly a light user of energy resources, has recently become a major consumer of petroleum as a source of energy to operate its growing economy.

Over time, the source of this energy has changed, with traditional water, wood, coal, and oil gradually being replaced with natural gas, wind, wave/tidal, nuclear, solar, geothermal, and fuel cell technologies. Each power source has its advantages and disadvantages for application in any particular industry. However, the cost per unit and availability of these power sources have always been prime determiners of the choices made. The development of these energy sources often uses valuable agricultural lands for the construction of extraction, storage, and transport of the energy they produce.

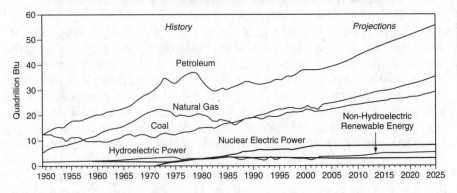

Energy Consumption History and Outlook, 1950–2025

Fossil Fuels

Fossil fuels, commonly used as an energy source in industrial processes for the past 200 years, include naturally occurring materials such as oil, coal, and natural gas. These fuels are drawn from underground deposits formed millions of years ago by biological and geological processes. Fossil fuel deposits are not being replenished by these same processes today. As a result, the supply of these fuels is finite and will someday be depleted. Fossil fuels are burned to power internal combustion engines or to produce steam for running electrical turbines. Their hydrocarbon chemistry results in the

231

emission of carbon dioxide (CO_2) and carbon monoxide (CO) gases when they are burned. Impurities in these fuels may also result in the emission of particulate matter, sulfur oxides, and nitrogen oxides. The drilling and mining operations used to obtain these fuels is also worthy of consideration, as the technology surrounding these operations can have negative consequences for the natural environment, as well as for humans.

Hydraulic fracturing, commonly known as **hydrofracking**, is a process that has been used, with variable success, to release natural gas from certain geologic shale formations (such as the Marcellus Formation underlying much of New York's southern tier). Because some of these shale formations are found in south/central New York State, hydrofracking has become an important point of discussion for many New Yorkers. The most prominent issue under discussion involves the trade-off between energy independence, economic development, and jobs vs. protection of the quality of aquatic ecosystems for native fish species and groundwater for human consumption. Are the short-term advantages of economic gain worth the long-term dangers of environmental disruption? This is one of many questions facing New Yorkers regarding the pros and cons of economic development vs. environmental quality.

The chemical by-products of fossil fuel combustion are released into the atmosphere and contribute to the worldwide air pollution problem as well as a phenomenon known as acid precipitation, or acid rain. **Acid precipitation** results from the combining of sulfur and nitrogen oxide ions with atmospheric water to produce sulfuric and nitric acids. When rain, snow, and sleet containing these acids falls on aquatic ecosystems they can significantly alter the acid/base chemistry of these ecosystems. This alteration often leads to stress on, or elimination of, acid/base sensitive species of fish and amphibians in those ecosystems. The burning of fossil fuels has also been linked to the phenomenon known as global warming (see page 235).

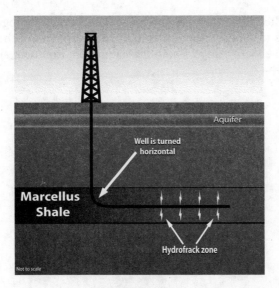

Hydrofracking

Nuclear fuel, in the form of plutonium, is a highly refined and concentrated form of the radioactive mineral uranium. In a containment vessel specially designed to promote it, plutonium emits high-energy particles that collide with atoms, forcing them to emit neutrons that collide with other atoms in a chain reaction. This reaction produces a tremendous quantity of heat, which is used to boil water to make steam for running electrical turbines. The wastes of this process are heat released into aquatic ecosystems and radioactive emissions released into the atmosphere from depleted fuel rods. Disposal of these wastes can be disruptive to natural environments and dangerous to human health. Waste heat can disrupt water environments by raising the temperature of those ecosystems to unnatural levels, killing fish and other aquatic life in the process. Radiation can contaminate soils, water, and air and can kill or cause mutations and cancers in living organisms including humans. For example, radioactive waste is currently being stored in outdoor containment areas in New York at Lake Ontario's nuclear power plants; its radioactivity is likely affecting communities downwind (southeast) of these facilities.

Alternative energy sources, such as wind, wave/tidal, solar, fuel cell, and geothermal, are in their developmental stages. Developed properly, these alternative energies have the potential to provide energy far in excess of the current annual energy needs of the United States, a fact that could lead to energy-independence, new jobs, and an improved economy. Because these energy sources are renewable, they have the advantage that they will not be depleted as will fossil fuels, allowing them to carry our energy needs far into the future. These technologies are also relatively pollution-free, making them a safer alternative to fossil and nuclear energy technologies. It benefits us as a species to support research into the development of these alternative energies, for the well-being of both the natural environment and ourselves.

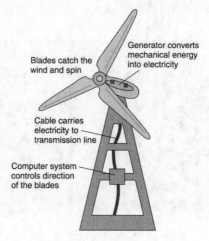

Wind Energy

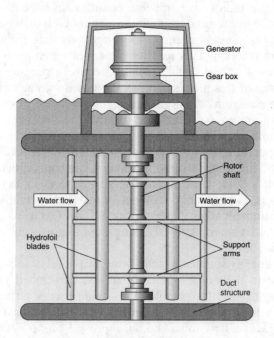

Tidal Energy

Global Warming

The negative effects of 200 years of worldwide industrial and technological development have recently manifested themselves as the phenomenon known as **global warming**. Although the signs of this phenomenon have been recognized by competent scientists for many years, it has taken industrialists, politicians, and the general public much longer to recognize the threat to the living environment posed by its effects. In the mid-20th century, atmospheric scientists and environmental biologists pointed out that increasing concentrations of carbon dioxide and methane, known collectively as **greenhouse gases**, had the effect of trapping solar radiation in Earth's atmosphere in the form of heat. These scientists warned that, if left unchecked, this process would lead to global warming sufficient to melt polar ice caps and glaciers, expand deserts, raise ocean levels, and significantly disrupt global weather patterns. They further warned that an increase of only a few degrees in average temperature would disrupt ecological communities in virtually all parts of the world, accelerating the pace of extinction of plant and animal species and threatening human survival. Recent estimates of the rate of global warming predict that Earth's average temperatures will rise by as much as 11°F (−12°C) by the year 2100. That much warming will create environmental effects that will devastate the entire human population.

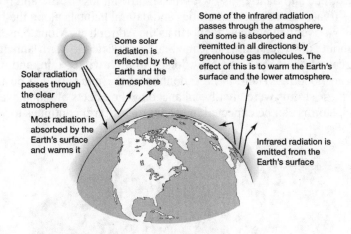

These scientists urged governments worldwide to impose controls on industries and technologies known to emit greenhouse gases into the atmosphere, but little was done at the time to curb these emissions. Recently, the serious effects of global warming have begun to capture worldwide attention. In 2007, the United Nations scientific panel on climate change noted that the evidence of global warming was "unequivocal" and that human technological oversight was the "likely" culprit. Still, the international community has been slow to respond with meaningful environmental controls of greenhouse gas emissions. Unfortunately for all of us, the actions taken may be "too little, too late." Adding to the problem is the repeated failure of vocal and influential politicians and corporate leaders to acknowledge that the phenomenon exists at all or, more specifically, to acknowledge that it is most likely a human-made problem requiring a human-made solution.

Reports of accelerated glacial melting, shrinking polar caps, epic droughts, intense and destructive weather patterns, catastrophic crop failures, disappearing fisheries, and rising ocean level are more and more frequent. If immediate steps are not taken to reduce greenhouse gas emissions, it is estimated that **ocean levels** will rise as much as 2 feet by the year 2050. This amount of sea level rise will drown coastal habitats and subject cities such as New York, Boston, and Philadelphia to massive flooding. It will also push salt water far up tidal estuaries (such as the Hudson River from New York to Albany), destroying freshwater habitats and flooding riverside agricultural lands and communities along the way. It will have a profound negative effect on the world's economic situation and will only get worse over the next several decades.

Weather patterns that in the past have been mild and predictable have become increasingly violent and unpredictable. Just weeks apart in 2005, hurricanes Katrina and Rita devastated the Gulf coast, taking many lives and destroying billions of dollars of property in Texas, Louisiana, Alabama, and Mississippi. In 2011 and 2012, respectively, New Yorkers felt the effects of hurricanes Irene and Sandy, also with the significant loss of life and property. Hurricane Irene destroyed communities and natural habitats along the Hudson and Mohawk river valleys to as far north as the Adirondack Mountains, as well as throughout New England. Hurricane Sandy subjected lower Manhattan and coastal areas of New York and New Jersey to high winds, rain, and massive storm surge flooding. There is little doubt that the intensity and destructive power of these storms were greatly enhanced by the effects of global warming. More such storms can be expected in the future as Earth continues to warm.

Another serious example of global warming's effect on weather patterns involves significant shifts in **temperature and rainfall patterns** across the world's most productive agricultural zones. Between the years 2000 and 2012, increasing temperatures in the "breadbasket" of the United States have had a significant impact on the production of wheat and other grains. An increase of just 1°F (0.56°C) in average temperature during this period has resulted in a 5% decrease in wheat production in North Dakota alone. Throughout the breadbasket, these changes have significantly shifted the areas where wheat can be grown at all, gradually pushing them northward and westward. Grains such as wheat, corn, and oats are essential crops needed to feed the country's population and/or its domestic meat-producing animals. Drastically decreased rainfall in this same area has resulted in record drought conditions that further threaten the production of food crops in the future. Similar effects have been noted in agricultural zones of Canada and Australia, as well as throughout Europe and Asia. Continued increases in temperature and decreases in rainfall threaten to seriously reduce or completely eliminate agricultural activities in these areas. The implications for human survival may be disastrous if the situation is not reversed.

In a perverse feedback effect of global warming, the melting of **polar ice and arctic permafrost** caused by global warming has been shown to be responsible for the release into the atmosphere of huge quantities of methane, a greenhouse gas that has been trapped in frozen soils and sea bottom muck for tens of thousands of years. The rapid addition of this methane to Earth's atmosphere is having the effect of compounding the global warming problem, which will have the effect of releasing still more methane into the atmosphere as increased temperatures promote still more ice/permafrost thawing.

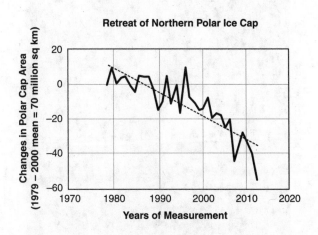

Retreat of Northern Polar Ice Cap

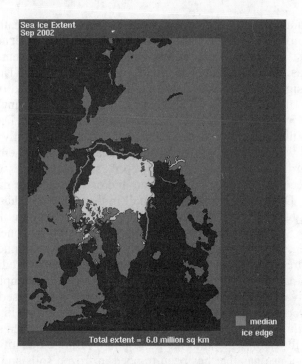

Polar Cap Extent 2002

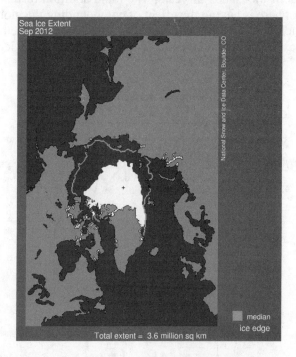

Polar Cap Extent 2012

While alternating **cycles of global warming and cooling** are known to have occurred many times over the four billion years of Earth's history, the current shift is happening with unprecedented speed. Whereas past temperature shifts have taken place over hundreds of thousands or millions of years, this one is taking place over mere decades or centuries. This means that we will see changes in our lifetimes that in past cycles would only have occurred over hundreds of human generations. Though difficult to observe from the perspective of a single human lifetime, these changes have been studied and predicted by competent scientists throughout the world and must be addressed by the present generation if we hope to survive as a species.

Need for Action

It is imperative that we, as a people, act together to make the regulatory and behavioral changes necessary to turn back the tide of global warming. Each of us has a personal obligation to reduce the amount of carbon dioxide we add to the atmosphere. We can do this by reducing personal energy consumption, buying locally produced foods and products that do not need to be shipped long distances, and minimizing the use of synthetic materials that require energy to produce. This is sometimes referred to as "**reducing our carbon footprint**." But we also must demand that political leaders enact effective legislation to enforce reduction of greenhouse gas emissions from vehicles, factories, power plants, and public buildings. We must insist that the U.S. government agree to enforce international treaties aimed at reducing energy use and environmental pollution, as well as those designed to protect fragile ecosystems and endangered species. To do less will only serve to hasten our extinction.

Individual Choices and Social Actions

Essential Questions:

- *How can humans take positive action to reverse the degradation of the natural environment caused by human technology?*
- *What trade-offs must be considered when making economic and political decisions that affect the environment?*

Performance Indicator 4.7.3 *The student should be able to explain how individual choices and social actions can contribute to improving the environment.*

Societies must develop procedures that ensure that we can prove the safety and effectiveness of new technologies before implementing them. This concept is sometimes referred to as the "**precautionary principle**," and is meant to help protect human health from reckless technological implementation. Individual voters need to make personal decisions that will help them assess the risks, costs, benefits, and trade-offs of such new technologies. Until it can be proven that a particular technology poses no significant risk to the natural environment and to human health, it should not be allowed to operate, no matter what the economic benefit to individuals, corporations, or governments. Developing the political will to do so will help ensure the survival of our environment, and of our species as well.

We must constantly keep in mind that we share this small planet with diverse species that together with us make up the living community. We must acknowledge that the **maintenance of biodiversity** promotes the general health and welfare of both human society and of the ecosystem. We cannot afford to lose a single species to extinction caused by environmental abuse, lest we threaten our own existence. We must remain vigilant that we do not revert to the dangerous and destructive practices of the past but continue to improve the ways in which we care for and understand our natural environment.

This requires that we become informed about environmental issues so we can apply the precautionary principle by asking critical questions about the potential impact of new technologies on these issues. We must also support research aimed at studying the effects of these technologies on human health as well as the health of other species. This information should then be weighed against the benefits that this technology would bring to society as a whole. This thought process is sometimes referred to as **examining the trade-offs**. If the benefits outweigh the risks, then the technology should be implemented with appropriate precautions to guard against unanticipated negative effects. If, on the other hand, the risks outweigh the benefits, then the technology should be set aside until adequate safeguards are developed.

Recently, some forward-thinking groups, governments, communities, and companies have made efforts to establish procedures that allow development without damaging the environment. Examples of such **positive environmental actions** include:

- Local governments requiring that developers replace wetlands destroyed by commercial and residential development activities;
- State and county governments requiring that logging industries replant native species in clear-cut tracts;
- State and federal governments requiring that coal and oil-burning power plants install smokestack scrubbers to remove greenhouse emissions and acid-producing ions from stack gases;
- Commercial companies and schools making the decision to build "green," energy-efficient buildings to reduce their carbon footprint;
- Auto manufacturers designing and marketing electric, hybrid, and other fuel-efficient vehicles that help conserve fossil fuels and reduce greenhouse gases released into the atmosphere;
- Schools, businesses, individuals, and governments installing geothermal, solar panel, and/or wind turbine energy as alternatives to traditional fossil fuel energy sources;
- Communities and schools developing and participating in recycling efforts to reduce energy use and materials depletion;

- States and local governments reducing the amount of road salt applied to highways during winter months in environmentally sensitive areas;
- State and local laws requiring that public landfills be properly constructed so as to minimize the likelihood of groundwater contamination and methane release into the atmosphere;
- Private institutions and foundations creating preservation lands for the conservation of habitat for endangered species;
- Federal and state governments setting aside wilderness areas for the preservation of natural ecosystems;
- Federal and state governments requiring industrial companies to clean PCBs and other toxic pollutants from ecosystems contaminated by the activities of these companies;
- Citizens' cooperatives being founded to promote the use of locally grown agricultural foods and other products to reduce the carbon footprint of the local economy;
- Environmental activists demanding that state and local governments investigate incidents of environmental contamination and prosecute violators of environmental and public health laws;
- Voters demanding that state governments require that energy companies prove the safety and effectiveness of hydrofracking and other energy-extraction technologies prior to implementation.

Environmental Responsibility

The decisions of one generation provide both the limit and range of possibilities open to the next generation. The technological advances of human beings have often had a negative impact on the environment, but we need not accept these past decisions going forward. Increasing awareness of the role of humans and all the other organisms in the ecosystem has begun to reverse the negative trend of the past. Only through continued efforts to protect wild species, conserve resources, preserve natural habitats, control human population growth, and value all life forms as essential contributors to the maintenance of a healthy environment will we, as a global community, survive and provide suitable living conditions for future generations.

As it becomes increasingly clear that the solution to pollution is to avoid creating hazardous wastes in the first place, today's generation is trying to open up more possibilities for those that follow. We are examining manufacturing processes more closely, demanding that industry use as few toxic inputs as possible and produce few or no toxic wastes. Grassroots activists are making it so difficult and expensive for industry to dump its hazardous wastes that cleaning up the production processes becomes a more cost-effective option for industries to consider. Consumers are beginning to show a preference for products that create less waste by demanding local manufacture, small carbon footprint, minimal packaging, and reusable vs. disposable products. In today's global economy, these ideals are sometimes hard to come by, but are an essential component in our economic and ecological recovery.

How will you respond to this environmental imperative? Will you shop at "big box" stores whose products have been manufactured and shipped from halfway around the world, giving them a large carbon footprint, or will you shop at local stores for products that have been produced locally? Will you insulate your home and set its programmable thermostat at a minimal level to conserve energy, or will you set it to a higher level without considering insulation merely to provide physical comfort? Will you allow local governments to issue building permits in your town without regard to their environmental impact, or will you become involved by demanding environmental review of such decisions? Will you be a citizen, only, or will you be an active voter, participant, and leader in the governance of your community? These are only a few of the questions that you should ask yourself as an emerging adult member of your town, county, state, nation, and world community. It will be your responsibility to make the best choices for the furtherance of our environment, our economy, and our species.

QUESTION SET 2.10—HUMAN IMPACT ON ECOSYSTEMS (ANSWERS EXPLAINED, P. 332)

1. The use of ladybugs and preying mantises to consume insect pests in gardens is an example of
   (1) biological control of insect pests
   (2) exploitation of insect pests
   (3) abiotic control of insect pests
   (4) use of biocides to control insect pests

2. The creation of wildlife refuges and the enforcement of game laws are conservation measures that promote increased
   (1) use of biocides
   (2) preservation of species
   (3) use of biological controls
   (4) exploitation of species

3. One practice that has successfully increased the number of bald eagles in the United States is the
   (1) protection of natural habitats
   (2) importation of food to their nesting sites
   (3) preservation of other eagle species that occupy the same niche
   (4) increased use of pesticides

4. Which human activity would be more likely to have a negative impact on the environment than the other three?
   (1) using reforestation and cover cropping to control soil erosion
   (2) using insecticides to kill insects that compete with humans for food
   (3) developing research aimed toward the preservation of endangered species
   (4) investigating the use of biological controls for pests

5. Which human activity would most likely result in the addition of an organism to the endangered species list?
   (1) cover cropping
   (2) use of pollution controls
   (3) use of erosion controls
   (4) habitat destruction

6–10. Base your answers to questions 6 through 10 on the passage below and on your knowledge of biology.

The Mystery of Deformed Frogs

Deformities, such as legs protruding from stomachs, no legs at all, eyes on backs, and suction cup fingers growing from sides, are turning up with alarming frequency in North American frogs. Clusters of deformed frogs have been found in California, Oregon, Colorado, Idaho, Mississippi, Montana, Ohio, Vermont, and Quebec.

Scientists in Montreal have been studying frogs in more than 100 ponds in the St. Lawrence River Valley for the past four years. Normally, less than 1 percent of frogs are deformed. In ponds where pesticides are used on surrounding land, as many as 69 percent of the frogs were deformed. A molecular biologist from the University of California believes that the deformities may be linked to a new generation of chemicals that mimic growth hormones. The same kind of deformities found in the ponds have been replicated in laboratory experiments.

Some scientists have associated the deformities with a by-product of retinoid, which is found in acne medication and skin rejuvenation creams. Retinoids inside a growing animal can cause deformities. For this reason, pregnant women are warned not use skin medicines that contain retinoids. Recent laboratory experiments have determined that a pesticide can mimic a retinoid.

A developmental biologist from Hartwick College in Oneonta, New York, questioned whether a chemical could be the culprit because no deformed fish or other animals were found in the ponds where the deformed frogs were captured. He believes parasites are the cause. When examining a three-legged frog from Vermont, the biologist found tiny parasitic flatworms packed into the joint where a leg was missing. In a laboratory experiment, he demonstrated that the invasion of parasites in a tadpole caused the tadpole to sprout an extra leg as it developed. Scientists in Oregon have made similar observations.

6. Why are pregnant women advised not to use skin medicines containing retinoids?
   (1) Retinoid by-products may cause fetal deformities.
   (2) Retinoid by-products cause parasites to invade developing frogs
   (3) Retinoid by-products mimic the effects of pesticides on fetal tissue.
   (4) Retinoid by-products reduce abnormalities in maternal tissue.

7. Some scientists argue that pesticides may not be the cause of the frog deformities because
   (1) pesticide use has decreased over the last four years
   (2) new pesticides are used in skin-care products
   (3) other animals in the ponds containing deformed frogs did not have abnormalities
   (4) laboratory experiments have determined that a pesticide can mimic retinoids

8. A possible reason for the absence of deformed fish in the ponds that contained deformed frogs is that
   (1) fish can swim away from chemicals introduced into the pond
   (2) parasites that affect frogs usually do not affect fish
   (3) fish cannot develop deformities
   (4) frogs and fish are not found in the same habitat

9. Which inference can be made from the information in the passage?
   (1) Only a few isolated incidents of frog deformities have been observed.
   (2) If frog parasites are controlled, all frog deformities will stop.
   (3) Deformities in frogs are of little significance.
   (4) Factors that affect frogs may also affect other organisms.

10. By using one or more complete sentences, describe how pesticides could cause deformities in frogs.

11. Habitat destruction is an environmental problem that affects our own generation and will affect future generations if not solved. Write an essay in which you identify a habitat that is being destroyed and explain how the destruction of this habitat relates to humans and the overall ecosystem. Your essay must include at least:

   - *two* human activities that contribute to the destruction of this habitat
   - *three* ways the destruction of this habitat has affected plants, humans, and other animals
   - *two* ways to limit further destruction of this habitat

12–13. Base your answers to questions 12 and 13 on the information below and on your knowledge of biology.

   In July 1997, about 25,000 *Galerucella pusilla* beetles were released at Montezuma National Wildlife Refuge in western New York State. These beetles eat purple loosestrife, a beautiful but rapidly spreading weed that chokes wetlands. Purple loosestrife is native to Europe, but here it crowds out native wetland plants, such as cattails, and does not support wildlife the way that native plants do. Purple loosestrife grows too thick to allow birds to nest. Most native insects do not eat it, leaving little for insect-eating birds to feed on. Bernd Blossey, a professor at Cornell University, spent 6 years in Europe trying to find out what limited the loosestrife population there.

12. By using one or more complete sentences, explain why the introduction of the beetle is an advantage over the use of herbicides to control the purple loosestrife population.

_____

13. By using one or more complete sentences, describe one possible environmental problem that may result from the introduction of this beetle.

_____

14–16. Base your answers to questions 14 through 16 on the information below.

In a rural area is a wetland with a large population of mosquitoes. Nearby residents are concerned because the mosquitoes are always annoying and occasionally carry diseases. The community decides to have an insecticide sprayed from an airplane onto the area during the prime mosquito season. Whenever they stop spraying, the mosquito population quickly rebounds to a higher level than existed before the spraying program began. After 10 years, the spraying became much less effective at reducing the mosquito population. Higher doses of insecticide were required to accomplish the same population decreases.

14. State *one* possible *disadvantage* of spraying the insecticide from the airplane.

_____

_____

15. State *one* alternative method of mosquito control that may have a more lasting impact on the mosquito population.

_____

_____

16. Give *one* positive effect or *one* negative effect, other than killing mosquitoes, of the alternative method of mosquito control you stated in question 15.

_____

APPENDIX

Answer Key

UNIT ONE

Question Set 1.1

1. 2
2. 2
3. 1
4. 1
5. 2
6. 3
7. 4
8. 2
9. 4
10. see Answers Explained
11. 3
12. 1
13. 2
14. 1
15. 1

Question Set 1.2

1–4. see Answers Explained
5. 4
6. 1
7. 4
8–10. see Answers Explained
11. 1
12. 4
13. 2
14. see Answers Explained
15. 3

Question Set 1.3

1. see Answers Explained
2. 1
3. 3
4. 1
5. 4
6. 2
7. see Answers Explained
8. see Answers Explained
9. see Answers Explained
10. 4
11. 4
12. see Answers Explained

Question Set 1.4

1. 2	17. 1	30. see Answers
2. B	18. 3	Explained
3. C	19. 2	31. see Answers
4. D	20. **8.9 cm**	Explained
5. 4	21. 1	32. see Answers
6. 3	22. 4	Explained
7. see Answers	23. see Answers	33. see Answers
Explained	Explained	Explained
8. 1	24. see Answers	34. 4
9. 4	Explained	35. 1
10. 2	25. 2	36. 2
11. 2	26. 3	37. 3
12. 4	27. 4	38. 3
13. 3	28. see Answers	39. 3
14. 1	Explained	40. 4
15. 4	29. see Answers	41. 4
16. 1	Explained	42. 2

Reading, Writing, and Current Events in Science

1. 3	8. see Answers	12. see Answers
2. 1	Explained	Explained
3. 1	9. see Answers	13. see Answers
4. 2	Explained	Explained
5. 3	10. see Answers	14. see Answers
6. see Answers	Explained	Explained
Explained	11. see Answers	
7. see Answers	Explained	
Explained		

UNIT TWO

Question Set 2.1

1. **1**	5. **1**	9. **2**
2. **4**	6. **4**	10. **1**
3. **4**	7. **3**	11. **4**
4. **4**	8. **1**	

Question Set 2.2

1. **4**	5. **1**	9. **4**
2. **1**	6. **4**	10. **2**
3. **2**	7. **3**	
4. **2**	8. **2**	

Question Set 2.3

1. **1**	5. **1**	10. **4**
2. **3**	6. **3**	11. **2**
3. **3**	7. **2**	12. **1**
4. **see Answers Explained**	8. **3**	
	9. **3**	

Question Set 2.4

1. **1**	6. **3**	11. **2**
2. **2**	7. **3**	12. **1**
3. **3**	8. **3**	13. **4**
4. **4**	9. **2**	
5. **3**	10. **3**	

Question Set 2.5

1. **1**	11. **2**	21. **1**
2. **4**	12. **3**	22. **3**
3. **4**	13. **1**	23. **2**
4. **1**	14. **2**	
5. **4**	15. **4**	
6. **3**	16. **3**	
7. **3**	17. **3**	
8. **2**	18. **3**	
9. **1**	19. **1**	
10. **4**	20. **2**	

Question Set 2.6

1. **4**	6. **2**	11. **2**
2. **3**	7. **3**	12. **see Answers Explained**
3. **1**	8. **1**	13. **see Answers Explained**
4. **1**	9. **3**	
5. **4**	10. **4**	

Question Set 2.7

1. 1	7. 3	13. 1
2. 4	8. 4	14. 3
3. 1	9. 2	15. 1
4. 4	10. 1	16–19. see Answers
5. 3	11. 3	Explained
6. 2	12. 4	

Question Set 2.8

1. 2	12. 2	21. 3
2. 2	13. 4	22. 4
3. 4	14. 4	23. 2
4. 2	15. 1	24–30. see Answers
5. 2	16. 4	Explained
6. 1	17. 1	
7. 2	18. see Answers	
8. 2	Explained	
9. 4	19. see Answers	
10–11. see Answers	Explained	
Explained	20. 2	

Question Set 2.9

1. 4	10. 3	19. 2
2. 2	11. 2	20. 2
3. 1	12. 3	21. 4
4. 6	13. 2	22. 2
5. 5	14. 3	23. 4
6. 2	15. 4	24. 1
7. 1	16. 4	25. 3
8. 2	17. 2	26. see Answers
9. 4	18. 3	Explained

Question Set 2.10

1. 1	5. 4	9. 4
2. 2	6. 1	10–16. see Answers
3. 1	7. 3	Explained
4. 2	8. 2	

Answers Explained

UNIT ONE

Question Set 1.1 (p. 16)

1. **2** The biologist should observe *a large number of frogs in their natural habitat*. The large sample will help to ensure the probability that the sample will be representative of the species's behavior. Observing the frogs in their natural habitat will ensure that no unusual conditions exist that may influence how the frogs react to each other.

Wrong Choices Explained
  (1) Observing *a small number of frogs in their natural habitat* will not ensure that a representative sample has been taken. If a few of the frogs in a small sample behave in an unusual way, they will represent a larger percentage of a small sample than they would of a large sample.
  (3), (4) Observing *several groups of frogs maintained in different temperatures in the laboratory* or *several groups of frogs maintained on different diets in the laboratory* will introduce variables that interfere with the natural behaviors of the frogs. Changes in temperature or diet may affect the frogs' behavior greatly, making the observations dependent on these experimental conditions instead of the natural factors that affect the frogs' behavior.

2. **2** Counting the number of lines on the cylinder between the meniscus at A and the meniscus at B yields the number 4. Since a total of 8 milliliters was removed, the graduations on the cylinder must be *2 milliliters* each.

Wrong Choices Explained
  (1) Graduations of *1 milliliter* would yield eight lines with the removal of 8 milliliters from the cylinder.
  (3) Graduations of *8 milliliters* would yield one line with the removal of 8 milliliters from the cylinder.
  (4) Graduations of *4 milliliters* would yield two lines with the removal of 8 milliliters from the cylinder.

3. **1** The statement *environmental conditions affect germination* represents a hypothesis. This statement contains a belief about the outcome of an investigation in which a dependent variable (germination) is changed by an experimental variable (environmental conditions). Although very general, the statement still predicts an outcome and therefore meets the definition of a hypothesis.

Wrong Choices Explained

(2) The statement *boil 100 milliliters of water, let it cool, and then add ten seeds to the water* represents an experimental procedure, not a hypothesis.

(3) The statement *is water depth in a lake related to available light in the water?* represents an experimental question, not a hypothesis.

(4) The statement *a lamp, two beakers, and elodea plants are selected for the investigation* represents a materials list for an experiment, not a hypothesis.

4. **1** The *millimeter* is the unit of measure most appropriate to perform this measurement. About 25.4 millimeters make up one inch.

Wrong Choices Explained

(2) The micrometer is a unit of measure used to describe the sizes of very small objects, such as cell organelles. It is not an appropriate unit of measure for an object of this size.

(3), (4) The foot and the meter are too large to be used to measure an object of this size, since the measurements would be fractional.

5. **2** A millimeter is made up of 1,000 micrometers. The number of whole millimeters shown in the diagram is 3 (3,000 micrometers). The amount of distance left to the right of the third millimeter is a little larger than half of a millimeter (about 70 micrometers). The total distance across the diameter, therefore, is approximately *3,700 micrometers*. (HINT: remember to count as 1,000 micrometers the distance from the left edge of one millimeter mark to the left edge of the next millimeter mark.)

Wrong Choices Explained

(1), (3), (4) Each of these distracters has been chosen to catch the unwary student. Count carefully and mark on the test booklet if that will assist you.

6. **3** *It differs in the one variable being tested* is the way that the control setup differs from the experimental setups in the same experiment. In all other respects, it is identical to the experimental setups. For example, in an experiment testing how varying concentrations of enzymes affect the rate of hydrolysis of a substrate, an appropriate control would be a setup containing water, only, but identical to the experimental tubes in all other respects. This control assures that the water in which the enzyme is dissolved has no effect on the hydrolytic action.

Wrong Choices Explained

(1), (2), (4) The statements *it tests a different hypothesis, it has more variables*, and *it utilizes a different method of data collection* are not true of a control group in a properly designed experiment. The experimental design should seek to eliminate the differences between the experimental setups and the control setup. The more variables and differences between the control and experimental setups, the less reliable the results of, and the conclusions drawn from, the experiment.

7. **4** The experimental variable is the condition changed (varied) by the experimenter. In this experiment, the condition varied is the *distance of the plant from light*.

Wrong Choices Explained

(1) No measure is being made of the *concentration of gas in the water*. If this condition is changing, it is incidental to the experiment.

(2) The *aquatic plant type*, in this case elodea, is a *constant* in the experiment as described.

(3) No measure is made of the *amount of water in the test tube*. If this condition is changing, it is incidental to the experiment.

8. **2** *Lugol's solution* will test for the presence of starch in foods. It turns blue black in the presence of starch.

Wrong Choices Explained

(1) *pH paper* is used to test how acidic or alkaline a solution is.

(3) *Methylene blue* is a commonly used stain in preparing wet-mount microscope slides.

(4) *Bromthymol blue* is used to detect the presence of carbon dioxide in a solution.

9. **4** Before a scientist begins to study a natural phenomenon, he/she must first pose a question to be answered or a *problem to be solved*. If this step is skipped, much time is wasted in the laboratory studying factors that may have no direct bearing on the object of study.

Wrong Choices Explained

(1), (2), (3) An appropriate sequence for conducting a scientific investigation is as follows:

— state the problem
— *formulate a hypothesis* (educated guess)
— define the experimental method
— *perform the experiment*
— gather experimental data
— *analyze the experimental data*
— draw inferences

10. *Goggles should be worn.* OR *Gloves should be worn.* OR *The student should take care not to touch the sharp scalpel to his skin.*
(NOTE: The preceding answers represent acceptable responses. Other complete-sentence responses are acceptable as long as they give correct information.)

11. **3** The *control* group in an experiment duplicates all aspects of the experiment except the independent variable (the variable being tested). In this case, the drug is the independent variable, so the group given pills without the drug, but with glucose, is considered the control group.

Wrong Choices Explained
(1) The *experimental group* is the group that receives pills containing the drug rather than pills containing glucose.
(2) The term *limiting factor* relates to environmental conditions that limit the survival of individuals or populations.
(4) An *indicator* is a chemical that helps to determine the chemical characteristics of biological samples. Examples of indicators include bromthymol blue and litmus.

12. **1** Of those given, the hypothesis *light intensity affects the growth of algae* best explains the observations reported. The intensity of natural light affects the depth to which it can penetrate the lake water. Algae can photosynthesize only in the presence of adequate light intensity. Therefore, we are led to the conclusion that the presence (or growth) of algae depends on the degree to which light is available (its intensity) in the environment.

Wrong Choices Explained
(2) The hypothesis *wind currents affect the growth of algae* has no basis for support in the reported observations. Not only is wind not mentioned as a variable, but logical analysis of the situation would indicate that wind is unlikely to directly affect algae 1 to 6 meters below the surface of the lake.
(3) The hypothesis *nitrogen concentration affects the growth of algae* has no basis for support in the reported observations. Nitrogen concentration is not mentioned as a variable in this experiment.
(4) The hypothesis *precipitation affects the growth of algae* has no basis for support in the reported observations. Precipitation is not mentioned as a variable in this experiment.

13. **2** *Diagram 2* illustrates how the hydra will appear to move under the compound light microscope. The optics of this instrument invert the image of what is seen in the field of view, so that everything appears upside down. As the hydra moves right, its image will appear to move left. If the hydra moves away from the observer, its image will appear to move toward the observer, and so on.

Wrong Choices Explained
(1), (3), (4) These diagrams all represent incorrect illustrations of how the image will move in the field of view under the conditions stipulated.

14. **1** The approximate length of a nucleus of one of these cells is *100 μm*. Two nuclei are visible as dark ovals within each of the two cells illustrated. Care should be taken to read the question carefully so that the correct object is measured.

Wrong Choices Explained
(2) *500 μm* is the approximate length of one cell.
(3) *1,000 μm* is the approximate length of both cells end to end.
(4) *1,500 μm* is the approximate diameter of the field of view (actually closer to 1,600 μm).

15. **1** The difference in length between leaves *A* and *B* is closest to *20 mm*. Leaf *A* is approximately 55 mm long, while leaf *B* is approximately 33 mm long. The difference, 22 mm, is closest to 20 mm.

Wrong Choices Explained
(2) *20 cm* is ten times the correct measurement. Students should take care that the correct units are used in determining the lengths of these specimens.
(3) *0.65 m* is a measurement equal to 650 cm or 6,500 mm (more than 2 feet) in length. These specimens are much smaller than this measurement.
(4) *1.6 μm* is a measurement equal to 0.0016 mm or 0.00016 cm. These specimens are much larger than this measurement.

Question Set 1.2 (p. 26)

1–3.

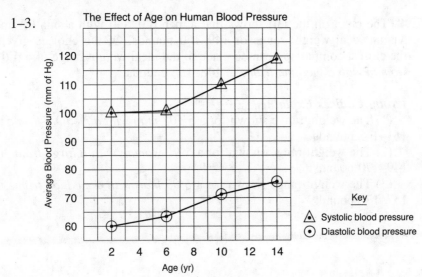

The Effect of Age on Human Blood Pressure

255

4. *Systolic pressure is higher than diastolic pressure.* OR *Both systolic pressure and diastolic pressure increase between the ages of 2 and 14.* (NOTE: Other correct, complete-sentence responses are acceptable.)

5. **4** Brook trout growth is represented by the lighter solid line in the graph. By tracing along the horizontal axis to 16°C and then tracing up to the point at which the brook trout line is encountered, we see that the growth rate is between 80% and 100% but closer to 100%. Estimate about *95%*.

Wrong Choices Explained
   (1) Brook trout growth rate is at 30% when the water temperature is about 18°C. Largemouth bass growth rate is about 30% at 16°C.
   (2) None of the species indicated have growth rates of 55% at 16°C. Brook trout growth rate is at 55% at about 17.5°C.
   (3) Brook trout growth rate falls to 75% at about 17°C. The growth rate of northern pike is at 75% at about 16°C.

6. **1** The diagram represents an experiment in cloning. Since the nucleus (which contains the genetic material) used in this experiment comes from a tadpole, the egg will produce new cells that have *genetic characteristics identical to those of the tadpole*.

Wrong Choices Explained
   (2) It is unclear from this diagram how frog *A* came into being.
   (3) This conclusion is refuted by the results of this experiment; a new frog was created without the use of fertilization.
   (4) This conclusion is refuted by the results of this experiment. The tadpole nucleus was taken from an intestinal cell, which is a body cell.

7. **4** The chart on the left shows that a man 6'0" tall with a small frame has an ideal weight range of 149–160 pounds. Of the choices given, the closest comparison can be made to the ideal weight range of a *6'0" woman with a medium frame* at 148–162 pounds.

Wrong Choices Explained
   (1) The weight range shown for a *5'10" man with a medium frame* is 151–163 pounds.
   (2) The weight range shown for a *5'9" woman with a large frame* is 149–170 pounds.
   (3) The weight range shown for a *6'0" man with a medium frame* is 157–170 pounds.

8.

Color of Light	Wavelength of Light (nm)	Percent Absorption by Spinach Extract
Violet	412	49.8
Blue	457	49.8
Green	533	17.8
Yellow	585	25.8
Orange	616	32.1
Red	674	41.0

9–10.

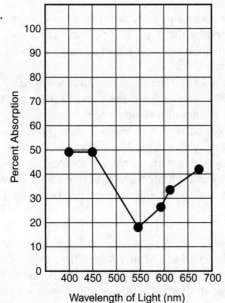

11. **1** The statement *photosynthetic pigments in spinach plants absorb blue light and violet light more efficiently than red light* is a valid conclusion that can be drawn from the data. The high point of the chart/graph data is very clearly shown to be over the blue and violet wavelengths of light.

Wrong Choices Explained
(2) The statement *the data would be the same for all pigments in spinach plants* is not supported by the results of this experiment. The chart/graph data show considerable variation in the experimental results as the wavelength of light varies.

(3) The statement *green and yellow light are not absorbed by spinach plants* is not supported by the results of this experiment. Although the chart/graph data show a lower absorption rate in these wavelengths, there is still some absorption in this range.

(4) The statement *all plants are efficient at absorbing violet and red light* is not supported by the results of this experiment. The experimental data is limited to the absorption of light by pigments found in one type of plant. This data cannot be extended to all plants unless all other types of plants are tested under the same experimental conditions and the results are found to be similar.

12. **4** Of those given, *refrigeration will most likely slow the growth of these bacteria* is the most reasonable inference that can be made from the graph data. The graph clearly shows a slower rate of growth (reproduction) at 5°C than at either 10°C or 15°C.

Wrong Choices Explained
(1) The inference that *temperature is unrelated to reproductive rate of bacteria* is not supported by the data. Temperature is the independent (experimental) variable in this study. It clearly has an influence on the bacterial reproductive rate.

(2) The inference that *bacteria cannot grow at a temperature of 5°C* is not supported by the data. The graph clearly shows that growth at this temperature, while slow, occurs at a steady pace.

(3) The inference that *life activities in bacteria slow down at high temperatures* is not supported by the data. The data indicates that, if anything, bacterial activity increases with increasing temperature. No data is shown for bacterial growth under temperatures above 15°C, so we cannot draw any inference about what will happen to the rate of bacterial growth at these extremes.

13. **2** These researchers should *make sure the conditions are identical to those in the first study*. The validity of any scientific experiment can be verified only if the same results are obtained under the same experimental conditions. Any change in these conditions will invalidate the results of the verification study.

Wrong Choices Explained
(1) If the researchers *give growth solution to both groups*, there will be no control group against which to compare the experimental group. The results of the verification study will be invalid since experimental conditions will have changed.

(3) If the researchers *give an increased amount of light to both groups of plants*, the original experimental method will be altered. The results of the verification study will be invalid since experimental conditions will have changed.

(4) If the researchers *double the amount of growth solution given to the first group*, the original experimental method will not be followed. The results of the verification study will be invalid since experimental conditions will have changed.

14. *130. The number of turns in the waggle dance decreases as the distance of the food supply from the hive increases.* OR *The closer to the hive the food source is located, the more turns there are in the waggle dance.* (NOTE: Any correct, complete-sentence answer is acceptable.)

15. **3** Other biologists in other laboratories should be able to perform *the same experiment and obtain the same results* if the experimental results are valid. Any experimental results obtained by one scientist must be validated by independent research by other scientists following the same procedures.

Wrong Choices Explained

(1), (4) If different scientists perform *an experiment with a different variable and obtain the same results* or *an experiment under different conditions and obtain the same results*, they will have neither validated nor invalidated the results of the original experiment. All variables and conditions must be kept the same if the experimental results are to be properly tested.

(2) If different scientists perform *the same experiment and obtain different results*, the original experimental results will have been invalidated.

Question Set 1.3 (p. 37)

1. Two credits are awarded for a correct two-part response that correctly identifies the finch species most affected and the reason for this effect. Acceptable responses may include:

   - *The small ground finch [1] because the sharp-billed ground finch looks for the same kinds of food. [1]*
   - *The native finch species most probably strongly affected will be the small ground finch. [1] The reason for this is that both the small ground finch and the sharp-billed ground finch depend on the same food source (small seeds). [1]*
   - *My hypothesis is that the small ground finch will be most affected by the introduction of the sharp-billed ground finch. [1] The chart shows that both species are ground-feeding seedeaters and have similar bill shapes, indicating that they will compete for the same types of available seeds. [1]*

2. **1** The *large ground finch* will have the greatest adaptive advantage under the changed environmental conditions described. According to the chart, the large ground finch is well-adapted for crushing the shells of large, thick-walled seeds. Since these seeds will be in abundance in the changed environment, it is expected that the large ground finch will thrive compared to the other species native to the island.

Wrong Choices Explained

(2) The *small ground finch*, with its small beak, is specifically adapted for picking up and eating small seeds. It is poorly adapted for eating large, thick-walled seeds. It is likely that this species will not thrive under the conditions of environmental change described.

(3), (4) According to the chart, the *large tree finch* and the *small tree finch* are adapted to feed on insects in an environment different from that of the other native finch species. It is unlikely that this environmental change will either significantly advantage or disadvantage insect-eating finch species.

3. **3** *Interspecies competition* occurs when one species tries to gain an advantage over another species in the same environment. Interspecies (between-species) competition may occur when any shared resource, including food, territorial area, nesting sites, or other habitat, is in short supply.

Wrong Choices Explained

(1) *Variation within a species* is not represented by one finch species gaining an advantage over other species in the same habitat. This term refers to the concept that each naturally occurring species contains differences (variations) among its individual members. Such variations are thought to be the basis of the ongoing evolution in all species by means of natural selection.

(2) *Environmental change* is not represented by one finch species gaining an advantage over other species in the same habitat. This term refers to any alteration (such as rainfall or temperature) in the set of environmental conditions that exist in a particular habitat.

(4) *Mutagenic agents* are not represented by one finch species gaining an advantage over other species in the same habitat. This term refers to forces in nature (such as radiation and chemicals) that are able to alter the genetic makeup of living things.

4. **1** *CUU-GUU-ACU-GAA-CAU-CAU-CCC-GUU* is the mRNA codon sequence that will result from this DNA sequence. In the formation of mRNA, ribonucleotides position themselves on the DNA template such that complementary nitrogenous bases link together. DNA bases *C, G, A*, and *T* link with RNA bases *G, C, U*, and *A*, respectively.

Wrong Choices Explained

(2), (3), (4) *GAA-CAA-UGA-CUU-GUA-GUA-GGG-CAA, CTT-GTT-ACT-GAA-CAT-CAT-CCC-GTT*, and *UGG-AGG-CUG-ACC-UCG-UCG-UUU-AGG* are not mRNA codon sequences that will result from this DNA sequence. Each of these sequences violates the rules of complementarity referenced above or includes the nitrogenous base *T* (thymine) instead of *U* (uracil) in its structure.

5. **4** *Leu-Val-Thr-Glu-His-His-Pro-Val* is the amino acid sequence that will result from this DNA sequence. In the synthesis of a protein strand on the ribosome, each amino acid is positioned according to the sequence of codons on the mRNA strand (which in turn derived its sequence from the DNA template in the nucleus). For example, the amino acid Leu (leucine) attaches at the mRNA codon *CUU* or *CUG* (which in turn is encoded by the DNA template codon *GAA* or *GAC*). Following this example through the entire DNA codon sequence leads to the sequencing of amino acid units as shown above.

Wrong Choices Explained

(1), (2), (3) *Leu-Glu-His-Pro-Val-Val-Pro-Val, Leu-Val-Thr-Leu-Glu-Glu-His-Pro*, and *Leu-Pro-His-Val-Thr-Thr-Glu-Pro* are not amino acid sequences that will result from this DNA sequence. Each of these sequences violates the rules of amino acid encoding referenced above and data shown in the Partial Genetic Code Chart.

6. **2** *Species X* produces a protein segment most similar to that produced by *S. hunta*. Because each of the amino acids in this segment can be encoded by more than one DNA code, the slight variations in this gene between the two species still result in the same amino acid sequence (*Leu-Val-Thr-Glu-His-His-Pro-Val*).

Wrong Choices Explained

(1), (2), (3) *Leu-Glu-His-Pro-Val-Val-Pro-Val, Leu-Val-Thr-Leu-Glu-Glu-His-Pro*, and *Leu-Pro-His-Val-Thr-Thr-Glu-Pro* are not amino acid sequences that will result from this DNA sequence. Each of these sequences violates the rules of amino acid encoding referenced above and data shown in the Partial Genetic Code Chart.

7. One credit is awarded for a correctly completed histogram, as shown below:

Number of Students
10
9
8
7
6
5
4
3
2
1
<51
51–60
61–70
71–80
81–90
>90
Average Pulse Rate Range (beats per minute)

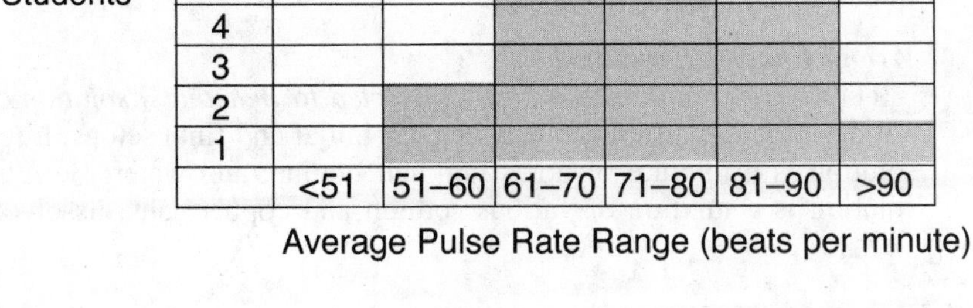

8. One credit is awarded for a complete sentence that correctly describes one pattern that is evident in the data that would help someone to understand how pulse rate is distributed among these 20 students. Answers may include:

- *Nearly two-thirds (65%) of this class has resting heart rates between 61 and 80 beats per minute.*
- *The average (mean) pulse rate for this class is 73.35 beats per minute.*
- *The distribution of heart rates for this class follows a bell curve that skews slightly to the left.*
- *Generally, the female members of this class have lower pulse rates than the males.*

9. One credit is awarded for a complete sentence that correctly states an experimental question about pulse rate that could be answered by further study of these 20 students. Answers may include:

- *What is the average pulse rate for males versus females in this class?*
- *How do the distribution curves differ for resting heart rate in females versus males?*
- *What is the relationship between resting pulse rate and body mass index (BMI)?*
- *What is the relationship between heart rate and amount of daily exercise among these students?*
- *What are the three-trial average pulse rates of these students after 5 minutes of moderate exercise?*

10. **4** The difference between the fluids in the initial and final setups is that *glucose has diffused from the fluid inside the dialysis tubing, across the membrane, and into the beaker fluid*. In the initial stage chemical test with Benedict's solution, the dialysis tubing was shown to contain glucose, whereas the beaker water was not. When the same test was performed on the setup in the final stage, glucose was detected in both the dialysis tubing and the beaker water. Since the dialysis tubing is a semipermeable membrane, it can be concluded that the small molecular size of glucose allowed it to diffuse through the membrane from a region of higher concentration (inside the tubing) to a region of lower concentration (outside the tubing).

Wrong Choices Explained

(1) *Lugol's solution has been converted to Benedict's solution* does not describe the difference between the initial and final setups. Lugol's solution is a mixture of potassium salt, iodine, and water. Benedict's solution is a mixture of various sodium and copper salts dissolved in

water. It is chemically impossible for Lugols's solution to be converted to Benedict's solution.

(2) *Starch has been converted to glucose by the indicator solutions* does not describe the difference between the initial and final setups. The conversion of starch to glucose requires the catalytic action of the enzymes amylase and maltase. Lugol's and Benedict's solutions contain no such enzymes.

(3) *Starch has diffused from the fluid inside the dialysis tubing, across the membrane, and into the beaker fluid* does not describe the difference between the initial and final setups. Starch is a polymer whose molecules are too large to diffuse through the dialysis tubing. Had this occurred, the final setup beaker water would have tested positive for the presence of starch (turned blue-black) in the presence of Lugol's solution.

11. **4** *More water moving out of the cell than into the cell* is the most likely cause of the change in appearance illustrated in diagram *B*. Under normal conditions, the cell cytoplasm and the watery environment have nearly identical salt concentrations, allowing water molecules to diffuse into and out of the cytoplasm at equal rates, as shown in *A*. When salt is added to the watery medium, the relative concentration of water decreases compared to the water concentration inside the cell. Under these conditions, water molecules diffuse out of the cell more rapidly than they diffuse into the cell, causing the cytoplasm to shrink, as shown in *B*.

Wrong Choices Explained
(1), (2) *More salt moving out of the cell than into the cell* and *more salt moving into the cell than out of the cell* are not probable causes of the condition shown in diagram *B*. Salt does not normally diffuse into or out of the cell. Rather, water is the substance that moves by osmosis in response to the concentration gradient.

(3) *More water moving into the cell than out of the cell* is not a probable cause of the condition shown in diagram *B*. If this were the case, the cytoplasm inside the cell would increase, rather than decrease, in volume.

12. Four credits are awarded for a paragraph, containing complete sentences, that correctly describes the biological explanation for the farmer's observations of the corn plants in field *B*. Sample paragraph:

When the farmer irrigated field B with the 10% salt solution on day 1, the ground around the roots became saturated with it. This created a situation in which the concentration of water inside the corn root cells was higher (>99%) than the concentration of water in the ground outside the corn root cells (90%). [1] As a result, water moved by osmosis (diffusion of water from a region of high relative concentration toward a region of low relative concentration) [1] from the inside to the outside of the corn plants.

On day 2, the rainfall washed the salt away from the soil around the roots of the corn plants sufficiently to allow the equilibrium (equal concentrations of water both inside and outside the cell) [1] to become reestablished, enabling the corn plant root cells to absorb water and the plants to regain their healthy appearance. [1]

Question Set 1.4 (on Laboratory Skills) (p. 49)

1. **2** The student's suggestion is considered to be *a hypothesis*. A hypothesis is an "educated guess" as to the scientific reason behind an observed phenomenon. The student is making this suggestion based on a hunch rather than on direct observation. To test the hypothesis, the student would need to design an experiment, collect data, analyze the data, and draw inferences and conclusions concerning the results.

Wrong Choices Explained
(1) The student's suggestion is not considered to be *a control*. A control is a laboratory setup created by the researcher in a scientific experiment that duplicates the experimental setup in every way except for the variable being tested.

(3) The student's suggestion is not considered to be *an observation*. An observation is an action taken by the researcher during a scientific experiment that normally results in the generation of data.

(4) The student's suggestion is not considered to be *a variable*. A variable is a condition that changes in a scientific experiment either because it is manipulated by the researcher or because it changes in response to the manipulated variable.

2. **B** A student testing for the presence of a carbohydrate in food should select set *B*. This laboratory setup contains two indicators (Benedict's solution and iodine) that are specialized to indicate the presence of specific carbohydrates (simple sugars and complex starches, respectively) in foods. The assorted glassware is needed to hold the foods being tested, and the heat source is used to heat the Benedict's solution in the presence of simple sugars. Goggles protect the eyes from accidental exposure to heated liquids.

Wrong Choices Explained
(A) The student should not select set *A*, which would probably be best used to determine the effect of different colors (wavelengths) of light on the rate of photosynthesis in a water plant.

(C) The student should not select set *C*, which would probably be best used to dissect a biological specimen in order to facilitate a study of gross anatomical features.

(D) The student should not select set D, which would probably be best used to study the microscopic features of very small biological specimens such as the organelles of a cell.

3. **C** A student determining the location of the aortic arches of the earthworm should select set C. This laboratory setup contains dissection instruments and a wax pan in which dissection of an earthworm can be carried out. The stereomicroscope is used to magnify the small anatomical features of the earthworm (in this case, aortic arches) for closer study. Goggles protect the eyes from accidental exposure to preservative chemicals and sharp instruments.

Wrong Choices Explained
(A) The student should not select set A, which would probably be best used to determine the effect of different colors (wavelengths) of light on the rate of photosynthesis in a water plant.
(B) The student should not select set B, which would probably be best used to determine the presence of various carbohydrates in foods.
(D) The student should not select set D, which would probably be best used to study the microscopic features of very small biological specimens such as the organelles of a cell.

4. **D** A student observing the chloroplasts in elodea should select set D. This laboratory setup contains a compound light microscope used to magnify the microscopic features of cells for closer study. The slide and water are used to provide a transparent viewing surface, and the forceps are used to manipulate the very small elodea leaf.

Wrong Choices Explained
(A) The student should not select set A, which would probably be best used to determine the effect of different colors (wavelengths) of light on the rate of photosynthesis in a water plant.
(B) The student should not select set B, which would probably be best used to determine the presence of various carbohydrates in foods.
(C) The student should not select set C, which would probably be best used to dissect a biological specimen in order to facilitate a study of gross anatomical features.

5. **4** The corresponding control group would most likely consist of truck drivers who drove *daily for 8 hours in very light traffic*. A control is a setup created by the researcher in a scientific experiment that duplicates the experimental setup in every way except for the variable being tested. Because the variable being tested is the stress of driving for 8-hour periods in heavy traffic, the stress factors that would contrast with these would be either reduced exposure to time driving in heavy traffic (not offered as a choice) or reduced exposure to traffic stresses for the same time duration (choice 4).

Wrong Choices Explained

(1) Truck drivers who drove *daily for 12 hours in very heavy traffic* is not the likely control group for this experiment. Because the experimental group is already considered by the researcher to be stressed with 8 hours in very heavy traffic, the control group would not be set up to receive even more time exposure to this stressful situation.

(2), (3) Truck drivers who drove *every third day for 8 hours in very heavy traffic* or who drove *every other day for 12 hours in very light traffic* are not the likely control groups for this experiment. The introduction of a third variable (intermittent exposure) in these groups would tend to confuse the research and raise questions about which variable of the three is really responsible for the stresses being observed in the experimental group.

6. **3** *Temperature* is a variable in this experiment. A variable is a condition that changes in a scientific experiment either because it is manipulated by the researcher or because it changes in response to the manipulated (experimental) variable. Of the conditions listed, only temperature changes in the experiment. The researcher places each of the five test tubes in a different temperature-controlled environment (0°C, 10°C, 20°C, 30°C, and 40°C), and so temperature is the experimental variable. The other variable in the experiment is the amount of egg white digested in each tube.

Wrong Choices Explained

(1), (2), (4) *Gastric fluid*, *length of glass tubing*, and *time* are not variables (conditions that change) in this experiment. These conditions remain constant throughout the experiment for all five laboratory setups.

7. (1) *The coarse adjustment knob is used to provide general focus on the specimen being viewed.*

(2) *The fine adjustment knob is used to adjust focus on an object more precisely and in finer detail than can be obtained under coarse adjustment.*

(3) *The objective contains optical lenses that permit magnification of the image of the specimen.*

(4) *The diaphragm is used to adjust the amount of light passing through the specimen to the objective.*

8. **1** The student should have focused under *low power using the coarse and fine adjustments, and then under high power using only the fine adjustment*. Following this procedure will allow the student to locate the portion of the specimen to be studied and then to focus on and study a small part of the specimen in detail. A compound light microscope should never be focused with the coarse adjustment when the high-power objective is in place. Even minor adjustments of the coarse adjustment knob under these conditions could crush the specimen and break the slide.

Wrong Choices Explained

(2) The student should not have focused under *high power first, then under low power using only the fine adjustment*. The high power objective should not be used to locate the general area of the specimen to study, but rather the low-power objective should be used first for this purpose.

(3) The student should not have focused under *low power using the coarse and fine adjustments, and then under high power using the coarse and fine adjustments*. A compound light microscope should never be focused with the coarse adjustment when the high-power objective is in place. Even minor adjustments of the coarse adjustment knob under these conditions could crush the specimen and break the slide.

(4) The student should not have focused under *low power using the fine adjustment, and then under high power using only the fine adjustment*. The fine adjustment is designed to move the objective very small distances and is not effective when used on low power unless the coarse adjustment has been used previously to gain general focus at this level of magnification.

9. **4** The diameter of the low-power field of vision is *2,000* micrometers (μm). By definition, a micrometer is one/one millionth (1/1,000,000) of a meter (m). Because there are 1,000 millimeters (mm) in a meter, each millimeter must contain 1,000 micrometers (1,000 mm × 1,000 μm/mm = 1,000,000 μm = 1 m). Because there are 2 millimeters that span the low-power field of view illustrated in this diagram, the total low-power-field diameter must be 2,000 micrometers (2 mm × 1,000 μm/mm = 2,000 μm).

Wrong Choices Explained

(1), (2), (3) The diameter of the low-power field in the illustration is not *1* micrometer, *2* micrometers, or *1,000* micrometers. See explanation above.

10. **2** The diameter of the high-power field of vision in this illustration would be closest to *0.5 mm* (millimeters). The ratio of magnification between the low-power and high-power objectives is 1:4. With the formula below, the high-power-field diameter can be calculated to be one fourth of the 2-mm low-power-field diameter, or 0.5 mm:

$$\frac{\text{Low-power objective magnification } (10\times)}{\text{High-power objective magnification } (40\times)} = \frac{\text{High-power-field diameter } (D \text{ mm})}{\text{Low-power-field diameter } (2 \text{ mm})}$$

$$40D = 2 \text{ mm} \times 10 = 20 \text{ mm}$$

$$D = 20 \text{ mm}/40 \text{ mm} = 1/2 \text{ mm} = 0.5 \text{ mm}$$

Wrong Choices Explained

(1), (3), (4) The diameter of the low-power field is not *0.05 mm*, *5 mm*, or *500 mm*. Each of these answers is a multiple of the correct answer. See explanation on page 267.

11. **2** *Iodine solution* is the substance that, when added to a wet mount containing starch grains, would react with the starch grains and make them more visible. Also known as Lugol's solution, this tan indicator turns blue-black when it contacts a substance containing starch.

Wrong Choices Explained

(1) *Litmus solution* would not react with the starch grains by changing color. Litmus changes color in reaction to acidic and basic solutions, not to solutions containing starch. Litmus turns blue in the presence of bases and red/pink in the presence of acids.

(3) *Distilled water* would not react with the starch grains by changing color. Distilled water is a neutral substance that normally does not react with any substance to produce a color change.

(4) *Bromthymol blue* would not react with the starch grains by changing color. Bromthymol blue is an indicator that turns brick red when heated in the presence of simple sugars such as glucose.

12. **4** The formation of bubbles on slide *A* could have been prevented by *bringing one edge of the coverslip into contact with the water and lowering the opposite edge slowly*. This technique allows air to escape from under the coverslip as it is lowered onto the slide, preventing the formation of air bubbles.

Wrong Choices Explained

(1) *Using a thicker piece of potato and less water* would not prevent the formation of air bubbles on slide *A*. The thicker specimen would tend to suspend the coverslip off the water surface and could actually increase the likelihood of bubble formation.

(2) *Using a longer piece of potato and a coverslip with holes in it* would not prevent the formation of air bubbles on slide *A*. The length of the specimen would not affect the formation of bubbles. However, a coverslip with holes in it would allow air to enter the area between the coverslip and slide, leading to the formation of air bubbles.

(3) *Holding the coverslip parallel to the slide and dropping it directly onto the potato* would not prevent the formation of air bubbles on slide *A*. This technique would not allow trapped air to escape from under the coverslip, increasing the likelihood that air bubbles will form and remain trapped.

13. **3** *It allows the stain to penetrate the potato tissue without the removal of the coverslip* is the statement that best describes the purpose of the procedure. This technique takes advantage of capillary forces to pull the liquid stain under the coverslip as water is being removed by the absorbent paper on the opposite side of the coverslip. When the stain comes into contact with the potato tissue, some of it is absorbed by the cells of the tissue, staining it for study.

Wrong Choices Explained
   (1) *It prevents the stain from getting on the ocular of the microscope* is not the statement that best describes the purpose of the procedure. The ocular lens should never be exposed to chemical stains, as a stain can coat the lens, obscuring visual observation of the specimen by the researcher. This procedure should be conducted away from the proximity of the ocular lens.
   (2) *It prevents the water on the slide from penetrating the potato tissue* is not the statement that best describes the purpose of the procedure. By drawing the stain under the coverslip, this technique helps to ensure that this penetration will occur.
   (4) *It helps to increase the osmotic pressure of the solution* is not the statement that best describes the purpose of the procedure. Although the active ingredient of many stains is a form of salt, and although this salt can have the effect of altering the osmotic gradient in a system, the main purpose of the procedure is to introduce the stain into proximity of the tissues to be stained, not to increase the osmotic pressure of the solution.

14. **1** *Onion epidermal cells* most likely make up the tissue represented in the drawing. The blocklike arrangement of the cells and the wall-like divisions between them are typical of a plant tissue. Iodine stain would tend to darkly stain the starch in vacuoles and the cellulose in the cell walls of these cells.

Wrong Choices Explained
   (2) *Ciliated protists* do not most likely make up the tissue represented in the drawing. Most ciliated protests are not colonial but free-living, whereas this tissue clearly shows a structure of contiguous blocks. Cilia are not apparent on any of the cells in the drawing.
   (3) *Cardiac muscle cells* do not most likely make up the tissue represented in the drawing. Such cells are shaped like interlocking spindles and show a prominent striped pattern within each cell. Muscle cells lack cell walls, whereas the cells in the diagram clearly contain cell walls.
   (4) *Blue-green algae* do not most likely make up the tissue represented in the drawing. Blue-green algae are among the most primitive of life forms, lacking organized nuclei. Structure *A* in the diagram is most likely an organized nucleus.

15. **4** The organelle labeled *B* in the diagram is most likely a *cell wall*. The thick, well-defined, rigid cell boundary illustrated in the diagram is typical of the cell walls of typical plant cells.

Wrong Choices Explained
(1) The organelle labeled *B* is not likely a *mitochondrion*. Mitochondria are too small to be clearly visible under typical laboratory high-power ($400\times$–$430\times$) magnification.

(2) The organelle labeled *B* is not likely a *centriole*. Centrioles are too small to be clearly visible under typical laboratory high-power ($400\times$–$430\times$) magnification. Centrioles are rod-shaped and are found only in animal cells.

(3) The organelle labeled *B* is not likely a *lysosome*. Lysosomes are specialized vacuoles that contain digestive enzymes used to break down complex molecules into small, soluble molecules. Lysosomes are irregularly spherical in shape.

16. **1** *The tissue is composed of more than one layer of cells* is the best conclusion to be made from these observations. In diagram *A*, the microscope is focused on the cell walls and nuclei of one cell layer, whereas the second layer remains out of focus in the foreground. As the fine adjustment knob is turned slowly, the second layer is brought into sharp focus, while the first layer is put out of focus in the background. This is illustrated in diagram *B*.

Wrong Choices Explained
(2) *The tissue is composed of multinucleated cells* is not the best conclusion to be made from these observations. A characteristic of a multinucleated cell is that it contains two or more nuclei. None of the cells in the diagrams have more than one nucleus.

(3) *The cells are undergoing mitotic cell division* is not the best conclusion to be made from these observations. In tissues undergoing mitotic cell division, additional cells are created. The diagrams do not show additional cells being created.

(4) *The cells are undergoing photosynthesis* is not the best conclusion to be made from these observations. The cells in the diagrams may or may not be carrying on photosynthetic activity, but this activity would not be responsible for the different views illustrated in diagrams *A* and *B*.

17. **1** The presence of carbon dioxide would be indicated by *a color change in the solution*. Bromthymol blue is blue in a slightly basic solution but turns yellow in a slightly acidic solution. When the student exhales through a straw into the bromthymol blue solution, carbon dioxide molecules in the exhaled air combine with water to form weak carbonic acid. As the acidity of the solution increases, the indicator reacts by changing from blue to yellow.

Wrong Choices Explained

(2) *A change in atmospheric pressure* would not indicate the presence of carbon dioxide in the exhaled air. This phenomenon occurs for meteorological reasons independent of the laboratory.

(3) *The formation of a precipitate in the solution* would not indicate the presence of carbon dioxide in the exhaled air. This is not a characteristic of the indicator bromthymol blue.

(4) *The release of bubbles from the solution* would not indicate the presence of carbon dioxide in the exhaled air. Bubbles of any gas that is less dense than water will rise to the surface when released in the water. This is not a characteristic of the indicator bromthymol blue.

18. **3** *Benedict's solution* is the indicator used to detect the presence of glucose and other simple sugars. When heated in the presence of glucose, blue Benedict's solution changes to brick red.

Wrong Choices Explained

(1) *Iodine solution* would not indicate the presence of glucose in the container. Iodine (Lugol's) solution is used to detect the presence of starch in foods.

(2) *Bromthymol blue* would not indicate the presence of glucose in the container. Bromthymol blue is used to detect the presence of carbon dioxide in solution.

(4) *pH paper* would not indicate the presence of glucose in the container. pH paper is used to detect the presence of acids or bases.

19. **2** The total volume of water indicated in the diagram is *11 mL* (milliliters). When reading the volume of liquid in a graduated cylinder, the observer should note the mark that corresponds to the bottom of the meniscus. In the diagram, the meniscus falls on the 11-mL mark.

Wrong Choices Explained

(1), (3), (4) The total volume indicated is not *10 mL*, *12 mL*, or *13 mL*. Each of these readings falls either below or above the 11-mL mark indicated at the bottom of the meniscus in the diagram.

20. *8.9 cm* (or *89 mm* or *0.089 m*) In the illustration, the earthworm is laid out along the metric ruler so that the proboscis is positioned at the 2.0-cm mark. The posterior of the specimen is shown to extend to the 10.9-cm mark on the ruler. Subtracting 2.0 cm from 10.9 cm provides the specimen's total length of 8.9 cm.

21. **1** *Micrometer, millimeter, centimeter, meter* is the group of measurement units correctly arranged in order of increasing size. A meter contains 1,000,000 micrometers, 1,000 millimeters, or 100 centimeters.

Wrong Choices Explained

(2), (3), (4) The groups *millimeter, micrometer, centimeter, meter; meter, micrometer, centimeter, millimeter;* and *micrometer, centimeter, millimeter, meter* do not correctly arrange these units in order of increasing size. See explanation on page 271.

22. **4** The *nerve cord* is ventral to the esophagus in earthworms and grasshoppers. The ventral surface of a bilaterally symmetrical animal is the surface normally oriented downward when the animal is in a horizontal position (except in humans and other upright-walking animals where it is located on the "front" of the body). This question requires a knowledge of the anatomy of these organisms gained through dissection or other means.

Wrong Choices Explained

(1) The *gizzard* of the earthworm and grasshopper is located posterior, not ventral, to the esophagus.

(2) The *brain* of the earthworm and grasshopper is located anterior and dorsal, not ventral, to the esophagus.

(3) The *intestine* of the earthworm and grasshopper is located posterior, not ventral, to the esophagus.

23. *Anther* The stamen is the male part of the flower and is made up of the filament and the anther. The anther contains tissues that produce pollen. This question requires a knowledge of the anatomy of flowers gained through dissection or other means.

24. A scientifically accurate, complete-sentence response is required. Examples of acceptable responses include:

- *This structure functions in the formation of pollen.*
- *The anther contains tissues that produce pollen.*

25. **2** The student should *discontinue heating and report the defect to the instructor.* A crack in any piece of laboratory glassware is a sign of imminent failure. If the beaker had shattered while the student was near it, she could have been spattered and burned by the hot water in the beaker and burned or cut by hot broken glass. The instructor should remove and discard the beaker from the student's lab station and replace it with a new one for her use.

Wrong Choices Explained

(1), (3), (4) The student should not *discontinue heating and attempt to seal the crack, discontinue heating and immediately take the beaker to the instructor,* or *continue heating as long as fluid does not seep from the crack.* These actions are unsafe for both the student and others in the laboratory, as each could result in injury from hot broken glass and hot water.

26. **3** *Rinse the eyes with water; then notify the teacher and ask for advice* is the proper laboratory procedure to follow if some laboratory chemical splashed into a student's eyes. Because of the sensitivity of the eyes and skin surface to chemical burns, the student should not wait for direction from the teacher but should immediately flood the exposed areas with clean water. Each school science laboratory should have an eyewash station for this purpose. The student should then immediately report the incident to the teacher. The teacher in turn should immediately refer the student to the school nurse for examination and treatment as needed.

Wrong Choices Explained
(1) The student should not *send someone to find the school nurse* before thoroughly rinsing the exposed area and informing the teacher of the incident. This delay could cause severe damage to the eyes that could result in blindness.

(2) The student should not *rinse the eyes with water and do not tell the teacher because he or she might become upset*. Chemical exposures should always be treated by qualified medical personnel. If the exposed tissues are not thoroughly rinsed and examined, the chemical could cause severe damage to the eyes that could result in blindness.

(4) The student should not *assume that the chemical is not harmful and no action is required*. Until it is determined otherwise, all chemicals used in the laboratory should be considered dangerous. Chemical exposures should always be treated by qualified medical personnel. If the exposed tissues are not thoroughly rinsed and examined, the chemical could cause severe damage to the eyes that could result in blindness.

27. **4** When heating a liquid in a test tube, the student should always keep the mouth of the tube *aimed away from everybody*. Doing so will minimize the risk that someone will be injured by the hot liquid in the test tube. If heated too quickly, the liquid in the tube could splash or boil from the open end of the tube and burn anyone close enough to be spattered by it.

Wrong Choices Explained
(1) The student should not keep the mouth of the tube *corked with a rubber stopper*. This action could result in the formation of hot, high-pressure gases inside the test tube that could eject the stopper at high speed or shatter the tube. Either of these events could cause injury to the student or someone standing nearby.

(2) The student should not keep the mouth of the tube *pointed toward the student*. This action may result in injury to the student. If heated too

quickly, the liquid in the tube could splash or boil from the open end of the tube and burn the student.

(3) The mouth of the tube should not be *allowed to cool*. Good laboratory practice suggests that the student should heat the tube evenly along its entire length so as to avoid uneven expansion of the glass walls of the tube that could result in cracking. This practice also helps to ensure that the liquid in the tube will heat evenly, minimizing the probability of uncontrolled boiling and splashing.

28. One credit is allowed for appropriately labeling column III of the chart. Acceptable responses include:

   - *Height of Bean Plants (cm)*
   - *Bean Plant Height in Centimeters*

29. One credit is allowed for appropriately organizing data on the chart. Students should complete the chart so that the values entered in column II are arranged in increasing order from top to bottom and that the corresponding bean plant measurement is entered in column III.

DATA TABLE

I	II	III
Group	Concentration of Substance X (%)	Height of bean plants (cm)
B	0	28.7
C	2	29.4
E	4	31.5
A	6	32.3
D	8	37.1
F	10	30.7

30. One credit is allowed for marking an appropriate scale on the axis labeled "Number of Members of Each Population." Students should be sure to start numbering from zero at the lower left corner of the grid and enter values along the vertical axis in equal increments up to 150, a value that just exceeds the highest recorded moose population.

31. One credit is allowed for marking an appropriate scale on the axis labeled "Year." Students should be sure to start numbering from 1970 at the lower left corner of the grid and enter values along the horizontal axis in equal increments up to 1980.

32. One credit is allowed for correctly plotting the data for the wolf population, surrounding each point with a small triangle, and connecting the points.

33. One credit is allowed for correctly plotting the data for the moose population, surrounding each point with a small circle, and connecting the points.

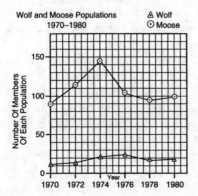

34. **4** According to the chart, *lake trout* can tolerate the highest level of acidity in their water environment. The chart shows that the survival rate for lake trout is strong at pH values ranging from 7.0 (neutral) to 5.5 (moderately acidic). At a pH value of 5.5, the survival rate for this species begins to decrease; at pH values below 5.0, the survival rate drops off drastically and reaches zero at a pH value of approximately 4.4.

Wrong Choices Explained
(1) *Mussels* do not tolerate the highest levels of acidity in their water environment. The chart shows that survival rates for this species begin to decrease at a pH value of 6.5 and reach zero at a pH value of approximately 5.9.

(2) *Smallmouth bass* do not tolerate the highest levels of acidity in their water environment. The chart shows that survival rates for this species begin to decrease at a pH value of 5.8 and reach zero at a pH value of approximately 5.3.

(3) *Brown trout* do not tolerate the highest levels of acidity in their water environment. The chart shows that survival rates for this species begin to decrease at a pH value of 5.5 and reach zero at a pH value of approximately 4.9.

35. **1** According to the chart, *mussels* would have been the first species eliminated from these lakes during the period from 1880 to 1980. The chart shows that survival rates for this species begin to decrease at a pH value of 6.5 and reach zero at a pH value of approximately 5.9. Assuming that lake acidification occurred at a constant rate throughout the period, by about 1930 a pH value of 5.9 would have been reached and mussels would have been eliminated from these lakes.

Wrong Choices Explained

(2) *Smallmouth bass* would not have been the first species to be eliminated from these lakes. The chart shows that survival rates for this species reach zero at a pH value of approximately 5.3. Assuming that lake acidification occurred at a constant rate throughout the period, by about 1960 a pH value of 5.3 would have been reached and smallmouth bass would have been eliminated.

(3) *Brown trout* would not have been the first species to be eliminated from these lakes. The chart shows that survival rates for this species reach zero at a pH value of approximately 4.9. Assuming that lake acidification occurred at a constant rate throughout the period, by about 1980 a pH value of 4.9 would have been reached and brown trout would have been eliminated.

(4) *Lake trout* would not have been the first species to be eliminated from these lakes. The chart shows that survival rates for this species reach zero at a pH value of approximately 4.4. As of 1980, the pH value in these lakes was approximately 4.9, and lake trout had presumably not yet been eliminated.

36. **2** The total change in the pH value of rainwater between 1880 and 1980 was *1.7*. In 1880, the pH value was approximately 5.7; in 1980, the pH value was approximately 4.0. The numerical difference between these values is found by carrying out the mathematical operation $5.7 - 4.0 = 1.7$.

Wrong Choices Explained

(1), (3), (4) The total change in the pH value of rainwater between 1880 and 1980 was not *1.3*, *5.3*, or *9.7*. These changes are incorrect according to the method described above.

37. **3** The total change in mass of the apple piece in the 10% sugar solution was *a decrease of 0.30 gram*. We assume that the natural sugar concentration of the apple slices is approximately 3%, a point at which the gain/loss in mass is at 0.0 gram on the graph. At a sugar concentration of 10%, the mass change shown on the graph is –0.3 gram.

Wrong Choices Explained

(1) A mass change is not shown as *a decrease of 0.45 gram*. This degree of mass change is indicated on the graph as corresponding to a sugar concentration of approximately 15%.

(2) A mass change is not shown as *an increase of 0.30 gram*. This degree of mass change is indicated on the graph as corresponding to a sugar concentration of approximately 1%.

(4) A mass change is not shown as *an increase of 0.10 gram*. This degree of mass change is not indicated on the graph as corresponding to

any specific sugar concentration but can be inferred to occur at a sugar concentration of approximately 2%.

38. **3** The apple slices should neither lose nor gain weight at a sugar concentration of approximately *3%*, a point at which the gain/loss in mass is at 0.0 gram on the graph. This value can be determined by drawing a curved line connecting the points on the graph and approximating the sugar concentration at the point where this line crosses the 0.0 mass change line.

Wrong Choices Explained
(1), (2), (4) According to the data shown in the graph, at sugar concentrations of *6%*, *10%*, and *20%* the apple slices will lose total mass. All these values exceed the equilibrium concentration of 3%.

39. **3** Based on the graph, it can be correctly concluded that the pituitary extract *affects the growth of rats* over a 200-day period. The graph shows data on their growth in grams of two groups of rats: one group (the control group) untreated and the other group (the experimental group) treated with anterior pituitary extract. The two groups of rats appear to grow at the same rate until day 70, when the treated (experimental) group begins to show an increased growth rate compared to the untreated (control) group.

Wrong Choices Explained
(1) It cannot be correctly concluded that the pituitary extract *is essential for life*. It is obvious from the data that the untreated rats have survived and grown over the 200-day period of this experiment, a fact indicating that the lack of extract does not lead to the death of these individuals.
(2) It cannot be correctly concluded that the pituitary extract *determines when a rat will be born*. There is no data on the graph concerning a measurement of the birthrates of rats.
(4) It cannot be correctly concluded that the pituitary extract *affects the growth of all animals*. Only rats, not "all animals," are being tested in this experiment. Further experiments involving other animal species may be conducted to provide this information, but the data from this experiment provides information only about the effect of anterior pituitary extract on rats.

40. **4** Based on the graph, it can be correctly concluded about plant hormones that *they stimulate maximum root and stem growth at different concentrations*. The graph shows data on the effect of relative plant hormone concentration on the relative degree of growth of roots and stems. The two types of plant tissues respond differently to the same concentrations of hormones. In general, root tissues are stimulated by

low concentrations of hormones and inhibited by high concentrations of hormones, whereas stem tissues are stimulated by moderately high hormone concentrations and inhibited by very low and very high concentrations of hormones.

Wrong Choices Explained

(1) It cannot be correctly concluded that *they stimulate maximum root growth and stem growth at the same concentration*. It is obvious from the data that the relative degree of stimulation or inhibition is different in roots than in stems at any particular concentration (other than at the point where the graph lines cross).

(2) It cannot be correctly concluded that *they stimulate maximum stem growth at low concentrations*. The data in the graph show that moderately high hormone concentrations (up to a point) are needed to stimulate the highest degree of stem growth.

(3) It cannot be correctly concluded that *they most strongly inhibit root growth at low concentrations*. The data in the graph show that relatively low hormone concentrations are needed to stimulate the highest degree of root growth.

41. **4** Based on the data, it can be correctly concluded that *as temperature increases from 15°C to 30°C, the number of compound eye sections in male Drosophila with bar-eyes decreases*. The chart presents data on the effect of incubation temperature on the number of compound eye sections in male bar-eye *Drosophila*. The data show that, as the incubation temperature increases from 15°C to 30°C, the number of eye sections decreases from 270 to 74 in males of this specific species and variety.

Wrong Choices Explained

(1) *The optimum temperature for culturing Drosophila is 15°C* is not the best conclusion to be drawn from analysis of the data. The chart contains no general information concerning the effect of temperature on the culturing of this or other varieties of *Drosophila*.

(2) *Drosophila cultured at 45°C will show a proportionate increase in the number of compound eye sections* is not the best conclusion to be drawn from analysis of the data. The chart contains no information concerning the effect of an incubation temperature of 45°C on the number of eye sections in this or other varieties of *Drosophila*.

(3) *Temperature determines eye shape in Drosophila* is not the best conclusion to be drawn from analysis of the data. The chart contains no information concerning the effect of incubation temperature on the expression of eye shape in varieties of *Drosophila* other than bar-eye.

42. **2** Based on the diagram, it can be correctly concluded that *nuclei are more dense than mitochondria*. At the very high spin rates of an ultracentrifuge, suspended organelles from broken cells are forced to separate into

layers based on their relative densities. The more dense the organelle, the lower in the tube it concentrates. The diagram shows nuclei concentrated in a lower layer than mitochondria, indicating that nuclei are more dense than mitochondria.

Wrong Choices Explained

(1), (3) *Ribosomes are more dense than mitochondria* and *mitochondria and ribosomes are equal in density* are not the best conclusions to be drawn from analysis of the data. The diagram shows that ribosomes concentrate in a higher layer than mitochondria, indicating that ribosomes are less dense than mitochondria.

(4) *The cell consists of only solid components* is not the best conclusion to be drawn from analysis of the data. The diagram shows that the highest (least dense) layer in the ultracentrifuge tube contains cell fluid, indicating that cells contain a liquid component.

Question Set on Reading, Writing, and Current Events in Science (p. 63)

1. **3** *There is not sufficient information given in the paragraph.* The first section of the paragraph mentions that scientists are now able to splice human genes into bacteria, but there is no indication of the technical difficulties involved in this process.

2. **1** *This statement is true according to the paragraph.* The first sentence of the second section of the paragraph states that "bacteria reproduce rapidly under certain conditions."

3. **1** *This statement is true according to the paragraph.* The second sentence of the third section of the paragraph states that "continued use of [animal] insulin can trigger allergic reactions in some humans."

4. **2** *This statement is false according to the paragraph.* The first sentence of the fourth section of the paragraph states that "the bacteria used for these experiments are . . . . common to the digestive systems of many humans."

5. **3** *There is not sufficient information given in the paragraph.* The fourth section of the paragraph mentions that scientists have used *E. coli* for these experiments, but no information is given concerning the use or nonuse of other bacterial species.

6. One credit is allowed for a scientifically correct, complete-sentence response that describes a difference between gradualism and punctuated equilibrium. Acceptable responses include:

- *Gradualism proposes that evolutionary change is continuous and slow, whereas punctuated equilibrium proposes that species undergo long periods without change and then undergo considerable change quickly.*
- *The theory of gradualism involves slow, gradual change, whereas the theory of punctuated equilibrium involves rapid, infrequent change.*
- *Gradualism is supported by the presence of transitional fossils, and punctuated equilibrium is supported by the lack of transitional fossils.*

7. One credit is allowed for a scientifically correct, complete-sentence response that states what may result from the accumulation of small variations. Acceptable responses include:

- *Gradualism theorizes that new species evolve as a result of the accumulation of many small genetic changes.*
- *The theory of gradualism states that the accumulation of many small genetic changes will eventually result in the formation of new species.*

8. One credit is allowed for a scientifically correct, complete-sentence response that states what fossil evidence supports the theory of gradualism. Acceptable responses include:

- *Gradualism is supported by fossil evidence that shows transitional forms.*
- *The theory of gradualism is supported by fossil evidence that includes transitional forms within an evolutionary line.*

9. Three credits are allowed for listing three human structures and describing in a scientifically correct response the structure's adaptive value to the human body. Acceptable responses include:

- *SWEAT GLAND—This structure regulates body temperature by releasing moisture from the blood to the skin surface where it can evaporate.*
- *PANCREAS—This organ produces digestive enzymes used in the digestive function.*
- *LIVER—This organ filters the blood of dead blood cells and toxic materials.*
- *EPIGLOTTIS—This structure prevents swallowed food from entering the trachea.*
- *CAPILLARY—This structure enables the blood to reach all body tissues.*
- *VILLUS—This structure permits the absorption of digested food from the intestine into the bloodstream.*
- *KIDNEY—This organ filters certain toxic materials from the blood and excretes them from the body.*
- *PLATELET—This structure promotes blood clotting by releasing enzymes into the blood.*

10. **4** Acid rain has a pH closest to that of *vinegar*. The second paragraph of the selection states that acid rain has a pH value between 3.0 and 5.0. The scale presents data on the pH of several common substances, one of which is vinegar with a pH value of approximately 3.0.

Wrong Choices Explained
   (1) *Ammonia* does not have a pH closest to that of acid rain. According to the scale, ammonia has a pH value of approximately 13.0.
   (2) *Tap water* does not have a pH closest to that of acid rain. According to the scale, tap water has a pH value of approximately 6.5.
   (3) *Baking soda* does not have a pH closest to that of acid rain. According to the scale, baking soda has a pH value of approximately 8.2.

11. **2** The most likely source of acid rain in New York State is the *midwestern United States*. The first paragraph of the selection states that this acid rain's principal sources are smokestack gases released by coal-burning facilities located in the midwestern states.

Wrong Choices Explained
   (1), (3), (4) The areas of the *far western United States, far eastern Canada*, and *far western Europe* are not the most likely sources of acid rain in New York State. Although smokestack gases are produced in these regions to a greater or lesser extent than in the midwestern United States, the prevailing westerly winds would be unlikely to carry them from these regions to the atmosphere over New York State.

12. **2** The food chain *algae → aquatic insect → trout → otter* would most immediately be affected by acid rain. These organisms all live in or near water and depend on clean, unpolluted water for survival.

Wrong Choices Explained
   (1), (3), (4) The food chains *grass → rabbit → fox → decay bacteria, shrub → mouse → snake → hawk*, and *tree → caterpillar → bird → lynx* are all terrestrial food chains. Although all would eventually be affected by acid rain, they would not be the most immediately affected.

13. **3** Acid rain can most appropriately be characterized as a type of *technological oversight*. The technologies of steam and electrical power generation have had a positive effect on the U.S. economy and the growth of the nation but have had an unintended negative impact on the natural environment. The gases produced by coal and oil-fueled power plants and factories have become a problem in that the acid precipitation that results has killed fish and aquatic insects that form an important part of the aquatic ecosystem. The destruction of these ecosystems in turn has profound negative implications for human health and survival.

Wrong Choices Explained

(1) Acid rain is not most appropriately characterized as a type of *biological control*. A biological control is an environmentally sound method for controlling unwanted insects and other pests by using their natural biological enemies to exert this control.

(2) Acid rain is not most appropriately characterized as *conservation of resources*. Conservation of resources involves using the minimum amount of our scarce natural resources, as well as recycling materials to reclaim these resources for reuse.

(4) Acid rain is not most appropriately characterized as a type of *land-use management*. Land-use management involves using the minimum amount of our dwindling land resources, preserving open space and arable land for enjoyment by future generations.

14. **3** A strain of fish that could survive under conditions of increased acidity could best be obtained by *selective breeding*. Fish of all species are organisms that reproduce by means of sexual reproduction, the foundation of selective breeding techniques. These techniques allow breeders to mate pairs of organisms that display the traits desired for breeding (in this case, fish with a tolerance for elevated levels of acidity). The offspring of these breeding pairs have an increased likelihood of displaying the desired trait and forming a self-reproducing variety of the species over time.

Wrong Choices Explained

(1), (2), (4) *Binary fission*, *vegetative propagation*, and *budding* are not reproductive processes that would be likely to produce a strain of fish that could survive under conditions of increased acidity. These reproductive processes are all examples of asexual reproduction that are not used by any species of fish as its primary method of reproduction. In addition, as a reproductive process, asexual reproduction does not normally result in a wide variety of traits. Rather, it tends to maintain genetic traits intact and unchanging over many generations.

Question Set 2.1 (p. 77)

1. **1** The human digestive system is characterized by *A, B,* and *C*. Peristalsis (*A*) is a wave of muscular contractions that moves food through the food tube. In humans, this occurs in the esophagus, stomach, and intestines. In humans, smooth (involuntary) muscle tissues (*B*) are responsible for peristaltic movement. Accessory organs (*C*) present in humans include the liver, gallbladder, and pancreas, which are not part of the food tube but that produce or store substances important to the digestive process.

Wrong Choices Explained

(2), (3), (4) The combinations of characteristics *A only*, *B only*, and *B and C only* do not occur in humans or in other commonly studied animals.

2. **4** *It has two atria and two ventricles, and it pumps blood directly into arteries* is a correct statement about the human heart. The human heart, like that of other mammal species, is a four-chambered vessel whose design helps to separate oxygenated and deoxygenated blood. These chambers include two thin-walled atria, which receive blood from veins, and two muscular ventricles, which pump blood out to the body through arteries.

Wrong Choices Explained

(1) *It has two atria and one ventricle, and it pumps blood directly into veins* is not a correct statement about the human heart. A three-chambered heart is characteristic of amphibians such as frogs but not humans. Blood is not pumped into veins from the heart.

(2) *It has one atrium and one ventricle, and it is composed of cardiac muscle* is not a correct statement about the human heart. Cardiac muscle is the muscle type found in human blood vessels, but the heart in humans is not two-chambered.

(3) *It has one atrium and two ventricles, and it is composed of visceral muscle* is not a correct statement about the human heart. This is a nonsense distracter since no known organism has this combination of heart chambers. Visceral muscle is not found in the human heart.

3. **4** *Filtering of bacteria from the lymph* is a process that would be affected by a malfunction of the lymph nodes. This process is one of the essential functions of the lymphatic system.

Wrong Choices Explained

(1), (2), (3) These are all nonsense distracters, since none of these processes is carried out in the lymph. *Release of carbon dioxide* and *oxygen* are associated with the respiratory system. *Filtering of glucose* may be carried out by the excretory system under certain circumstances.

4. **4** Of the choices given, *decreased consumption of complex carbohydrates* (for example, starches) is least likely to have the effect of promoting heart attack. By decreasing the consumption of starchy foods, there may even be a positive effect on the body's ability to ward off heart disease.

Wrong Choices Explained

(1), (2), (3) *An increase in arterial blood pressure, oxygen deprivation of cardiac muscle*, and *narrowing of the arteries transporting blood to the heart muscle* are all potential contributors to heart disease and may help to set up the conditions for heart attack in certain people. All place additional strain onto the heart muscle, making failure under stress more likely.

5. **1** Structure *A* (mouth) is the site in which the initial hydrolysis of carbohydrates occurs. Saliva secreted into the mouth contains digestive enzymes that hydrolyze starch into disaccharides.

Wrong Choices Explained

(2) Structure *E* (small intestine) is the organ where the digestion of all food types is completed and the end products of digestion are absorbed into the blood.

(3) Structure *C* (stomach) is the site in which the chemical digestion of protein begins.

(4) Structure *D* (pancreas) is an accessory organ that produces a variety of digestive juices important for the complete hydrolysis of carbohydrates, proteins, and lipids.

6. **4** Structure *E* (small intestine) is the organ where the digestion of all food types is completed and the end products of digestion are absorbed into the blood.

Wrong Choices Explained

(1) Structure *F* (large intestine) is where solid wastes accumulate and water is reabsorbed.

(2) Structure *H* (liver) is an accessory organ that manufactures bile and stores glycogen.

(3) Structure *C* (stomach) is the site in which the chemical digestion of protein begins.

7. **3** Structure *C* (stomach) is the site in which the chemical digestion of proteins begins.

Wrong Choices Explained

(1) Structure *G* (gallbladder) is the site where bile is stored after it is manufactured by the liver.

(2) Structure *B* (esophagus) is the structure that moves food from the mouth to the stomach.

(3) Structure *E* (small intestine) is the organ where the digestion of all food types is completed and the end products of digestion are absorbed into the blood.

8. **1** *An artery*, with its muscular walls, is the blood vessel most likely to produce a detectable pulse. The pulse is a wave of muscular contractions that help blood to move through the circulatory system.

Wrong Choices Explained

(1) *A vein*, by lacking muscular walls, is not likely to produce a detectable pulse. Veins serve as a return system for blood to the heart.

(2) *A capillary* is only one-cell thick and contains blood under very low pressure. Pulses would not be detected in the capillary.

(3) *A lacteal* is a microscopic extension of the lymphatic system found within a villus in the small intestine. There is no blood pressure in a lacteal.

9. **2** Structures *B, F, and I* are those most likely to contain deoxygenated blood in the human heart. Deoxygenated blood enters the heart from the body in structure *I* (left atrium) and then passes to structure *F* (left ventricle), where it is pumped through structure *B* (pulmonary artery) to the lungs.

Wrong Choices Explained

(1) Structures *A, B, and C* is not the correct choice. Structure *A* is the aorta, which carries oxygenated blood to the body. Structures *C* are the pulmonary veins, which carry oxygenated blood from the lungs to the heart.

(3) Structures *C, D, and E* is not the correct choice. Structures *C* are the pulmonary veins, which carry oxygenated blood from the lungs to the heart. Structure *D* is the right atrium, which receives oxygenated blood from the pulmonary veins. Structure *E* is the right ventricle, which pumps oxygenated blood into the aorta.

(4) Structures *D, H, and I* is not the correct choice. Structure *D* is the right atrium, which receives oxygenated blood from the pulmonary veins. Structure *H* is the vena cava, which carries deoxygenated blood from the body to the heart.

10. **1** Structure *G* is the left atrioventricular (A-V) valve that opens when blood passes from structure *I* to structure *F*, but then closes to prevent the blood from flowing back into structure *I* (left atrium).

Wrong Choices Explained

(2) Structure *B*, the aorta, is the major artery leading from the heart to all parts of the body. It does not control the backflow of blood into an atrium.

(3) Structures *C* are the pulmonary veins. These blood vessels carry oxygenated blood from the lungs to the heart and do not control the backflow of blood into an atrium.

(4) Structure *H* is the vena cava, which carries deoxygenated blood from the body to the heart. It is not involved in the control of backflow of blood.

11. **4** *Lymph vessels* have specialized regions, known as nodes, where bacteria are attacked by phagocytic white blood cells and removed from the bloodstream.

Wrong Choices Explained

(1) *Arteries* are muscular blood vessels specialized for moving blood from the heart to the body extremities and internal organs.

(2) *Capillaries* are microscopic blood vessels that connect arteries to veins. Most molecular exchange occurs in the capillaries.

(3) *Veins* are valved blood vessels that carry blood back to the heart from all around the body.

Question Set 2.2 (p. 83)

1. **4** This question requires a knowledge of the structural design of the human respiratory system. Air passing out of the body during exhalation would start in the alveoli of the lungs and then proceed in order through the bronchioles, bronchi, trachea, *pharynx*, and oral/nasal cavity to the environment.

Wrong Choices Explained

(1), (2), (3) These distracters all reference parts of the human respiratory system. However, during exhalation, air passes through the *alveoli*, *bronchioles*, and *bronchi* before reaching the trachea.

2. **1** In humans, blood is cooled as it passes through capillaries surrounding *sweat glands* in the skin. Heat (along with excess salt and water) is absorbed from the blood and transferred to the glands as sweat. As sweat droplets evaporate from the skin surface, heat dissipates with it and the body cools.

Wrong Choices Explained

(2) The *nephron* is the functional unit of the human kidney. Although it is associated with removal of materials (urea, water, and salts) from the blood, it is not directly involved in temperature regulation.

(3) The *liver* is involved with filtering the blood; it is responsible for removing worn-out red blood cells and recycling their components in the body. The liver also removes a variety of toxic substances, including alcohol, from the blood. The liver is not directly involved in temperature regulation.

(4) The *urinary bladder* is responsible for collecting and storing urine passing from the kidneys. It is not directly involved in temperature regulation.

3. **2** This question requires a knowledge of the structural and functional design of the human excretory system. In humans, the urethra (*X*) is responsible for the *transport of urine out of the body* to the environment.

Wrong Choices Explained
(1) The *filtration of cellular wastes from the blood* is performed by the paired kidneys, shown at the top of the diagram.
(3) The *storage of urine* is a function performed by the urinary bladder, shown immediately above the urethra in the diagram.
(4) The *secretion of hormones* is not a function performed by the human excretory system. However, the adrenal glands, which are a part of the human regulatory system, are located near the kidneys (not shown in the diagram).

4. **2** The region labeled *X* is most likely an *alveolus* within the human lung. The diagram shows oxygen (O_2) diffusing into the blood from the alveolus and carbon dioxide (CO_2) diffusing out of the blood and passing into the alveolus.

Wrong Choices Explained
(1) *A glomerulus* is a portion of the nephron within the human kidney. Glomeruli are not associated with gas exchange in humans.
(3) *A villus* is a microscopic projection of the inner lining of the human small intestine. Villi are not associated with gas exchange in humans.
(4) *The liver* is an organ in humans specialized for the removal and storage of toxic materials from the blood. The liver is not associated with gas exchange in humans.

5. **1** *Anaerobic respiration in muscle cells, forming lactic acid* is a condition arising from the continual use of muscles during prolonged physical exercise such as that encountered in a marathon. In the absence of sufficient oxygen, the muscle cells carry on lactic acid fermentation to produce energy. The resulting lactic acid buildup produces a sensation of pain in the muscle, commonly called muscle fatigue.

Wrong Choices Explained
(2), (3), (4) *Aerobic respiration in muscle cells, generating glycogen*; *anaerobic respiration in liver cells, producing glucose*; and *aerobic respiration in liver cells, synthesizing alcohol* are nonsense distracters. No type of respiration in humans produces glycogen, glucose, or alcohol.

6. **4** The *nasal cavity* in humans is lined with a ciliated mucous membrane that warms, moistens, and filters air. The cilia move dust-laden mucus to

the outside, while tissues warm and moisten the air before it enters the lung interior.

Wrong Choices Explained

(1) The *pharynx* is a cavity at the back of the throat, where the mouth cavity and the nasal cavity join. No ciliated tissues are in the pharynx.

(2) The *alveolus* is the respiratory sac in the lung where gas exchange occurs between the environment and the blood. The air reaching the alveolus must already be warmed and moistened to prevent drying of the lung tissues.

(3) The *epiglottis* is a small flap of tissue that closes the trachea while food is being swallowed. The epiglottis helps to prevent choking.

7. **3**   The ureter transports urine from the *kidney to the urinary bladder*. From the urinary bladder, the urine passes through the urethra to the environment.

Wrong Choices Explained

(1) From the *blood to the kidney* is incorrect. This is accomplished in the nephron within the kidney.

(2) From the *liver to the kidney* is incorrect. This is accomplished only by the interaction of the bloodstream, as the waste products of protein deamination are removed for filtering in the kidney.

(4) From the *urinary bladder to outside the body* is incorrect. This is accomplished by way of the urethra.

8. **2**   *Constriction of the bronchial tubes and wheezing* are symptoms commonly associated with asthma. In its acute state, asthma can be life threatening, as the victim's breathing tubes narrow and close.

Wrong Choices Explained

(1) *Enlargement and degeneration of the alveoli* may be a symptom of the human disorder known as emphysema.

(3) *Inflammation and swelling of the epiglottis* is not associated with any known human disorder.

(4) *Constriction of the nasal cavity and watery eyes* may be a symptom of an allergic reaction in humans.

9. **4**   Structure *D*, the diaphragm, is a sheet of muscle tissue that contracts, causing a pressure change in the chest cavity during breathing. When the diaphragm presses downward, pressure in the chest cavity is reduced and air is forced into the lungs by atmospheric pressure. When the diaphragm presses upward, pressure in the chest cavity is increased and air is expelled to the atmosphere.

Wrong Choices Explained

(1) Structure *A* is the trachea, which is a rigid tube that conducts air from the mouth and nasal cavities downward toward the lung.

(2) Structure *B* is the bronchus, one of two extensions of the trachea leading deep into the lung tissues.

(3) Structure *C* is the lung, which contains thousands of small air sacs (alveoli) that facilitate the absorption of oxygen from the atmosphere.

10. **2** The statement *it regulates the chemical composition of the blood* is true of the main function of the nephron. Blood passes through capillary networks where materials diffuse out of the blood fluid and into the nephron interior. Metabolic wastes, such as urea, salts, and water, are removed from the blood. However, beneficial materials, such as dissolved food molecules, are returned to the blood. With this mechanism, the chemical composition of the blood is maintained.

Wrong Choices Explained

(1), (3) The statements *it breaks down red blood cells to form nitrogenous wastes* and *it forms urea from the waste products of protein metabolism* are true of the liver but not of the nephron.

(4) The statement *it absorbs digested food from the contents of the small intestine* is true of the villus but not of the nephron.

Question Set 2.3 (p. 92)

1. **1** This process is an example of *maintenance of homeostasis*. As the runner exercises, an increase in metabolism produces heat. This heat is absorbed by the blood and is carried to sweat glands in the skin, where it is transferred to droplets of perspiration. As these droplets evaporate from the skin surface, the blood is cooled, helping to maintain a stable body temperature.

Wrong Choices Explained

(2) *Antigen-antibody reactions* involve the body's defense mechanism against disease. By itself, it is an example of maintenance of homeostasis. However, the description given does not match this particular process.

(3) *Acquired characteristics* is a concept contained in the evolutionary theories of Jean Lamarck. These theories have been largely discredited by modern science.

(4) *Environmental factors affecting phenotype* is a concept discussed in modern genetic theory. It states that the expression of certain genetic traits can be altered by particular environmental conditions.

2. **3** *Neurotransmitters*, such as acetylcholine, are secreted by the terminal branches of neurons. These chemical neurotransmitters diffuse across the synapse and cause subsequent neurons to fire nerve impulses. In this manner, nerve impulses are transferred from neuron to neuron until the impulse reaches the central nervous system.

Wrong Choices Explained

(1) *Antibodies* are produced by the blood in response to the detection of foreign proteins (antigens) in the blood.

(2) *Antigens* are foreign proteins that may enter the blood and stimulate the production of antibodies.

(4) *Lipids* are biochemicals that take on a wide variety of particular shapes and characteristics. Lipids are composed of a glycerol molecule bonded chemically to three fatty acid molecules.

3. **3** *C* represents the most logical choice of those given. The statement should read as follows: "The *endocrine* glands produce *hormones*, which are transported by the *circulatory* system." Endocrine glands lack ducts and so depend on the bloodstream to carry hormones throughout the body.

Wrong Choices Explained

(1), (2), (4) These nonsense distracters indicate combinations of glands, products, and transport systems that do not exist in nature.

4. **Definitions** (acceptable sample responses):

- Coordination—regulates the body's activities through the use of nerve impulses and hormones
- Excretion—gets rid of cellular wastes from the body
- Digestion—breaks complex foods down into soluble subunits that can be absorbed by the cell
- Circulation—moves materials through the body from one place to another
- Synthesis—manufacture of complex molecules from simpler subunits

Explanations of interaction (acceptable sample responses):

- Circulation/excretion—The circulatory system moves cell wastes to places where they can be removed from the body.
- Coordination/digestion—The nervous system coordinates the peristaltic movement of food through the food tube so that digestion and absorption of the food can occur.

5. **1** In humans, muscle is anchored to bones by connective tissues known as tendons. *Cartilage* does not perform this function in humans. (HINT: Always pay close attention to exam questions that use the italicized word *not*.)

Wrong Choices Explained

(2), (3), (4) These distracters all reference functions characteristic of human cartilage.

6. **3** This question requires a knowledge of the structural and functional design of the human nervous system. An *interneuron* is shown at *X* lying within the spinal cord and linking a sensory neuron (*below*) with a motor neuron (*above*). The motor neuron in turn enervates a muscle effector. A reflex arc typically involves this structural pattern, designed to provide rapid responses (reflexes) to potentially dangerous situations in nature.

Wrong Choices Explained

(1) The *effector* in this diagram is the muscle, which enables a movement response.

(2) The *motor neuron* is shown linking the interneuron (*X*) with the muscle.

(4) The *receptor* (not shown) could be any of the body's sense organs or internal receptors.

7. **2** *Production of blood cells* is a function of the human skeletal system. Both red and white blood cells are produced in bone marrow.

Wrong Choices Explained

(1) *Transmit impulses* is a function of the regulatory system in humans.

(3) *Produce lactic acid* occurs in human muscle under anaerobic conditions.

(4) *Store nitrogenous wastes* is a function of the excretory system in humans.

8. **3** The *pituitary gland* is responsible for the elongation of structures *A* and *B* in the photograph. Structures *A* and *B* are long bones of the arm and leg. Pituitary growth hormone produced by the pituitary gland is responsible for the elongation of these bones.

Wrong Choices Explained

(1) The *islets of Langerhans* in the pancreas are responsible for producing the hormone insulin; insulin regulates the concentration of sugar in the blood.

(2) The *liver* acts as a reservoir for glycogen produced when the body stores excess sugar. The liver produces no hormones and does not directly influence the growth of bone tissue.

(4) *Striated muscles* produce no hormones and do not directly affect the elongation of these bones.

9. **3** *Both the central and peripheral nervous systems* are brought into play in the example given. When the child hears the clap of thunder, both the central and peripheral nervous systems are involved in the initial startle reflex (reflex arc). The central nervous system is involved in the child's decision to find his/her mother for comfort (cerebrum). The peripheral nervous system is involved in the child's feelings of fear (endocrine response) and the movement of muscles needed to allow the child to run to his/her mother (locomotion). Coordination of those muscles is a central nervous system function (cerebellum).

Wrong Choices Explained

(1) *The central nervous system only* is not correct. The peripheral nervous system is needed to relay central commands to the muscles and glands.

(2) *The peripheral nervous system only* is not correct. Central control of a complex response is necessary for survival.

(4) *Neither the central nor the peripheral nervous system* is not correct. Both systems are needed to effect a coordinated response.

10. **4** *Ligaments*, strong elastic connective tissues, are responsible for attaching bone to bone at joints in the skeleton, including the shoulder and elbow joints.

Wrong Choices Explained

(1) *Cartilage* is a type of connective tissue found in many flexible parts of the human body. Cartilage acts as a cushion, not a connector, between bones.

(2) *Tendons* are another form of elastic connective tissue in the body. Tendons are specialized for connecting muscle to bone, not bone to bone.

(3) *Extensors* are muscles that extend joints. They are not associated with linkages between bones.

11. **2** The *adrenal* gland secretes the hormone adrenaline in times of emergency. Adrenaline speeds up metabolic rate, enabling the body to respond to the emergency.

Wrong Choices Explained

(1) The *thyroid* gland secretes the hormone thyroxin, which regulates the body's general metabolic rate.

(3) The *islets of Langerhans*, embedded within the pancreas, secrete the hormone insulin, which regulates the concentration of sugar in the blood.

(4) The *parathyroid* gland secretes the hormone parathormone, which regulates the metabolism of calcium in the body.

12. **1** The *thyroid* gland secretes the hormone thyroxin, which regulates the body's general metabolic rate. Dietary iodine is necessary for the proper synthesis of thyroxin.

Wrong Choices Explained

(2), (3), (4) The *adrenal* gland, *islets of Langerhans*, and *parathyroid* gland do not require dietary iodine to produce their respective hormones.

Question Set 2.4 (p. 115)

1. **1** The *compound light microscope* typically has a low-power magnification of 100× and relatively limited resolution. At this level of magnification and resolution, viewing cells and some of the larger cell organelles (such as *cell walls, chloroplasts*, and *nuclei*) is posssible, especially if stains are added to enhance the color or contrast of these features.

Wrong Choices Explained

(2), (3), (4) The items mentioned in these distracters are generally too small or too difficult to stain to be viewed under a compound light microscope on low power. Under conditions of higher magnification and specialized specimen preparation (such as those used in electron microscopy), viewing *ribosomes, endoplasmic reticula, lysosomes,* and *mitochondria* would be possible. *Genes* and *nucleotides*, being molecular in size, would be difficult or impossible to view using known methodologies.

2. **2** In a living cell, *ribosomes* provide the site for the manufacture of new polypeptide molecules (such as enzymes and structural proteins) from free amino acid molecules in the cytoplasm. While the precise role of the ribosomes in protein synthesis is not completely understood, they are known to be necessary for its successful completion. If ribosomes were to become disabled in the cell, the synthesis of essential enzymes would theoretically cease.

Wrong Choices Explained

(1) *Uncontrolled mitotic cell division* is a characteristic of many cancers. It is associated with changes that are thought to occur within the cell's nucleus. Ribosomes are not directly involved in this process.

(3) *Antibodies* are substances produced by specialized cells in response to the antigens of a foreign invading cell or virus. Antibodies are thought to combine to specific antigens in the body, rendering them harmless. Ribosomes are essential to the production of antibodies, which are proteins. If the ribosomes were not functioning, antibody production would cease, not increase.

(4) The *transport of glucose* and other small molecules in a cell depends on the cell membrane and cytoplasm of the cell. Ribosomes are not directly involved in this process.

3. **3** The diagram most likely represents *a protozoan ingesting food during heterotrophic nutrition.* The protozoan illustrated is most likely an ameba carrying out the process of phagocytosis to ingest a food particle found in its water environment. The food particle is first detected (illustration 1), then engulfed by cytoplasmic extensions known as pseudopods (illustration 2), and finally enclosed in a food vacuole for digestion (illustration 3).

Wrong Choices Explained

(1) The diagram does not represent *a virus destroying a cell by extracellular digestion.* An illustration of this process would show a crystalline virus particle attaching to the exterior of a cell, injecting its genetic material into the cell interior, producing a number of new viruses from the cell's genetic material, and bursting the cell in order to release the new virus particles.

(2) The diagram does not represent *a member of the bryophyte phylum performing intercellular digestion.* Members of the bryophyte phylum include mosses, which are autotrophic and whose cells are surrounded by cell walls. The cell illustrated is clearly not a plant cell carrying on photosynthesis.

(4) The diagram does not represent *a lysosome egesting a food particle into the cytoplasm.* Lysosomes are specialized vacuoles in the cell that contain digestive enzymes. During the nutritional process, lysosomes combine with food vacuoles to facilitate chemical digestion. They are not directly involved with the process of egestion.

4. **4** The arrows in this diagram represent the movement of the indicated substances from the interior of a unicellular protist (a paramecium) to the environment. The substances indicated are carbon dioxide (CO_2) and ammonia (NH_3), which are both *metabolic wastes* produced in the cell. These substances are being *excreted* from the cell in this illustration.

Wrong Choices Explained

(1) Certain types of *anaerobic respiration* may result in the production of carbon dioxide, but this process is not carried on in paramecia. Ammonia is never a product of anaerobic respiration.

(2) Certain types of *autotrophic nutrition* may use (but not release) carbon dioxide. Ammonia is never a product of autotrophic nutrition. Paramecia do not carry on autotrophic nutrition.

(3) *Deamination of amino acids* may occur in a paramecia and may result in the production of ammonia as a metabolic waste. However, carbon dioxide is never a product of deamination.

5. **3** In this diagram, structure *A* illustrates a flagellum of an alga cell. Like the flagella of protists, these structures are primarily used to allow the organism to move about in its watery environment. This movement is classified as a type of *locomotion*.

Wrong Choices Explained
    (1), (2) The structure most closely associated with the life functions of *excretion* and *transport* in this organism is the cell membrane, which is not labeled.
    (4) The structure most closely associated with the life function of *reproduction* in this organism is the nucleus, which is not labeled.

6. **3** $C_{12}H_{22}O_{11}$ illustrates the organic compound maltose. The presence of the key elements carbon (C), hydrogen (H), and oxygen (O) clearly indicate that this compound is organic.

Wrong Choices Explained
    (1) $Mg(OH)_2$ is the chemical formula for magnesium hydroxide, an inorganic base. No carbon is present in the formula of this compound; therefore it cannot be organic.
    (2) *NaCl* is the chemical formula for sodium chloride, common table salt, which is an inorganic salt. No carbon is present in the formula of this compound, therefore it cannot be organic.
    (4) NH_3 is the chemical formula for ammonia, an inorganic material resulting from protein metabolism in certain bacteria. No carbon is present in the formula of this compound, therefore it cannot be organic.

7. **3** *Diffusion* is the process that results in the movement of molecules from a region of higher concentration (area *A*) to a region of lower concentration (area *B*). This net movement will occur until the concentrations of molecules have reached equilibrium between area *A* and area *B*.

Wrong Choices Explained
    (1), (2) *Phagocytosis* and *pinocytosis* are processes by which certain protists engulf their food and enclose it within a vacuole for digestion.
    (4) *Cyclosis* refers to the streaming of cytoplasm in the cell, a simple form of intracellular transport.

8. **3** This diagram illustrates a molecule of lipid. Lipids have *a high energy content*. The many carbon-hydrogen bonds present in this molecule indicate that a large amount of energy will be derived from its complete breakdown.

Wrong Choices Explained
    (1) The statement *it has the ability to control heredity* is false. A molecule that controls heredity is the nucleic acid DNA.

(2) The statement *it has the ability to control reactions* is false. Molecules capable of controlling cell reactions are known as catalysts and enzymes.

(4) The statement *it is involved in photosynthesis* is false. Photosynthesis is a chemical process involving principally carbon dioxide, water, ATP, PGAL, glucose, and oxygen, as well as a number of enzymes.

9. **2** The molecule shown, which may also be represented by the empirical formula $C_3H_5(OH)_3$, is a molecule of glycerol. Glycerol is one of two products that result from the chemical digestion of a triglyceride lipid, the other being fatty acid.

Wrong Choices Explained

(1) The molecule shown, which may also be represented by the empirical formula CO_2, is a molecule of carbon dioxide. Carbon dioxide is not an end product of any known chemical digestion but is an end product of aerobic respiration.

(3) The molecule shown, which may also be represented by the empirical formula H_2O, is a molecule of water. Water is a reactant, not an end product, of chemical digestion.

(4) The molecule shown, which may also be represented by the empirical formula CH_4, is a molecule of methane. Methane is not an end product of any known chemical digestion but is an end product of decomposition.

10. **3** *A carboxyl group* is illustrated in box *Y*. This group is common to organic acids, including amino acids, and is often written as –COOH.

Wrong Choices Explained

(1) *An amino group*, also common to amino acids, is written –NH$_2$.

(2) *A variable group*, more commonly known as a radical group, refers to the chemical composition of individual amino acid molecules. Each type of amino acid contains a unique chemical composition in the radical group position.

(4) *A peptide group* is a nonsense distracter. No known chemical group carries this name.

11. **2** There are two peptide bonds illustrated in this diagram. These bonds are the C–N bonds that link one amino acid to the next prior to hydrolysis.

Wrong Choices Explained

(1), (3), (4) These bond counts are not correct for the molecule illustrated.

12. **1** Diagram A illustrates a group of plant cells whose cell membranes are pressed tightly against the cell wall by cytoplasm containing a relatively high concentration of water. Diagram B illustrates the same cells after the addition of *salt water*. Note that the cytoplasm has shrunk in volume as water has moved out of the cell by osmosis (diffusion of water) from a region of high concentration of water (the cell interior) toward a region of lower concentration of water (the cell exterior).

Wrong Choices Explained
(2) The addition of *distilled water* to this system would have the opposite effect from that illustrated; the cytoplasm would swell with the rapid inward movement of water, perhaps to the point of bursting.

(3), (4) Most common sources of *pond water* or *tap water* contain dissolved mineral concentrations similar to that found in the natural habitats of most terrestrial or aquatic species. The addition of water from these sources would probably have little or no effect on the cells' appearance.

13. **4** *Structure X transports mateirals for metabolic activities* is the statement that is valid inference concerning structure *X* represented in the diagram. Structure *X* represents a vein of a maple tree leaf. The vein in a leaf contains vascular tissues that are specialized to transport water, sugar, and other materials resulting from, or needed for, the plant's metabolic activities.

Wrong Choices Explained
(1) *Structure X contains guard cells that regulate glucose intake* is *not* the statement that is valid inference concerning structure *X* represented in the diagram. Guard cells are specialized cells that flank the stomates of leaves and regulate the passage of atmospheric gases into and out of the leaf. Stomates and guard cells in plants such as the maple tree are located in the underside of flat areas, not in the veins, of their leaves.

(2) *Structure X carries out heterotrophic nutrition* is *not* the statement that is valid inference concerning structure *X* represented in the diagram. Heterotrophic nutrition is a feeding mechanism in which an organism consumes and metabolizes preformed organic matter for food. Green plants such as the maple tree are autotrophic, not heterotrophic, organisms.

(3) *Structure X produces gametes for asexual reproduction* is *not* the statement that is valid inference concerning structure *X* represented in the diagram. Gametes (pollen and ova) in flowering plants such as the maple tree are produced in their flowers, not in their leaves.

Question Set 2.5 (p. 136)

1. **1** The *Cell Theory* is one of the central principles of modern biology. The "units of structure and function" are the cells that make up the bodies

of all known living organisms, from amebas to humans. The "functions" referenced are the chemical reactions that constitute the cell's metabolic activities. These, in turn, enable the life functions of living organisms. A unique feature of living cells is their ability to produce new living cells like themselves (cells "arising from preexisting units").

Wrong Choices Explained

(2) The *lock-and-key model of enzyme activity* is a theoretical concept used to explain the specificity of enzymes involved in the cell's metabolism. This concept is much more limited in scope than cell theory. Enzymes are important in the functioning of a cell but are not structural units and cannot arise from preexisting enzymes.

(3) *Natural selection* is a term used to describe the mechanism by which favorable variations within a species are thought to be selected by natural forces to increase in frequency (and by which unfavorable traits are gradually reduced in frequency) in a species population. Natural selection deals with forces outside the organism and therefore cannot be described as making up the structure of living things.

(4) The *heterotroph hypothesis* is a theory proposed by scientists to explain the origin of life on Earth from primitive organic molecules floating in the oceans of the early Earth. Because the heterotroph hypothesis deals with events that theoretically took place before true cells existed, it is not used to describe the structure and function of modern life-forms.

2. **4** The statement "*in both plants and animals, the daughter cells are genetically identical to the original cell*" is true. Mitotic cell division ensures this continuity by replicating the genetic material prior to its separation into the two new daughter cells. This process occurs in all organisms employing mitotic cell division as a mechanism of growth and repair of body tissues.

Wrong Choices Explained

(1) The statement "*it is exactly the same in plant and animal cells*" is not correct since minor differences in the mechanics of cell division are evident when comparing plant and animal cells.

(2) The statement "*the walls of plant cells pinch in, but the membranes of animal cells do not*" is not correct. In plant cells, cell division is associated with the formation of the cell plate, which divides the cytoplasm and separates the daughter nuclei. In animal cells, the cell membrane pinches in to perform the same function.

(3) The statement "*most plant cells use centrioles, but most animal cells do not*" is not correct. Centrioles are commonly found in animal cells undergoing mitotic cell division. This process occurs in plant cells without the presence of centrioles.

3. **4** Sperm cells and egg cells are *monoploid* (*n*) gametes formed during the process of meiotic cell division. A zygote results from the fusion of two monoploid nuclei in fertilization, so it must be *diploid* (2*n*) in chromosome number.

Wrong Choices Explained
(1), (2), (3) Each of these distracters contains an incorrect combination of choices (see above).

4. **1** Gregor Mendel performed his experiments in the 1800s, *prior to modern scientific research findings concerning chromosome structure or function*. He described observed characteristics of and performed controlled breeding experiments with garden peas that led to the development of theories concerning the inheritance of characteristics. He postulated the existence of physical structures, which he referred to as factors (today's genes) that are passed from generation to generation in the reproductive process.

Wrong Choices Explained
(2) Mendel developed the principle of *dominance*, but he did so without reliance on or knowledge of the *gene-chromosome theory*.
(3) The modern compound light *microscope* was developed over time through the contributions of many scientists; Mendel is not known for his contributions in this area. He would not have studied *genes* with a microscope since he was unaware of the genes' existence except as theoretical factors.
(4) Mendel had no direct knowledge of the existence of *gene mutations* and so would not have been able to use such mutations as an explanation for variation.

5. **4** The X and Y chromosomes carry genes that determine maleness and femaleness in humans. Genes located on either of these two chromosomes are said to be *sex-linked*.

Wrong Choices Explained
(1) The term *hybrid* refers to the heterozygous condition of an allele pair.
(2) The term *codominant* refers to a type of intermediate inheritance.
(3) The term *autosomal* refers to chromosomes other than the X and Y chromosomes.

6. **3** Mutations can be passed on to offspring only if they occur in cells (*primary sex cells*) that give rise to gametes. Mutations in other (somatic) cells may cause cancerous conditions in their recipients, but these conditions cannot be passed on to succeeding generations.

Wrong Choices Explained

(1), (2), (4) *Skin cells*, *lung cells*, and *uterine* cells are all examples of somatic cells. Mutations in somatic cells cannot be passed on to offspring.

7. **3** Fur color in the Himalayan hare depends on the temperature of the skin during fur growth. The diagram illustrates an experiment that shows that artificial cooling stimulates the skin to produce dark fur. Since temperature is an environmental condition, this experiment shows that certain *genetic traits can be influenced by environmental conditions*.

Wrong Choices Explained

(1) Heredity always influences *gene expression*. The experiment is not focused on this area.

(2) Genes are arranged on *homologous chromosomes* in essentially the same order. The experiment is not focused on this area.

(4) It is unlikely that this experiment would have resulted in *gene mutations* so extensive as to be immediately observable as a change in hair color.

8. **2** Molecule 1 represents a DNA molecule. The presence of units labeled T (for thymine) confirms this fact. In the living cell, DNA is found in the *nucleus*, associated with the chromatin.

Wrong Choices Explained

(1) The *centriole* is an organelle found in animal cells that serves as a point of attachment for the spindle fibers during mitotic cell division.

(3) The *cell wall* is an organelle that bounds and gives mechanical support to animal cells.

(4) The *lysosome* is an organelle that houses hydrolytic enzymes used in the digestion of complex food molecules in the cell.

9. **1** Molecule 3 represents a polypeptide (protein) molecule composed of four units of *amino acid*. Each of the amino acids is coded by a unique combination of three nitrogenous bases in DNA, as translated by mRNA molecules (molecule 2).

Wrong Choices Explained

(2) *DNA* is a very large polymer composed of repeating units of nucleic acid. DNA does not serve as a building block for other molecules.

(3) *Fatty acids* serve as components of lipid molecules, not of protein molecules.

(4) *RNA*, like DNA, is a very large polymer composed of repeating units of nucleic acid. RNA does not serve as a building block for other molecules.

10. **4**   Molecule 2 represents a molecule of messenger RNA (mRNA). This molecule is manufactured from free nucleic acids, using DNA as a template. mRNA then migrates to the *ribosome* to perform the task of protein synthesis.

Wrong Choices Explained
   (1) Cell *vacuoles* are specialized for storing various materials, including ingested foods, water, and toxic wastes.
   (2) The *plasma membrane* provides the outer covering of the cell. It is specialized to regulate the entry and exit of materials into and out of the cell.
   (3) The *lysosome* is an organelle that houses hydrolytic enzymes used in the digestion of complex food molecules in the cell.

11. **2**   *Dehydration synthesis* (joining by removing water) is a process by which complex molecules are formed by the chemical combination of two or more simpler subunits. By using this reaction, two amino acid molecules might be joined to synthesize one molecule of a dipeptide. Water is a by-product of this reaction.

Wrong Choices Explained
   (1) *Deamination* is a process by which amino acids are broken down into their component parts for conversion into urea.
   (3) *Enzymatic hydrolysis* is sometimes referred to as splitting with water. A complex molecule is split into two simpler molecules by adding the elements making up a water molecule to the bond that used to join the molecules. It is the chemical opposite of dehydration synthesis.
   (4) *Oxidation* is a general biochemical term that relates to chemical reactions in which oxygen is chemically added to an element or compound.

12. **3**   In the first paragraph, reference is made to the fact that *all embryos develop female sex organs* during the first 35–40 days (5–6 weeks) after fertilization.

Wrong Choices Explained
   (1) Although these structures are formed during this period, they are not the only structures to be formed.
   (2), (4) According to the passage (second paragraph), sex differentiation occurs after 35–40 days, not within the first 5 weeks.

13. **1**   The passage identifies the *SRY* gene as the determiner of maleness.

Wrong Choices Explained
   (2) The *SRY* gene is a single locus on the *Y* chromosome; the remainder of the chromosome is evidently not critical to this process.

(3) The *X* chromosome does not carry the *SRY* gene, and therefore cannot determine maleness.

(4) The *MIS* gene is activated by the *SRY* gene therefore the *MIS* gene cannot activate the process.

14. **2** The *MIS* gene is activated by a *chemical* produced by the *SRY* gene, which stimulates it to cause the female organs to disappear.

Wrong Choices Explained
(1) The *X* chromosome is not involved in the production of maleness in humans.

(3), (4) The presence of reproductive structures, male or female, is not cited in this passage as having anything to do with the stimulation of the *MIS* gene.

15. **4** Both the *SRY and the MIS genes* are involved in this process, the *SRY* gene acting as a trigger factor and the *MIS* gene performing the task of tissue resorption.

Wrong Choices Explained
(1) The *X* chromosome is not involved in the production of maleness in humans.

(2) The *SRY* gene is a single locus on the *Y* chromosome; the remainder of the chromosome is evidently not critical to this process.

(3) The *MIS* gene is activated by the *SRY* gene, therefore the *MIS* gene cannot activate the process.

16. **3** *Messenger RNA* (mRNA) is manufactured from free nucleic acids, using DNA as a template to produce a specific coded message concerning the design of an enzyme molecule. mRNA then migrates to the enzyme-producing region (ribosome) to perform protein synthesis.

Wrong Choices Explained
(1) *Hormones* are chemical substances produced by endocrine glands that are responsible for the chemical regulation of the body. They are not directly involved in protein synthesis.

(2) *Nerve impulses* are carried along neurons as electrochemical changes on the cell membrane. They are not directly involved in protein synthesis.

(4) *DNA molecules* provide the code for protein synthesis. They do not directly participate in protein synthesis.

17. **3** *Genetic engineering* refers to a set of laboratory techniques that involve the artificial translocation of genes from one cell to the chromosomes of other host cells. The host cells may then be used to produce

specific enzymes or other materials unique to the translocated gene. The examples given are two of the economic benefits that might be derived from such activities.

Wrong Choices Explained

(1) *Natural selection* is a process occurring in nature by which favorable varieties of a species may be selected by natural forces for survival and then pass their favorable traits on to their offspring. It is unlikely that natural selection would lead to the kinds of traits referenced in the question.

(2) *Sporulation* is a reproductive process carried out by certain molds and fungi. As an asexual reproductive process, it is unlikely to result in unique new varieties such as those referenced.

(4) *Chromatography* is a laboratory technique used to separate and study complex organic compounds found in cells. It does not relate directly to the production of new genetic varieties.

18. **3** *One cell with two identical nuclei* would result from this situation. Mitosis is a process by which a cell nucleus first replicates its genetic material and then undergoes a nuclear division that results in two identical daughter nuclei. Cytoplasmic division normally accompanies mitosis, separating the daughter nuclei into two separate daughter cells.

Wrong Choices Explained

(1) *Two cells, each with one nucleus* would be the result of normal mitotic cell division.

(2) *Two cells, each without a nucleus* could not result since mitosis normally results in the formation of identical daughter nuclei. Also, two cells could not result in the absence of cytoplasmic division.

(4) *One cell without a nucleus* could not result since mitosis normally results in the formation of identical daughter nuclei.

19. **1** *Artificial selection and inbreeding* are techniques that may be used to develop animals with certain desirable characteristics. The breeder selects individuals with evidence of desirable traits to be used as breeding stock. Once a favorable line has been established, it is maintained in future generations by inbreeding related individuals within the breeding population.

Wrong Choices Explained

(2) *Grafting and hybridization* are techniques normally associated with plant reproduction. Grafting refers to a process by which a plant slip (scion) is forced to grow into a stem (stock) of a different plant of the same species.

(3) *Regeneration and incubation* are processes associated with repair of damaged or lost tissues (regeneration) or the care by parents of eggs in a nest (incubation).

(4) *Vegetative propagation and binary fission* are methods of asexual reproduction carried out by plants and ameba, respectively.

20. **2** *The environment influences wing phenotype in these fruit flies.* For this particular mutation, the expression of the curly wing phenotype depends not only on the homozygous recessive genotype but also on the presence of high environmental temperatures (25°C) during development. Since the same genotype expresses a different phenotype (straight wing) at lower temperatures (16°C), we say that the trait is influenced by environmental conditions.

Wrong Choices Explained
(1) The statement *fruit flies with curly wings cannot survive at high temperatures* is not supported by the information given. If anything, the question leads us to believe that high temperatures support the curly wing condition.

(3) The statement *high temperatures increase the rate of mutations* is not supported by the information given. There is no evidence derived from the question that this mutant condition was caused by the environmental condition of temperature but instead, only influenced by it.

(4) The statement *wing length in these fruit flies is directly proportional to temperature* is not supported by the information given. The question implies an all-or-none condition with respect to the expression of this trait. No gradations in wing length are referenced, so wing length cannot be described as being proportional.

21. **1** A *genetic code* is contained within structure *A*, which represents a chromosome inside the nucleus of an animal cell. This genetic code is specific for the production of one or more proteins in the cell.

Wrong Choices Explained
(2) A *single nucleotide only* is not found in structure *A*. A chromosome is made up of hundreds, it not thousands, of nucleotides for each strand of DNA contained within it.

(3) A *messenger RNA molecule* is not found in structure *A*. Messenger RNA molecules are formed next to DNA molecules but then migrate from the nucleus to the ribosome.

(4) A *small polysaccharide* is not found in structure *A*. Polysaccharides, such as starch, are not normally associated with chromosome structure.

22. **3** *Recombinant DNA* is represented by structure *B*. Genes from an animal chromosome (structure *A*) have been spliced into the genome of a bacterial cell, forming a new genetic combination. The bacterial cell will now be capable of producing the protein or proteins coded for by the animal DNA.

Wrong Choices Explained

(1) *A ribosome* is not represented by structure *B*. Ribosomes are small organelles containing RNA and are located on the endoplasmic reticulum of most cells.

(2) *Transfer RNA* is not represented by structure *B*. The diagram indicates that structure A is composed of DNA, not RNA.

(4) *A male gamete* is not represented by structure *B*. Male gametes are sperm cells. Although sperm cells contain DNA, they are not found in the nuclei of animal cells.

23. **2** The technique illustrated is *genetic engineering*. Genetic engineering involves using various laboratory procedures to move genes from one cell to another.

Wrong Choices Explained

(1) *Cloning* is a technique in which the undifferentiated cells of an organism are used to produce a new organism with the same set of characteristics.

(3) *Protein synthesis* is a natural cellular process involving DNA and RNA molecules where amino acids are joined in particular sequences to produce specific proteins for use by the cell.

(4) *In vitro fertilization* is a laboratory technique in which mature eggs are fertilized outside the mother's body and the resulting zygote is reintroduced into the mother's uterus for development.

Question Set 2.6 (p. 155)

1. **4** *Organism A was probably more primitive than organism B and organism C* is a correct statement about the relationship among these organisms. By assuming that the fossils illustrated in the diagram are of sequentially evolved species, it can be inferred from evolutionary theory that the most primitive fossil is in the lowest (oldest) rock layer (A), and that the most advanced fossil is in the upper (youngest) rock layer (C).

Wrong Choices Explained

(1) *Organism A was probably more structurally advanced than organism B and organism C* is incorrect because it is unlikely that an older form of organism is more structurally advanced than a more recent form of that organism.

(2) *Organism C probably gave rise to organism A and organism B* is not a correct statement about the relationship among these organisms. Since the rock layers are "undisturbed," the top layer must have been the last one laid down. For this reason, it would be impossible for an organism whose fossils are found in layer C to have lived before, and given rise to, organisms whose fossils are found in deeper strata (A or B).

(3) *All of these organisms probably evolved at the same time* is not a correct statement about the relationship among these organisms. Given the long time frame required for the deposition of sedimentary rock, it is unlikely that all three of these organisms could have lived at the same time.

2. **3** Comparative embryology is often used to assess the degree of similarity between two species. The more similar two organisms are in terms of their embryological development, the more closely related they are assumed to be. This assessment can, and often does, point to the likelihood that they *share a common ancestry*.

Wrong Choices Explained
(1) Birds and reptiles are assigned to different genera. Therefore, they cannot *belong to the same species*.
(2) Species do not have to be closely related to be *adapted for life in the same habitat*. An example of this phenomenon is cacti and snakes that live successfully in the desert.
(4) The embryos of birds and reptiles show marked similarities. Neither is an *animal-like protist*. Rather, they are members of closely related vertebrate genera.

3. **1** Modern evolutionary theory borrows a great deal from the work of Charles Darwin. In addition, it uses the discoveries of geneticists of the 20th century in the areas of genetic inheritance and gene mutation to shed light on the mechanisms of the evolutionary process. As a result of this work, we now know that *variations are the result of mutations and gene recombination*.

Wrong Choices Explained
(2) *Overproduction of organisms leads to extinction* is not a concept included in modern theories of evolution. Overpopulation can be a serious problem for species. However, by itself, it rarely results in the extinction of a species since natural checks and balances tend to keep species population growth under control.
(3) *Variations exist only in large populations* is not a correct statement. Variations are not limited to large populations but exist in all populations, small and large.
(4) *Competition occurs only between members of the same species* is not a correct statement. Competition for environmental resources is not limited to members of the same species but occurs both between different species and within a single species.

4. **1** The concept of *organic evolution* is illustrated by these diagrams. Organic evolution refers to the mechanism thought to govern the changes in living things over geologic time. The diagram illustrates the physical changes that have occurred within the horse family over 60 million years, based on the fossil record.

Wrong Choices Explained

(2) *Ecological succession* refers to the changes that occur in ecological communities over long periods of time. It is not used to describe evolutionary change within a species and is not referenced in this diagram.

(3) *Intermediate inheritance* is used to describe a pattern of genetic inheritance in which the contrasting traits controlled by an allele pair blend to produce an intermediate phenotype. No reference is made to this pattern of inheritance in the diagram.

(4) *Geographical isolation* is used to describe situations in which species populations are separated by physical barriers (such as oceans, mountains, or canyons). This separation tends to promote variation leading to speciation. No geographical isolation can be inferred from this diagram.

5. **4** When long periods of evolutionary stability (equilibrium) are interrupted by relatively short periods of rapid evolutionary change (punctuation), the pattern of evolution is referred to as *punctuated equilibrium*.

Wrong Choices Explained

(1) The term *use and disuse* refers to the evolutionary theory of Jean Lamarck, which has been largely discredited as a viable theory of evolutionary change.

(2) *Reproductive isolation* refers to behavioral mechanisms that can isolate two varieties of a species from interbreeding even though they may inhabit the same geographical area.

(3) The term *homologous structures* refers to the study of comparative anatomy that postulates that similarities in structure, regardless of function, signal common ancestry between two or more different species.

6. **2** The phenomenon illustrated in the diagram is *an increase in the adaptive value of gene a*. We are led to believe that selective pressures have been brought to bear on the test species over 10 generations such that the lighter phenotype is negatively selected. This selection pressure gradually eliminates organisms displaying the light phenotype, eliminating with them a large proportion of the dominant allele A. As the frequency of allele A diminishes, the frequency of allele a increases proportionately, resulting in more organisms displaying the darker phenotype controlled by genotype *aa*.

Wrong Choices Explained

(1) *A decrease in the adaptive value of gene a* would result in an increase in the proportion of light phenotypes in this population rather than a decrease.

(3) A quick comparison of the branch at generations 1 and 10 reveals the same relative density of this population. No evidence indicates *an increase in the population of this insect* in this diagram.

(4) No information is given in the diagram that would lead us to believe that *a decrease in the mutation rate of gene A* has occurred. This situation might be caused by an increased mutation rate of *A* but not a decreased rate.

7. **3** *Natural selection* is the centerpiece of Darwin's theory of evolution. Darwin theorized that natural conditions (selection pressures) favor some varieties of a species over other varieties of the same species. As the favorable variety survives these pressures at a higher rate, its members are more readily available to produce offspring expressing the same favorable variety. In this way, Darwin thought, new and favorable varieties of a species are selected by nature to become common in the species. When carried to an extreme, these varieties can become different species surviving under different sets of environmental conditions.

Wrong Choices Explained

(1), (2) Evolution by means of the use and *disuse of body structures* and *the transmission of acquired characteristics* were central ideas expressed in the evolutionary theory of Jean Lamarck. Lamarck's theories have been largely disproved by modern science.

(4) *Mutagenic agents* and gene mutation as a means of creating new varieties in species were concepts unknown at the time that Darwin developed his evolutionary theory. These concepts have been developed in the context of modern science.

8. **1** The *increasing need for new antibiotics* to kill new, antibiotic-resistant strains of bacteria is often used as a clear example of evolution at work in the present day. Bacteria, like other living species, express many genetic variations, including resistance to the effects of antibiotics. In an environment where the presence of antibiotics is common (such as a hospital), the antibiotics act as a selection pressure that eliminates most (but not all) members of the bacterial species. A single bacterium that contains a mutation for resistance to a particular antibiotic can reproduce millions of bacterial cells like itself in a very short time. New types of antibiotics must constantly be produced to combat these resistant strains.

Wrong Choices Explained

(2) *An increasing number of individuals in the human population* is an environmental problem not directly supporting the concept that evolution is occurring in the present.

(3) *A decreasing number of new fossils discovered in undisturbed rock layers* is not directly related to evolution occurring in living species. The discovery and study of fossils provides evidence that evolution occurred in the past.

(4) *Decreasing activity of photosynthetic organisms due to warming of the atmosphere* references an environmental problem not directly supporting the concept that evolution is occurring in the present.

9. **3** Similar nucleotide sequences in the genetic material of separate species provide evidence that these species *may have similar evolutionary histories*. Genetic material, and particularly gene sequences, are passed from generation to generation by reproduction. The more similar these sequences are, and the more sequences there are that are similar, the more closely related the species are considered to be.

Wrong Choices Explained

(1) It is extremely unlikely that these two species *are evolving into the same species*. The process of evolution tends to lead to greater, not less, diversity among living things.

(2) No information is presented in the question that would point to the possibility that these two species *contain identical DNA*. If this were the case, then they would be the same, not different, species.

(4) No information is presented in the question that would indicate that these two species *have the same number of mutations*. Mutations are thought to occur randomly in nature, so the likelihood that two separate species have the same number of mutations is small.

10. **4** *Gradualism* is a theory of evolutionary change that assumes that mutations occur at a predicable pace and the environmental pressures stay relatively constant over time. These two factors, then, result in a relatively constant rate of change of the gene pool of most species, leading to slow, gradual evolution.

Wrong Choices Explained

(1) The term *punctuated equilibrium* refers to a theory of evolutionary change in which long periods of relatively slow change are interrupted by relatively short periods of rapid evolution.

(2) *Geographic isolation* is a term relating to a condition (physical separation) that can promote the development of new species varieties as well as new species.

(3) *Speciation* is a term that relates to the change of separate varieties of a single species into two separate and distinct species.

11. **2** The relationship between the bird and moth is *predator-prey*. The bird is the predator of the moth and other insects that constitute its prey.

Wrong Choices Explained

(1) A *producer-consumer* relationship would have to involve a plant (the producer) and a herbivore (the consumer). No such relationship is indicated in the passage.

(3) A *parasite-host* relationship is a symbiosis between an organism (the parasite) that lives in or on another (the host) and harms it in the process. The bird and moth live separate existences in this case.

(4) An *autotroph-heterotroph* relationship would have to involve a plant (the autotroph) and a herbivore (the heterotroph). No such relationship is indicated in the passage.

12. *They introduced industry to the area.* OR *They built factories that produced pollution.*
    (NOTE: Other correct, complete-sentence responses are acceptable.)

13. *Mutations change DNA, resulting in new traits.* OR *Crossing-over during meiosis may produce new gene combinations.* OR *Fertilization involves the union of sex cells from each of two parents, resulting in offspring different from either parent.*
    (NOTE: Other correct, complete-sentence responses are acceptable.)

Question Set 2.7 (p. 171)

1. **1** Of those shown, diagram 1 most clearly illustrates the process of binary fission. The diagram shows an ameba-like organism undergoing the cytoplasmic division stage of mitotic cell division. Clearly visible in each of the two forming daughter cells is a newly formed nucleus.

Wrong Choices Explained

(2) Diagram 2 illustrates a yeast cell undergoing budding. Budding is a form of asexual reproduction characterized by mitosis followed by unequal cytoplasmic division.

(3) Diagram 3 illustrates a strawberry plant reproducing by runners. While mitotic cell division is involved in this form of reproduction, it is not binary fission.

(4) Diagram 4 illustrates a bread mold in the process of spreading spores. Spore formation is a form of asexual reproduction, but does not involve binary fission.

2. **4** *Four monoploid cells* normally form in a male when a primary sperm cell undergoes meiotic cell division. Meiotic cell division involves two distinct phases, the first of which reduces the number of chromosomes from diploid ($2n$) to monoploid (n). The second division separates homologous chromosomes into gamete cells known as sperm cells.

Wrong Choices Explained

(1), (2), (3) Each of these is a nonsense distracter. Normal cell division does not result in the combinations indicated.

3. **1** The sequence that represents the correct order of events in the development of sexually reproducing animals is *fertilization* → *cleavage* → *differentiation* → *growth*. Fertilization refers to the fusion of gametes to form a zygote. Cleavage is the rapid mitotic cell division of the zygote to form a cell mass. Differentiation is a process by which the new cells begin to specialize into embryonic tissues. Growth is the increase in cell number and organism size that leads to the formation of an adult organism.

Wrong Choices Explained

(2), (3), (4) These sequences are each out of order.

4. **4** Structures *E and G* secrete hormones that regulate the development of secondary sex characteristics. Structure *E* is the male testis; structure *G* is the female ovary. These structures produce hormones that regulate the development of secondary sex characteristics such as production of body hair, breast enlargement, and voice and muscle tone changes in the adult human.

Wrong Choices Explained

(1) Structures *A and J* represent the male urinary bladder and the female urinary bladder, respectively. These structures serve to store urine prior to its controlled elimination.

(2) Structures *D and H* represent the male vas deferens and the female oviduct, respectively. The primary role of these structures is to transport gametes to the site of fertilization.

(3) Structures *F and I* represent the male scrotum and the female uterus, respectively. The scrotum surrounds the testes and provides conditions conducive to the formation and storage of sperm cells. The uterus functions to receive a fertilized egg and provide a stable environment for its development.

5. **3** The pathway followed by sperm cells implanted inside the female is *K to I to H*. *K* represents the vagina, which serves as the site for the implantation of sperm during intercourse. *I* represents the uterus,

through which the sperm must travel to reach the egg. *H* represents the oviduct, within which the sperm meet the mature egg and where fertilization normally occurs.

Wrong Choices Explained
(1), (2), (4) These sequences are each out of order.

6. **2** Gametogenesis occurs within structures *E and G*. Structure *E* is the male testis; structure *G* is the female ovary. These structures produce sperm cells and egg cells, respectively, which are the male and female gametes involved in the process of sexual reproduction.

Wrong Choices Explained
(1) Structures *A and J* represent the male urinary bladder and the female urinary bladder, respectively. These structures serve to store urine prior to its controlled elimination.

(3) Structures *B and I* represent the male urethra and female uterus, respectively. In males, the urethra serves to guide the release of sperm cells into the female reproductive tract. The urethra also conducts urine from the urinary bladder to the environment. The uterus functions to receive a fertilized egg and provide a stable environment for its development.

(4) Structures *D and H* represent the male vas deferens and the female oviduct, respectively. The primary role of these structures is to transport gametes to the site of fertilization.

7. **3** Structures *G and I* are directly affected by hormones involved in the menstrual cycle. Structure *G* is the female ovary, which produces eggs throughout the period of sexual maturity in females. Structure *I* is the uterus, which functions to receive a fertilized egg and provide a stable environment for its development.

Wrong Choices Explained
(1) Structures *C and E* represent the male penis and the male testis, respectively. These structures are not affected by hormones produced in the female menstrual cycle.

(2) Structures *A and D* represent the male urinary bladder and the male vas deferens, respectively. These structures are not affected by hormones produced in the female menstrual cycle.

(4) Structures *I and J* represent the female uterus and the female urinary bladder, respectively. While the uterus is affected by the production of hormones, the urinary bladder is not similarly affected.

8. **4** When a human egg is released from the ovary, it passes directly into the *oviduct* for transport to the uterus.

Wrong Choices Explained

    (1) The *cervix* is a muscular structure at the base of the uterus that keeps the developing embryo and its associated membranes within the uterus during gestation.

    (2) The *vagina* serves as the site for the implantation of sperm during intercourse as well as the birth canal following gestation.

    (3) The *uterus* is the organ in which the embryo implants and the placenta forms during gestation.

9. **2**   The uterus periodically readies itself for implantation of a fertilized egg. The duration of the menstrual cycle is approximately 28 days. If no fertilized egg is received by the 28th day of the cycle, the uterine lining is shed in a process known as *menstruation*.

Wrong Choices Explained

    (1) *Ovulation* is the process by which a mature egg cell is released from the ovarian follicle once every 28 days in the menstrual cycle.

    (3) The *follicle* stage of the menstrual cycle is the 14-day period, immediately prior to ovulation, in which the egg undergoes maturation inside the ovarian follicle.

    (4) The *corpus luteum* stage of the menstrual cycle is the 8- to 10-day period, immediately following ovulation, in which the cells of the follicle transform into the corpus luteum for the production of progesterone.

10. **1**   The sequence $D \rightarrow B \rightarrow C \rightarrow A$ provides for follicle stage (D), followed by ovulation (B), followed by corpus luteum stage (C), followed by menstruation (A).

Wrong Choices Explained

    (2), (3), (4) These are nonsense distracters whose sequences do not coincide with any known reproductive cycle.

11. **3**   The fluid *acts as a transport medium for sperm*. This fluid is known as semen. Its primary function is to provide a protective watery medium for the sperm cells as they enter the female reproductive tract.

Wrong Choices Explained

    (1) *Removes polar bodies from the surface of the sperm* is a nonsense distracter. Polar bodies are not associated with sperm production.

    (2) *Activates the egg nucleus so that it begins to divide* is not an advantage of this fluid medium. The egg is stimulated to divide by the act of fertilization. Semen is not directly involved in this process.

    (4) *Provides currents that propel the egg down the oviduct* is not an advantage of this fluid medium. Cilia that line the oviduct are responsible for establishing fluid currents that both carry the egg downward

toward the uterus and carry sperm upward toward the ovary. Semen is not directly involved in this process.

12. **4** The *placenta* is the structure through which substances can diffuse from the mother's blood into the fetal blood. This structure forms in the uterus and establishes a physical link between the maternal and fetal tissues.

Wrong Choices Explained

(1) The *amnion* is a membrane that immediately surrounds the developing embryo and provides a watery environment during development.

(2) The *fallopian tube*, also known as the oviduct, is the site of fertilization and early development (cleavage).

(3) The *yolk sac* provides nutrition for the human embryo during the very early stages of development, before the placenta is formed. It deteriorates as soon as the placental connection is established.

13. **1** Stage *11* shows the first appearance of the mesoderm layer, which can be seen as a group of cells lying between the outer ectoderm layer and inner endoderm layer. The mesoderm will eventually give rise to many of the body's internal organs, including muscle, bone, and circulatory structures.

Wrong Choices Explained

(2) Stage *8* represents the blastula stage of embryonic development, characterized by a hollowing of the cell mass created by the cleavage divisions. No cell specialization is present at this stage.

(3) Stage *6* represents the cell mass resulting from five consecutive cleavage divisions. No cell specialization is present at this stage.

(4) Stage *4* represents the cell mass resulting from three consecutive cleavage divisions. No cell specialization is present at this stage.

14. **3** The *oviduct* is normally the site of the first several cleavage divisions. Cleavage follows fertilization and precedes implantation of the embryo in the uterus.

Wrong Choices Explained

(1) The *ovary* is the site of egg formation and the process of ovulation. The released egg is carried along the oviduct toward the uterus.

(2) The *vagina* is the site of sperm implantation during intercourse. Sperm cells must swim through the female tract until they contact the egg within the oviduct.

(4) The *uterus* is the site of embryo implantation and fetal development.

15. **1** *Gametogenesis and fertilization* most immediately precede the sequence of developmental stages represented in the diagram.

Gametogenesis is the process by which sperm cells and egg cells are produced in the testes and ovaries, respectively. Fertilization is the fusion of egg cell and sperm cell nuclei within the female tract.

Wrong Choices Explained

(2) *Menstruation and menopause* do not most directly precede this sequence of developmental stages. Menstruation is the shedding of the uterine lining that occurs when an egg is not fertilized. Menopause is the cessation of the menstrual cycle that occurs in most females at about age 50.

(3) *Prenatal development and gestation* do not most directly precede this sequence of developmental stages. Prenatal development refers to the entire developmental process, encompassing this sequence of stages up to birth. Gestation refers to the 9-month period during which prenatal development occurs in humans.

(4) *Placental formation and metamorphosis* do not most directly precede this sequence of developmental stages. Placental formation is an event that follows, rather than precedes, this sequence of stages. The placenta forms a connection between mother and fetus during gestation. Metamorphosis is a term relating to development in insects, not humans.

16. *The data in the tables do support the scientists' claim. The measurements of newborns of drinkers are uniformly lower than those of nondrinkers, including shorter gestation periods (4.7% shorter), lower birth weight (17.4% lower), less length (6.6% shorter), and smaller head size (7.0% smaller).*
    (NOTE: Any correct, complete-sentence response is acceptable.)

17. *The data are based on a relatively small sample size of 40 drinkers, so additional data is needed.*
    (NOTE: Any correct, complete-sentence response is acceptable.)

18. *Embryological tissues are very sensitive to chemical changes in the uterus. The younger and more highly undifferentiated fetal tissues are, the more they can be affected by chemicals in their environment.*
    (NOTE: Any correct, complete-sentence response is acceptable.)

19. Sample essay: *"Sexual and asexual reproduction are alike in that they both result in the production of new organisms from a parent or parents. They are unlike in that asexual reproduction results in offspring that have no genetic variation from the parent, while sexual reproduction allows a mixing of new traits each time an offspring is produced. This is due to the fact that in sexual reproduction, two monoploid sex cells fuse, each carrying half of each parent's genetic information. The resulting zygote contains a unique combination of traits.*

"Examples of organisms that reproduce asexually include the ameba, hydra, yeast, and planaria worm. Organisms that reproduce sexually include earthworms, humans, and flowering plants."

(NOTE: Any essay that correctly addresses the points required in the question is acceptable.)

Question Set 2.8 (p. 191)

1. **2** *Living matter is able to control chemical activities with organic catalysts.* The controlled chemical activity in living matter is one of the distinguishing characteristics of life. Enzymes (organic catalysts) are principally responsible for controlling this chemical activity, enabling reactions to occur under conditions found in the cell.

Wrong Choices Explained
(1) *Living matter is unable to diffuse materials* is not a true statement. Diffusion is a process by which molecules of a substance move from a region of higher concentration to a region of lower concentration of that substance. It occurs naturally in many living and nonliving systems and does not depend on living things for its operation.

(3) *Living matter is able to create energy* is not a true statement. Scientists theorize that the amount of energy in the universe is fixed and is neither created nor destroyed but only transferred from one form to another. Energy, although essential to the survival of living things, cannot be created by them.

(4) *Living matter is unable to use energy for metabolic activities* is not a true statement. The life function of respiration is involved with the transfer of chemical bond energy in foods (such as glucose) to molecules of ATP. The energy in ATP is subsequently used by the cell as a controlled source of energy to run cellular reactions (metabolism). Respiration is common to all known life-forms.

2. **2** The nutritional process illustrated by the equation $A + C \rightarrow A + B + D$ is photosynthesis. The carbon-fixation reactions combine the atoms of hydrogen, oxygen, and carbon found in water (A) and carbon dioxide (C) into molecules of glucose (B); Water (A) and oxygen (D) are by-products of this autotrophic nutritional process. Water (A) is both a reactant in and a product of this process.

Wrong Choices Explained
(1) The equation $B + D \rightarrow A + C$ illustrates the process of aerobic respiration, which involves the same substances as does photosynthesis but is its chemical opposite.

(3), (4) These equations, $B + C \rightarrow A + D$ and $A + B + D \rightarrow B + C$, are nonsense equations that cannot occur either in nature or in the laboratory.

3. **4** *B and D* represent chemicals needed for cellular respiration. Glucose (B) is combined with oxygen (D) to release energy for use in the cell. This is a process carried on by both animals and plants.

Wrong Choices Explained

(1) The substances water (A) and carbon dioxide (C) are reactants in the process of photosynthesis, the chemical opposite of aerobic respiration.

(2), (3) The combinations of substances in these distracters are nonsense combinations that would rarely, if ever, occur in nature.

4. **2** Respiration involves chemical reactions in which food molecules (such as glucose) are broken down and molecular energy is released to form molecules of adenosine triphosphate (*ATP*). This process is carried on by all known living organisms as a means of making energy available for cellular processes.

Wrong Choices Explained

(1) $C_6H_{12}O_6$ is the chemical formula for glucose, which is synthesized in the process of photosynthesis, not respiration.

(3) *Alcohol* is a by-product of fermentation, a type of anaerobic respiration. While it is a product of the respiratory process, it is not produced by all organisms but rather by a few specialized yeasts.

(4) Molecular *oxygen* is a by-product of photosynthesis, not respiration.

5. **2** When an organism reacts to a change in environmental conditions, its reaction is known as a *response*. The brine shrimp discussed in this paragraph respond to a change in light conditions by moving toward the light, a nervous response. Response to environmental stimuli is a characteristic of regulation.

Wrong Choices Explained

(1) *Negative feedback* is a concept normally associated with hormonal regulation in animals. Hormonal negative feedback is not normally associated with simple nervous responses such as the one discussed in the paragraph.

(3) A *stimulus* is defined as an environmental change that results in a response by an organism. In this paragraph, the stimulus is the change in light conditions.

(4) *Active transport* is a process by which materials may be actively moved into or out of a cell through the expenditure of energy. The life function associated with active transport is that of transport.

6. **1** An examination of graph A shows that, from a temperature of 0°C to a temperature of 38°C, enzyme action tends to increase at a fairly steady pace. Once the temperature of *40°C* is reached, however, the rate of enzyme action begins to decrease rapidly, indicating that the enzyme molecules in the system are being destroyed through the irreversible process of denaturation.

Wrong Choices Explained

(2) Graph A shows that the rate of enzyme action is approximately 50% of optimum at *23°C*.

(3), (4) Graph B shows that, at least for the human stomach enzyme pepsin, the rate of enzyme action is near optimum at *pH 2* and *pH 3*.

7. **2** The measure of acidity or basicity used in the biology laboratory is pH (concentration of hydrogen ion). Graph B is clearly labeled as a representation of the effect of pH on enzyme action.

Wrong Choices Explained

(1) Graph A represents the effect of temperature on the rate of enzyme action.

(3) Graph C represents the effect of enzyme concentration on the rate of enzyme action.

(4) Graph D represents the effect of substrate concentration on the rate of enzyme action.

8. **2** The equation represents those steps in photosynthesis known as the carbon-fixation reactions. Prior to these reactions, water (H_2O) is split into hydrogen and oxygen in the reaction known as photolysis, making water the most likely source of hydrogen for this equation.

Wrong Choices Explained

(1), (3) *PGAL* (phosphoglyceraldehyde) is the intermediate and glucose ($C_6H_{12}O_6$) the final molecular product resulting from this process. These molecules are the recipients, not the sources, of the hydrogen represented in the equation.

(4) *ATP* (adenosine triphosphate) is a specialized molecule that provides energy for many cellular reactions, including this one. However, it does not serve as a source of hydrogen for these reactions.

9. **4** The cell organelle indicated as structure *D* is a chloroplast. The chloroplast is the site of the photosynthetic reactions, including photolysis and carbon-fixation reactions.

Wrong Choices Explained

(1) Structure *A* represents a plant cell nucleus, which is involved in cell regulation.

(2) Structure *B* represents the cytoplasm, which provides a watery medium for cell activities.

(3) Structure *C* represents a food vacuole, which is involved in storage of manufactured starch.

10. In *active immunity*, the body is stimulated to mount its own defenses against foreign antigens in the blood. These foreign antigens can be acquired naturally, by contracting a disease, or introduced artificially by inoculation with a vaccine. Active immunity is usually long lasting.

Wrong Choices Explained
Passive immunity is acquired when antibodies produced in another organism are introduced into the blood via injection. Passive immunity is usually short-lived.

Allergies arise in response to specific foreign antigens. The body responds to these antigens by releasing chemicals known as histamines, which in turn affect the body by causing headaches, runny nose, watery eyes, and a variety of other symptoms.

Tissue rejection is used to describe the reaction of a host organism to a transplanted tissue or organ from a donor organism. If the antigens of the host and those of the donor do not closely match, the host may reject the transplanted tissues in a process similar to that associated with natural active immunity.

11. *Allergies* arise in response to specific foreign antigens. The body responds to these antigens by releasing chemicals known as histamines, which in turn affect the body by causing headaches, runny nose, watery eyes, and a variety of other symptoms.

Wrong Choices Explained
See question 10.

12. **2** The human disorder known as *anemia* is associated with insufficient amounts of hemoglobin, the oxygen-carrying compound in red blood cells. Anemia sufferers usually tire quickly due to insufficient amounts of oxygen reaching body tissues.

Wrong Choices Explained
(1) *Angina* pectoris refers to chronic chest pain caused by insufficient oxygen reaching the heart muscle. Angina is usually associated with narrowing of the coronary arteries due to cholesterol buildup on the artery walls.

(3) *Coronary thrombosis*, commonly known as heart attack, is an acute disorder of the circulatory system resulting from complete blockage of the coronary artery.

(4) *High blood pressure* is a chronic disorder of the circulatory system, which may result from the gradual narrowing of arterial vessels throughout the body.

13. **4** *A lack of iodine in the diet, which has caused the development of a goiter* is the correct response. Goiter is often associated with diets insufficient in iodine, an element essential to the operation of the thyroid gland located in the throat. Without iodine, the thyroid enlarges, producing the swelling described in the question.

Wrong Choices Explained
(1), (2), (3) *An excess of calcium in the diet, which has caused a muscle deformity*, *deposits of fat under the skin caused by a vegetable diet*, and *inherited neck deformities caused by elevated environmental temperatures* are all nonsense distracters that relate to no known human disorders.

14. **4** *Glucose* is the principal product of photosynthesis. Glucose may be *metabolized into more complex carbohydrates by dehydration synthesis*, a chemical process occurring in plant cells that produces starch, cellulose, and other complex carbohydrates.

Wrong Choices Explained
(1), (2) *Starch* and *protein* can result from dehydration synthesis of smaller molecular units. Dehydration synthesis results in more complex, not less complex, molecular by-products.
(3) *Glycerol*, a component of certain fat molecules, cannot be synthesized into complex carbohydrates.

15. **1** The graphs indicate that *this enzyme works best at a temperature of 35°C and a pH of 8*. The temperature graph clearly shows the high point of the graph directly at the midpoint between 30°C and 40°C, while the pH graph shows the optimum pH at 8.

Wrong Choices Explained
(2), (4) The statements *this enzyme works best at a temperature of 50°C and a pH of 12* and *this enzyme works best at a temperature above 50°C and a pH above 12* are not supported by the information in the graphs. The graphs show that the rate of enzyme activity drops off drastically above 40°C and above pH 9.
(3) The statement *temperature and pH have no effect on the action of this enzyme* is incorrect. The graphs clearly show that the rate of enzyme activity varies greatly when these two variables are manipulated.

16. **4** Of those given, the statement *it reduces the amount of raw material reaching the active site of the enzyme that produces prostaglandins* most

correctly describes how aspirin acts to reduce flu symptoms. In the second paragraph, the article describes how aspirin blocks a critical area of the enzyme that manufactures prostaglandins, preventing the raw materials used in its manufacture from reaching the active site of the enzyme.

Wrong Choices Explained

(1), (2), (3) These statements are not supported by the information given in the article. Close reading of the article is necessary to eliminate these distracters.

17. **1** Of those given, the statement *it interferes with the activity of an enzyme that helps to protect the stomach* most accurately describes how aspirin can irritate the stomach of some people. In the third paragraph, aspirin's effect on the stomach-protecting enzyme PGHS-1 is discussed. By completely blocking this enzyme, stomach-protecting substances cannot be manufactured, and stomach irritation can result.

Wrong Choices Explained

(2), (3), (4) These statements are not supported by the information given in the article. Close reading of the article is necessary to eliminate these distracters.

18. *The molecule is a crystal with a tube running up the middle of it.*
    (Other correct, complete-sentence responses are acceptable.)

Wrong Choices Explained

Students should take care to give their responses in biologically accurate, complete-sentence format, including a subject, predicate, and appropriate punctuation. Responses that do not meet these basic criteria cannot be considered for credit.

19. *The willow tree bark contains salicylic acid, which is similar to acetyl-salicylic acid, the active ingredient in aspirin.*
    (Other correct, complete-sentence responses are acceptable.)

Wrong Choices Explained

Students should take care to give their responses in biologically accurate, complete-sentence format, including a subject, predicate, and appropriate punctuation. Responses that do not meet these basic criteria cannot be considered for credit.

20. **2** *Cancer* is characterized by abnormal cells that suddenly begin to undergo cell division at a very rapid rate. A cancerous cell can produce a mass of abnormal cells in a relatively short time, crowding out normal tissues.

Wrong Choices Explained

(1) *Albinism* is a genetic condition, controlled by a single mutant somatic gene, in which the skin lacks the ability to produce pigments.

(3) *Hemophilia* is a genetic condition, controlled by a single mutant sex-linked gene, in which the blood lacks the ability to produce clotting factors.

(4) *Color blindness* is a genetic condition, controlled by a single mutant sex-linked gene, in which the eyes lacks the ability to produce proteins necessary for recognizing certain colors.

21. **3** *Asthma* is a disorder in humans characterized by constricted airways due to the presence of an irritant. Asthma can be life-threatening if not properly treated.

Wrong Choices Explained

(1) *Coronary thrombosis* is a human disorder in which the heart muscle becomes damaged due to blockage of the coronary artery.

(2) *Arthritis* is a human disorder in which skeletal joints become inflamed, swollen, and painful.

(4) *Emphysema* is a human disorder in which lung tissue deteriorates, leaving the lung inefficient at absorbing oxygen.

22. **4** *A lack of essential amino acids in the diet* is the most likely cause of the condition described in the passage. Because breast milk contains proteins produced by the mother's own body, it will contain the correct balance of amino acids needed by the infant for proper growth. When the diet is changed to cereal, mainly carbohydrates, these proteins are no longer available to the child. As a result, the child is not able to obtain essential amino acids for its own cells to synthesize proteins.

Wrong Choices Explained

(1) *Too many nucleic acids in the diet* is not the likely cause of these symptoms. Nucleic acids are manufactured in the cell nucleus from free nucleotides. They are not found in large quantities in cereal grains.

(2) *An overconsumption of complete protein foods* is not the likely cause of these symptoms. The disorder is described as a protein deficiency in the passage. Overconsumption of complete protein foods would lead to an abundance, not a deficiency, of protein in the body.

(3) *Not enough carbohydrates in the diet* is not the likely cause of these symptoms. Cereal grains contain large concentrations of carbohydrates, which provide an energy source but relatively small quantities of protein.

23. **2** *Cell type B would be fewer in number and lighter in appearance* is the statement that best describes the changes observed in the blood of a

person with anemia, assuming all other variables are kept constant. The cells labeled *B* are red blood cells, which contain the red oxygen-carrying pigment hemoglobin. A person with anemia cannot produce hemoglobin efficiently, so red blood cell production drops. The red blood cells that are produced may contain less hemoglobin than those of a person without this condition.

Wrong Choices Explained

(1), (4) *Cell type A would be fewer in number and larger in size* and *cell type A would be larger in size and darker in appearance* are not correct responses. The cell labeled *A* is a white blood cell. White blood cells lack hemoglobin and so would be unaffected by the anemic condition.

(3) *Cell type B would be larger in size and greater in number* is not the correct response. Cell size would be unaffected by anemia. The cells' appearance would be lighter, not darker, due to the lack of sufficient hemoglobin.

24. *The experimental group and control group should be set up identically, except that the control group should receive a placebo rather than a dose of Lowervil.* OR *All parts of the experiment have to be the same except that the control group does not receive the drug.*

(NOTE: Any correct, complete-sentence answer is acceptable.)

25. *Using a large number in the experiment in order to eliminate bias is important.* OR *If the experiment is run with a small number of subjects, then statistical significance will be difficult to measure.*

(NOTE: Any correct, complete-sentence answer is acceptable.)

26. *The researcher should carefully measure the blood pressure of both experimental and control subjects before and after the administration of the drug.* OR *The researcher should establish a baseline blood pressure for each participant in the study then monitor blood pressure changes after administration of the drug or placebo.*

(NOTE: Any correct, complete-sentence answer is acceptable.)

27. *Veins and arteries may be blocked and tissue damage may result.* OR *The body may reject the new organ.*

(NOTE: Any correct, complete-sentence answer is acceptable.)

28. *A transplant patient might take an immunosuppressant drug in order to prevent the rejection of the new organ or tissue.*

(NOTE: Any correct, complete-sentence answer is acceptable.)

29. *The drug may weaken the patient's ability to fight diseases.* OR *The drug may leave the patient less able to fight infection.*

(NOTE: Any correct, complete-sentence answer is acceptable.)

30. *Tissues or organs grown from stem cells of the patient would not be rejected by the patient's immune system.* OR *Organs produced by this process would not be foreign material and would not be attacked by the patient's immune system.*

(NOTE: Any correct, complete-sentence answer is acceptable.)

Question Set 2.9 (p. 217)

1. **4** From the information given in the diagram, the statement that best describes some organisms in the food web illustrated is *raccoons, fish, and ducks are secondary consumers*. Each of these organisms is shown consuming another organism that is either itself a primary or secondary consumer.

Wrong Choices Explained
(1) The statement *minnows and fish are primary consumers* is not supported by the information in the diagram. Fish are shown consuming aquatic crustaceans, which are primary consumers.
(2) The statement *algae and floating plants are decomposers* is not supported by information given in the diagram. Algae and floating plants are producers.
(3) The statement *aquatic crustaceans are omnivores* is unsupported by the information in the diagram. Aquatic crustaceans are shown consuming only plant matter.

2. **2** Of the statements given, *population B competed more successfully for food than population A did* best describes the data shown in the graph. We are led to believe that populations *A* and *B* inhabit the same environmental niche and compete for the same food supply in their range. Over time, these two stable populations undergo a significant change, perhaps in the face of a sudden reduction in available food supply. As a result of this change, population *A* is virtually eliminated, while population *B* increases in number.

Wrong Choices Explained
(1) The statement *all of the plant populations in this habitat decreased* is not supported by the data. While some of the plant populations likely decreased, assuming that all plant populations decreased is unreasonable.
(3) The statement *population A produced more offspring than population B did* is false. The data shows population *B* increasing in number, indicating that many offspring are being produced. At the same time, we see population *A* decreasing in number, indicating a reduced rate of reproductive activity.

(4) The statement *population A consumed the members of population B* is false. The question states that these populations are herbivores, so *A* would not consume members of *B*. Also, if *A* had consumed *B*, then we would expect that *B* would decrease in relation to *A*, but this is not the case.

3. **1** The *tundra* is a biome characterized by very cold average temperatures such that the *subsoil is permanently frozen*. Because of the lack of soil moisture, only surface plants such as *lichens and mosses* have a chance for survival.

Wrong Choices Explained
(2) The *taiga* is a biome characterized by cool average temperatures. Coniferous forests are common, as are caribou and moose.

(3) The *temperate deciduous forest* biome is characterized by moderate average temperatures, good rainfall, broadleaf (deciduous) forests, and index animals such as deer and squirrels.

(4) The *grassland* biome is characterized by moderate average temperatures, low rainfall, grassy prairies, and index animals including bison and prairie dogs.

(5) The *desert* biome is characterized by wide variations in daily temperature, little rainfall, sagebrush and cactus, and index animals such as roadrunners, coyotes, and rattlesnakes.

(6) The *tropical forest* biome is characterized by high average temperature, abundant rainfall, dense tropical forests, and index animals such as the parrot and monkey.

4. **6** The *tropical forest* biome is characterized by high average temperature, abundant rainfall, dense tropical forests, and index animals such as the parrot and monkey.

Wrong Choices Explained
(1), (2), (3), (4), (5) See question 3.

5. **5** The *desert* biome is characterized by wide variations in daily temperature, little rainfall, sagebrush and cactus, and index animals such as roadrunners, coyotes, and rattlesnakes.

Wrong Choices Explained
(1), (2), (3), (4), (6) See question 3.

6. **2** Because a number of different types of organisms are present, both plant and animal species, this diagram illustrates the level of ecological organization known as a *community*.

Wrong Choices Explained

(1) The term *biosphere* refers to the entire inhabited portion of the planet, a much broader level of organization than that illustrated in the diagram.

(3) The *habitat* is the natural environment for a species, including both biotic and abiotic factors. This diagram illustrates only biotic factors.

(4) The *biome* level of ecological organization is a major grouping of large, but similar ecosystems, a much broader level of organization than is illustrated in the diagram.

7. **1** As the *owl population increases*, interspecies and intraspecies competition for what is presumably a fixed food supply will increase dramatically. This food supply may be taxed beyond its ability to support the owl and hawk populations, and eventually many members of these populations would die of starvation.

Wrong Choices Explained

(2) An *increase in the mouse population* will benefit the hawks and owls since additional food resources will be made available. This should allow a slight increase in the owl and hawk populations over time.

(3) A *decrease in the hawk population* will also benefit the food supply since hawks compete with owls for a limited food supply.

(4) A *decrease in the owl population* will benefit the food supply since owls compete with hawks for a limited food supply.

8. **2** *All the* Homo sapiens *living in New York State* represent a population. The term "population" is used to refer to all the members of a single species inhabiting a given geographic area at a given time. The species *Homo sapiens* (humans) inhabiting the geographic area of New York State constitutes such a population.

Wrong Choices Explained

(1), (3), (4) All the *vertebrates*, *plant and animal species*, or *flowering plants* inhabiting New York State are much too broad groupings to represent a population since each of these represents multiple species.

9. **4** *Temperature and oxygen content* are examples of abiotic factors in the environment.

Wrong Choices Explained

(1), (2), (3) Each of these distracters references the measurement of living things, representing biotic factors. These biotic factors include *scavengers*, *green plants and snails*, and *fish*.

10. **3** When green plants are eaten by primary consumers that are in turn eaten by secondary consumers, energy is passed from level to level in the *food chain*.

   Wrong Choices Explained
   (1) *The process of photosynthesis* is involved in the food chain only to the extent that the process is responsible for capturing light energy and using it to manufacture the substances used as food by the consumers.
   (2) *Natural selection* describes the mechanism by which favorable variations within a species are thought to be selected by natural forces to increase in frequency (and by which unfavorable traits are gradually reduced in frequency) in a species population. The term does not relate to this question.
   (4) *Ecological succession* refers to the changes that occur in ecological communities over long periods of time. The term does not relate to this question.

11. **2** The *nitrogen cycle* depends primarily on various chemical and biological processes associated with the metabolism of nitrogen compounds. Plant and animal nitrogenous wastes are processed by a variety of soil bacteria specialized for this task. None of the processes mentioned deal primarily with the metabolism of nitrogen compounds.

   Wrong Choices Explained
   (1), (4) The *carbon-hydrogen-oxygen cycle* is very much associated with the processes of photosynthesis and respiration, which function to recycle these elements through the environment as complementary biological processes.
   (3) The *water cycle* depends on the physical processes of transpiration, evaporation, and condensation to operate in nature.

12. **3** The tapeworm's association with the human is harmful to the human because nutrients are removed by the tapeworm before they can be absorbed by the intestine. Therefore the relationship is one of *parasitism* (+, −).

   Wrong Choices Explained
   (1) *Commensalism* is a symbiotic relationship in which the host organism is not harmed but the other organism is benefited (0, +).
   (2) *Mutualism* is a symbiotic relationship in which both organisms in the relationship are benefited (+, +).

13. **2** The bacteria use the legume as a site from which to carry on their metabolic activities; the legume utilizes nutrients from the bacteria's activities. This relationship can be characterized as *mutualism* since both organisms in the relationship are benefited (+, +).

Wrong Choices Explained
   (1), (3) See question 12.

14. **3** The flea's association with the dog is harmful to the dog because blood is removed from the dog by the flea, making it unavailable for the dog to use. Therefore, the relationship is one of *parasitism* (+, –).

Wrong Choices Explained
   (1), (2) See question 12.

15. **4** *Lichens* → *grasses* → *shrubs* → *trees* represents a typical ecological succession from bare rock. Bare rock succession requires a specialized type of pioneer organism. Lichens are known for their ability to survive on bare rock and to begin the conversion of rock into soil. Lichens are succeeded by grasses that help soil building by adding organic matter to the soil. Shrubs with shallow root structures are next, followed by trees with deeper root structures.

Wrong Choices Explained
   (1), (2), (3) These distracters all contain incorrect sequences of succession communities.

16. **4** This ecosystem will be self-sustaining if *materials cycle between the organisms labeled A and the organisms labeled B*. Organisms labeled *A* include the animals illustrated in the diagram; organisms labeled *B* include the plants. In any balanced ecosystem, materials such as carbon, hydrogen, and oxygen cycle between these two types of organisms.

Wrong Choices Explained
   (1), (2) If *the organisms labeled A outnumber the organisms labeled B* or *the organisms labeled A are equal in number to the organisms labeled B* the entire ecosystem would be disrupted. In any stable ecosystem, the number and biomass of producer organisms must be greater than that of consumers.
   (3) If *the type of organisms represented by B are eliminated*, this would represent an extreme condition that would lead to the total collapse of the ecosystem.

17. **2** *Abiotic factors* (nonliving factors) in the environment are necessary for the survival of living things. The abiotic materials and energy sources mentioned in the question are all essential to the survival of the plant as it carries out its life functions.

Wrong Choices Explained

(1) *Biotic factors* important to plants might include other living plant and animal species in the plant's environment. No biotic factors are mentioned in the statement.

(3), (4) *Symbiotic relationships* and *carnivore-herbivore relationships* always involve other living things and, as such, are biotic relationships.

18. **3** The statement *decomposers release a material that is acted upon by other organisms* is best represented in the diagram. The diagram shows various types of bacteria taking in nitrogen-containing compounds and releasing others after having acted upon them internally. Some of these nitrogen-containing materials are in turn acted upon by other types of organisms, such as plants and animals.

Wrong Choices Explained

(1) The statement *respiration and photosynthesis are interrelated* is not supported by information given in the diagram. No reference to these life processes is made in the diagram nor are their chemical reactants or products shown.

(2) The statement *transpiration and condensation are related to the water cycle* is not supported by information given in the diagram. No reference to these abiotic processes is made in the diagram.

(4) The statement *predators and their prey are involved in many interactions* is not supported by information given in the diagram. No reference to these nutritional relationships is made in the diagram.

19. **2** The statement *stored energy decreases from consumer 2 to consumer 3* is correct. In any environment, the total available energy decreases with each successive step of the food chain, energy being lost as heat and motion at each exchange of food matter between levels. The biomass pyramid represents this phenomenon.

Wrong Choices Explained

(1) The statement *the producer organisms contain the least amount of stored energy* is not correct. In fact, the producer level of the biomass pyramid contains the most energy of any level.

(3) The statement *consumer 3 contains the greatest amount of stored energy* is not correct. In fact, the top consumer level of the biomass pyramid contains the least energy of any level.

(4) The statement *stored energy increases from the producer to consumer 1* is not correct. In fact, the total amount of stored energy decreases in the biomass pyramid from producer to any consumer level.

20. **2** The statement *it persists until the environment changes* is correct. The climax stage is the the final stage of succession. It is characterized

by a self-perpetuating community of plant and animal species. When the environment changes drastically and the climax community is disturbed, the process of succession may begin again.

Wrong Choices Explained

(1) The statement *it changes rapidly* is incorrect. In fact, the climax community displays a marked stability over long periods of time.

(3) The statement *it is the first community to inhabit an area* is incorrect. The first community to inhabit an area is known as the pioneer community.

(4) The statement *it consists entirely of plants* is not correct. Stable climax communities are characterized by the diversity of plant and animal life that inhabit them.

21. **4** Of the choices given, *an increase in sediment, fallen leaves, and tree limbs accumulating in the bottom of the pond* would most likely lead to terrestrial succession. As these materials build up on the pond bottom, pond water is displaced and becomes gradually shallower. Eventually, terrestrial plants will gain a foothold on this moist organic mat and terrestrial succession will begin.

Wrong Choices Explained

(1) *A decrease in the number of suspended particles in the pond water* would not tend to speed the rate of terrestrial succession of a pond. The sedimentation rate would decrease and water displacement would slow.

(2) *An increase in current velocity of the pond water* would not tend to speed the rate of sediment buildup in the pond. Suspended particles and accumulated sediment could be more readily washed away.

(3) *A decrease in the number of diverse organisms in the shallow water of the pond* would signal a decline in biological activity in the pond. This might lead to a decrease in the rate of organic matter deposition in the pond, thereby slowing the rate of filling and the rate of terrestrial succession.

22. **2** *Competition between the squirrels* is the most likely result when resources are limited. This may result in a reduction in the squirrel population over time to bring the natural community back into proper balance.

Wrong Choices Explained

(1) *An increase in the number of squirrels* is not a likely result of this situation. In fact, a decrease in the population is likely to occur.

(3) *Increased habitats for the squirrels* is not a likely result of this situation. The habitat is limited to the areas capable of supporting the population. These areas will be unlikely to increase in the short term.

(4) *A greater diversity of food for the squirrels* is not a likely result of this situation. The squirrels' food requirements are most likely limited to relatively few choices. It is unlikely that the squirrels would be capable of changing them at will.

23. **4** The *mouse, deer, and cricket* are the primary consumers shown in the diagram. Each of these organisms is shown drawing its nutrition from plant matter, which is the main characteristic defining a primary consumer.

Wrong Choices Explained
(1), (2), (3) The *mouse, snake, and hawk*, the *snake, hawk, and frog*, and the *cricket, frog,and deer* are all combinations that contain organisms that are not primary consumers. The diagram shows that snakes consume mice, hawks consume mice and frogs, and frogs consume crickets.

24. **1** *X—biotic factors; Y—abiotic factors* is the information that belongs in areas *X* and *Y*. Biotic factors are those relating to the presence and activities of living things in the environment. Abiotic factors are those relating to the nonliving portion of the environment.

Wrong Choices Explained
(2) *X—ecological relationships; Y—biotic relationships* is not a correct response. The subcategories of area *Y* are clearly abiotic, not biotic, factors.
(3) *X—abiotic factors; Y—interacting populations* is not a correct response. The subcategories of area *X* are clearly biotic, not abiotic, factors.
(4) *X—energy flow; Y—biotic factors* is not a correct response. The sub-categories of area *Y* are clearly abiotic, not biotic, factors.

25. **3** *Parasitism* best describes the relationship between the dogwood and the fungus. In this relationship, the parasite (fungus) is benefited (+) and the host organism (dogwood) is harmed (–).

Wrong Choices Explained
(1) *Commensalism* is a type of symbiosis in which one organism is helped (+) and the host organism is neither helped nor harmed (0).
(2) *Mutualism* is a type of symbiosis in which one organism is helped (+) and the host organism is helped as well (+).
(4) *Saprophytism* refers to plants and fungi that depend on decaying organic matter for their food supply.

26. Sample essay: *"The organisms in this food web interact on many different nutritional levels. Producers absorb the Sun's energy and convert it to food energy. The plants are consumed by herbivores, which are then consumed by carnivores, including the wolf (or worms or fish). The energy is absorbed by the grass, is passed to the deer when the deer eats grass, and eventually reaches the wolf when the wolf consumes the deer. Decomposers are important in the food web because they recycle the materials that make up the bodies of dead organisms so that new organisms can use them."*

"If the wolf population decreases, then the deer population will increase because no predator is present to limit it. As the deer population increases, its members will consume the grass at a much faster rate. After a year, the grass population will decrease as a result."

"When the farmer sprays pesticides near the pond, the pesticides will wash into the pond. Their presence there will pollute the water, killing beneficial insect and other species. The pesticide-contaminated insects will be consumed by the fish. As this occurs, the pesticide will build up in the bodies of the fish until they die or become inedible to humans."

(NOTE: Any essay that correctly addresses the points required in the question is acceptable.)

Question Set 2.10 (p. 242)

1. **1** The use of *biological controls of insect pests* is one of the positive effects humans have had on the environment in the past several years. In this example, the ladybugs and praying mantises represent nonharmful insects (biological controls) that prey on harmful ones.

Wrong Choices Explained

(2) This choice is a nonsense distracter since scientists do not refer to insect pests as being *exploited*.

(3) The term *abiotic control* has no explicit meaning in the context of controlling insect pests.

(4) *Biocides* are chemical substances toxic to living things. Their use in the environment to control insect pests can have unintended, severe negative consequences.

2. **2** *Preservation of species* is a major positive outcome of the measures listed in this question. Wildlife refuges enable wild animals to live in natural surroundings and to thrive without interference by humans. Game laws help to ensure that now common species will not be overhunted in the future.

Wrong Choices Explained

(1), (4) The *use of biocides* and *exploitation of species* are considered major negative aspects of human involvement in the environment. They would not result from wildlife refuges and game laws.

(3) The *use of biological controls* is a positive aspect of human involvement in the environment. However, it would not result from wildlife refuges and game laws.

3. **1** *Protection of natural habitat* is a positive effect that humans have had on their environment in the past several years. This practice has brought several species, including the American bald eagle, back from the brink of extinction.

Wrong Choices Explained

(2) *Importation of food into their nesting sites* is an activity that would have little positive value for the preservation of a wild species. In fact, the less interference by humans in the lives of wild species, the better for the species' ultimate survival.

(3) *Preservation of other eagle species that occupy the same niche* would have a negative impact on the survival of this species since interspecies competition would be made much more significant by this change.

(4) *Increased use of pesticides* would have a negative impact since many pesticides can interfere with the reproductive cycles of eagles and other bird species.

4. **2** *Using insecticides to kill insects that compete with humans for food* is a human activity that would be more likely to have a negative impact on the environment than the other three. Insecticides contain chemicals that may be toxic to beneficial insects, amphibians, fish, and birds, among other animal species. If these chemicals enter the food chain, they may build up in concentration and eventually pose a threat to top-level feeders, including humans.

Wrong Choices Explained

(1) *Using reforestation and cover cropping to control soil erosion* is generally considered a positive human activity since its goal is the stabilization of soils that might otherwise erode into streams and be lost as a resource.

(3) *Developing research aimed toward the preservation of endangered species* is generally considered a positive human activity since its goal is the maintenance of biodiversity in the environment. Biodiversity is thought to be of vital importance in maintaining the overall health of the environment.

(4) *Investigating the use of biological controls for pests* is generally considered a positive human activity since its goal is using the natural enemies of these pests instead of a chemical biocide to control the pests' numbers.

5. **4** *Habitat destruction* is most likely to endanger species. Some organisms depend on the conditions available in relatively limited habitat zones. When these zones are disturbed or destroyed, the organisms cannot find suitable conditions elsewhere, cannot breed, and may be eliminated altogether in a very short time.

Wrong Choices Explained

(1) *Cover cropping* is a positive human effect on the environment. It involves planting crops on cultivated land in order to control soil erosion.

(2) *Use of pollution controls* is a positive human effect on the environment. It involves using techniques to limit the amount of pollution that enters our air, water, and soil resources.

(3) *Use of erosion controls* is a positive human effect on the environment. It involves using a variety of techniques to limit the amount of soil lost to erosion.

6. **1** *Retinoid by-products may cause fetal deformities* is the reason pregnant women are advised not to use medicines containing retinoids. This information is found in the third paragraph of the passage.

Wrong Choices Explained

(2) *Retinoid by-products cause parasites to invade developing frogs* is not supported by the passage. Parasites are mentioned in the fourth paragraph of the passage as being a possible alternate cause for the frog deformities referenced in the first and second paragraphs of the passage. The passages makes no reference to retinoids causing parasites to invade developing frogs.

(3) *Retinoid by-products mimic the effects of pesticides on fetal tissue* is not supported by the passage. The key word "mimic" is found in the second and third paragraphs of the passage. In the third paragraph, the passage notes that certain pesticides can mimic retinoids, but no direct reference to the effects of pesticides on fetal tissue is made.

(4) *Retinoid by-products reduce abnormalities in maternal tissue* is not supported by the passage. Nowhere in the passage is this statement made. In fact, the third paragraph states that pregnant women are warned not to use retinoids because of their possible deforming effect on fetal tissues.

7. **3** *Other animals in the ponds containing deformed frogs did not have abnormalities* is the reason some scientists argue that pesticides may not be the cause of the frog deformities. Instead, some scientists believe that the deformities may be caused by parasites of the frogs. This information is found in the fourth paragraph of the passage.

Wrong Choices Explained

(1) *Pesticide use has decreased over the last four years* is not supported by the passage. No statement to this effect is made in the passage.

(2) *New pesticides are used in skin-care products* is not supported by the passage. No statement to this effect is made in the passage.

(4) *Laboratory experiments have determined that a pesticide can mimic retinoids* is supported by the passage in the third paragraph. However, this fact is not used in the passage as supporting evidence that pesticides may not be the cause of the frog deformities.

8. **2** *Parasites that affect frogs usually do not affect fish* is a possible reason for the absence of deformed fish in the ponds that contained deformed frogs. While this is not stated explicitly in the passage, an inference to this effect may be made from the information supplied in the fourth paragraph of the passage.

Wrong Choices Explained

(1) *Fish can swim away from chemicals introduced into the pond* is not a possible reason for the absence of deformed fish in the ponds that contained deformed frogs. Any water-soluble chemical introduced into an aquatic environment will diffuse through the water until it reaches equal concentration everywhere in the system. Therefore, fish cannot swim away from it.

(3) *Fish cannot develop deformities* is not a possible reason for the absence of deformed fish in the ponds that contained deformed frogs. Any organism whose embryonic cells are assaulted by a mutagenic chemical or parasite can develop deformities.

(4) *Frogs and fish are not found in the same habitat* is not a possible reason for the absence of deformed fish in the ponds that contained deformed frogs. In fact, fish and frogs are frequently found inhabiting the same aquatic habitats such as ponds, streams, and marshes.

9. **4** *Factors that affect frogs may also affect other organisms* is an inference that can be made from the information in the passage. The fact that pregnant women are warned against using retinoids is evidence that scientists understand that chemical substances can be harmful to all living things, especially in their developmental stages.

Wrong Choices Explained

(1) *Only a few isolated incidents of frog deformities have been observed* is not an inference that should be made from the information in this passage. In fact, the passage leaves the impression that many such deformities have been noted and studied over the past 4 years.

(2) *If frog parasites are controlled, all frog deformities will stop* is not an inference that should be made from the information in this passage. Parasites as a possible cause of frog deformities has not been firmly established by scientific research. Several possible causes for these deformities are being studied.

(3) *Deformities in frogs are of little significance* is not an inference that should be made from the information in this passage. In fact, scientists are sufficiently concerned about this phenomenon that significant resources are being put into researching the possible causes of the frog deformities and their relationship to human birth defects. This research has led scientists to warn pregnant women not to use substances containing retinoids or their by-products.

10. *Some pesticides mimic retinoids, which can cause deformities inside growing animals such as frogs.*

    (NOTE: Any correct, complete-sentence response is acceptable.)

11. Sample essay: *"The Brazilian Amazon River rain forest is being destroyed by logging and agriculture. Loggers come in and strip the forest of trees, which are sold for timber. Farmers then follow and plant crops in the cleared land, selling the crops they produce. The soil in these areas is too thin and unproductive to serve as farmland for long, so new areas are continually being cleared."*

    "When this happens to an area of the rain forest, the environment is changed drastically. Trees are removed, which exposes the soil to drying. Understory plants cannot survive the changed conditions and die out rapidly. Animals that depend on the deep forest habitat for food and shelter are forced to migrate to other forest habitats or die from exposure or lack of food. Humans are affected because the soils in these areas quickly erode and cause silting of the rivers."

    "Steps can be taken to preserve these areas from further destruction. Laws can be passed to limit the amount of forest that can be destroyed in this way in any one year. Parks and preserves can be established to ensure that sizable tracts of this habitat are never destroyed."

    (NOTE: Any essay that correctly addresses the points required in the question is acceptable.)

12. *Chemical herbicides may cause air or water pollution that could kill species other than the loosestrife.* OR *The beetle is a natural enemy of the loosestrife and so less likely to cause habitat disruption than herbicides.*
    (NOTE: Any correct, complete-sentence response is acceptable.)

13. *The beetles might multiply and compete with native insect species.* OR *It is possible that the beetles could also consume native species of plants such as the cattail.*
    (NOTE: Any correct, complete-sentence response is acceptable.)

14. *The spray may reach other areas and harm people, pets, or other animals in that area.* OR *The spray could kill beneficial insects.*
    (NOTE: Any correct, complete-sentence response is acceptable.)

15. *Predators or parasites of the mosquitoes could be released into the area to control them.* OR *Swamp areas could be drained to reduce the breeding areas.* OR *Sterilized male mosquitoes could be released to mate with females to reduce the rate of reproduction.*
    (NOTE: Any correct, complete-sentence response is acceptable.)

16. Positive: *Biological control (or draining or sterilization) means that no pesticide is released that could harm other species.*
    Negative: *Habitat modifications could be detrimental to other species.* OR *Predators or parasites released could harm other species or spread disease.*
    (NOTE: Any correct, complete-sentence response is acceptable.)

GLOSSARY OF BIOLOGICAL TERMS _____

abiotic factor Any of several nonliving, physical conditions that affect the survival of an organism in its environment.

absorption The process by which water and dissolved solids, liquids, and gases are taken in by the cell through the cell membrane.

accessory organ In human beings, any organ that has a digestive function but is not part of the food tube. (See **liver**; **gallbladder**; **pancreas**.)

acid A chemical that releases hydrogen ion (H^+) in solution with water.

acid precipitation A phenomenon in which there is thought to be an interaction between atmospheric moisture and the oxides of sulfur and nitrogen that results in rainfall with low pH values.

active immunity The immunity that develops when the body's immune system is stimulated by a disease organism or a vaccination.

active site The specific area of an enzyme molecule that links to the substrate molecule and catalyzes its metabolism.

active transport A process by which materials are absorbed or released by cells against the concentration gradient (from low to high concentration) with the expenditure of cell energy.

adaptation Any structural, biochemical, or behavioral characteristic of an organism that helps it to survive potentially harsh environmental conditions.

addition A type of chromosome mutation in which a section of a chromosome is transferred to a homologous chromosome.

adenine A nitrogenous base found in DNA and RNA molecules.

adenosine triphosphate (ATP) An organic compound that stores respiratory energy in the form of chemical-bond energy for transport from one part of the cell to another.

adrenal cortex A portion of the adrenal gland that secretes steroid hormones that regulate various aspects of blood composition.

adrenal gland An endocrine gland that produces several hormones, including adrenaline. (See **adrenal cortex**; **adrenal medulla**.)

adrenal medulla A portion of the adrenal gland that secretes the hormone adrenaline, which regulates various aspects of the body's metabolic rate.

adrenaline A hormone of the adrenal medulla that regulates general metabolic rate, the rates of heartbeat and breathing, and the conversion of glycogen to glucose.

aerobic phase of respiration The reactions of aerobic respiration in which two pyruvic acid molecules are converted to six molecules of water and six molecules of carbon dioxide.

aerobic respiration A type of respiration in which energy is released from organic molecules with the aid of oxygen.

aging A stage of postnatal development that involves differentiation, maturation, and eventual deterioration of the body's tissues.

air pollution The addition, due to technological oversight, of some unwanted factor (for example, chemical oxides, hydrocarbons, particulates) to our air resources.

albinism A condition, controlled by a single mutant gene, in which the skin lacks the ability to produce skin pigments.

alcoholic fermentation A type of anaerobic respiration in which glucose is converted to ethyl alcohol and carbon dioxide.

allantois   A membrane that serves as a reservoir for wastes and as a respiratory surface for the embryos of many animal species.

allele   One of a pair of genes that exist at the same location on a pair of homologous chromosomes and exert parallel control over the same genetic trait.

allergy   A reaction of the body's immune system to the chemical composition of various substances.

alveolus   One of many air sacs within the lung that function to absorb atmospheric gases and pass them on to the bloodstream.

amino acid   An organic compound that is the component unit of proteins.

amino group   A chemical group having the formula $-NH_2$ that is found as a part of all amino acid molecules.

ammonia   A type of nitrogenous waste with high solubility and high toxicity.

amniocentesis   A technique for detecting genetic disorders in unborn human beings in which a small amount of amniotic fluid is removed and the chromosome content of its cells analyzed. (See **karyotyping**.)

amnion   A membrane that surrounds the embryo in many animal species and contains a fluid to protect the developing embryo from mechanical shock.

amniotic fluid   The fluid within the amnion membrane that bathes the developing embryo.

amylase   An enzyme specific for the hydrolysis of starch.

anaerobic phase of respiration   The reactions of aerobic respiration in which glucose is converted to two pyruvic acid molecules.

anaerobic respiration   A type of respiration in which energy is released from organic molecules without the aid of oxygen.

anal pore   The egestive organ of the paramecium.

anemia   A disorder of the human transport system in which the ability of the blood to carry oxygen is impaired usually because of reduced numbers of red blood cells.

angina pectoris   A disorder of the human transport system in which chest pain signals potential damage to the heart muscle due to narrowing of the opening of the coronary artery.

animal   One of the five biological kingdoms; it includes multicellular organisms whose cells are not bounded by cell walls and that are incapable of photosynthesis (for example, human being).

Annelida   A phylum of the animal kingdom whose members (annelids) include the segmented worms (for example, earthworm).

antenna   A receptor organ found in many arthropods (for example, grasshopper), which is specialized for detecting chemical stimuli.

anther   The portion of the stamen that produces pollen.

antibody   A chemical substance produced in response to the presence of a specific antigen that neutralizes that antigen in the immune response.

antigen   A chemical substance, usually a protein, recognized by the immune system as a foreign invader and that is neutralized by a specific antibody.

anus   The organ of egestion of the digestive tract.

aorta   The principal artery carrying blood from the heart to the body tissues.

aortic arches   A specialized part of the earthworm's transport system that serves as a pumping mechanism for the blood fluid.

apical meristem   A plant growth region located at the tip of the root or tip of the stem.

appendicitis   A disorder of the human digestive tract in which the appendix becomes inflamed as a result of bacterial infection.

aquatic biome   An ecological biome composed of many different water environments.

artery A thick-walled blood vessel that carries blood away from the heart under pressure.

arthritis A disorder of the human locomotor system in which skeletal joints become inflamed, swollen, and painful.

Arthropoda A phylum of the animal kingdom whose members (arthropods) have bodies with chitinous exoskeletons and jointed appendages (for example, grasshopper).

artificial selection A technique of plant/animal breeding in which individual organisms displaying desirable characteristics are chosen for breeding purposes.

asexual reproduction A type of reproduction in which new organisms are formed from a single parent organism.

asthma A disorder of the human respiratory system in which the respiratory tube becomes constricted by swelling brought on by some irritant.

atrium In human beings, one of the two thin-walled upper chambers of the heart that receive blood.

autonomic nervous system A subdivision of the peripheral nervous system consisting of nerves associated with automatic functions (for example, heartbeat, breathing).

autosome One of several chromosomes present in the cell that carry genes controlling "body" traits not associated with primary and secondary sex characteristics.

autotroph An organism capable of carrying on autotrophic nutrition. Self-feeder.

autotrophic nutrition A type of nutrition in which organisms manufacture their own organic foods from inorganic raw materials.

auxin A biochemical substance, a plant hormone, produced by plants that regulates growth patterns.

axon An elongated portion of a neuron that conducts nerve impulses, usually away from the cell body of the neuron.

base A chemical that releases hydroxyl ion (OH^-) in solution with water.

bicarbonate ion The chemical formed in the blood plasma when carbon dioxide is absorbed from body tissues.

bile In human beings, a secretion of the liver that is stored in the gallbladder and that emulsifies fats.

binary fission A type of cell division in which mitosis is followed by equal cytoplasmic division.

binomial nomenclature A system of naming used in biological classification that consists of the genus and species names (for example, *Homo sapiens*).

biocide use The use of pesticides that eliminate one undesirable organism but that have, due to technological oversight, unanticipated effects on beneficial species as well.

biological controls The use of natural enemies of various agricultural pests for pest control, thereby eliminating the need for biocide use—a positive aspect of human involvement with the environment.

biomass The total mass of living material present at the various trophic levels in a food chain.

biome A major geographical grouping of similar ecosystems, usually named for the climax flora in the region (for example, Northeast Deciduous Forest).

biosphere The portion of the earth in which living things exist, including all land and water environments.

biotic factor Any of several conditions associated with life and living things that affect the survival of living things in the environment.

birth In placental mammals, a stage of embryonic development in which the baby passes through the vaginal canal to outside of the mother's body.

blastula In certain animals, a stage of embryonic development in which the embryo resembles a hollow ball of undifferentiated cells.

blood The complex fluid tissue that functions to transport nutrients and respiratory gases to all parts of the body.

blood typing An application of the study of immunity in which the blood of a person is characterized by its antigen composition.

bone A tissue that provides mechanical support and protection for bodily organs and levers for the body's locomotive activities.

Bowman's capsule A cup-shaped portion of the nephron responsible for filtering of soluble blood components.

brain An organ of the central nervous system responsible for regulating conscious and much unconscious activity in the body.

breathing A mechanical process by which air is forced into the lung by means of muscular contraction of the diaphragm and rib muscles.

bronchiole One of several subdivisions of the bronchi that penetrate the lung interior and terminate in alveoli.

bronchitis A disorder of the human respiratory system in which the bronchi become inflamed.

bronchus One of the two major subdivisions of the breathing tube; the bronchi are ringed with cartilage and conduct air from the trachea to the lung interior.

Bryophyta A phylum of the plant kingdom that consists of organisms lacking vascular tissues (for example, moss).

budding A type of asexual reproduction in which mitosis is followed by unequal cytoplasmic division.

bulb A type of vegetative propagation in which a plant bulb produces new bulbs that may be established as independent organisms with identical characteristics.

cambium The lateral meristem tissue in woody plants responsible for annual growth in stem diameter.

cancer Any of a number of conditions characterized by rapid, abnormal, and uncontrolled division of affected cells.

capillary A very small, thin-walled blood vessel that connects an artery to a vein and through which all absorption into the blood fluid occurs.

carbohydrate An organic compound composed of carbon, hydrogen, and oxygen in a 1:2:1 ratio (for example, $C_6H_{12}O_6$).

carbon-fixation reactions A set of biochemical reactions in photosynthesis in which hydrogen atoms are combined with carbon and oxygen atoms to form PGAL and glucose.

carbon 14 A radioactive isotope of carbon used to trace the movement of carbon in various biochemical reactions, and also used in the carbon dating of fossils.

carbon-hydrogen-oxygen cycle A process by which these three elements are made available for use by other organisms through the chemical reactions of respiration and photosynthesis.

carboxyl group A chemical group having the formula —COOH and found as part of all amino acid and fatty acid molecules.

cardiac muscle A type of muscle tissue in the heart and arteries associated with the rhythmic nature of the pulse and heartbeat.

cardiovascular disease In human beings, any disease of the circulatory organs.

carnivore A heterotrophic organism that consumes animal tissue as its primary source of nutrition. (See **secondary consumer**.)

carrier An individual who, though not expressing a particular recessive trait, carries this gene as part of his/her heterozygous genotype.

carrier protein A specialized molecule embedded in the cell membrane that aids the movement of materials across the membrane.

cartilage A flexible connective tissue found in many flexible parts of the body (for example, knee); common in the embryonic stages of development.

catalyst Any substance that speeds up or slows down the rate of a chemical reaction. (See **enzyme**.)

cell plate A structure that forms during cytoplasmic division in plant cells and serves to separate the cytoplasm into two roughly equal parts.

cell theory A scientific theory that states, "All cells arise from previously existing cells" and "Cells are the unit of structure and function of living things."

cell wall A cell organelle that surrounds and gives structural support to plant cells; cell walls are composed of cellulose.

central nervous system The portion of the vertebrate nervous system that consists of the brain and the spinal cord.

centriole A cell organelle found in animal cells that functions in the process of cell division.

centromere The area of attachment of two chromatids in a double-stranded chromosome.

cerebellum The portion of the human brain responsible for the coordination of muscular activity.

cerebral hemorrhage A disorder of the human regulatory system in which a broken blood vessel in the brain may result in severe dysfunction or death.

cerebral palsy A disorder of the human regulatory system in which the motor and speech centers of the brain are impaired.

cerebrum The portion of the human brain responsible for thought, reasoning, sense interpretation, learning, and other conscious activities.

cervix A structure that bounds the lower end of the uterus and through which sperm must pass in order to fertilize the egg.

chemical digestion The process by which nutrient molecules are converted by chemical means into a form usable by the cells.

chemosynthesis A type of autotrophic nutrition in which certain bacteria use the energy of chemical oxidation to convert inorganic raw materials to organic food molecules.

chitin A polysaccharide substance that forms the exoskeleton of the grasshopper and other arthropods.

chlorophyll A green pigment in plant cells that absorbs sunlight and makes possible certain aspects of the photosynthetic process.

chloroplast A cell organelle found in plant cells that contains chlorophyll and functions in photosynthesis.

Chordata A phylum of the animal kingdom whose members (chordates) have internal skeletons made of cartilage and/or bone (for example, human being).

chorion A membrane that surrounds all other embryonic membranes in many animal species, protecting them from mechanical damage.

chromatid One strand of a double-stranded chromosome.

chromosome mutation An alteration in the structure of a chromosome involving many genes. (See **nondisjunction**; **translocation**; **addition**; **deletion**.)

cilia Small, hairlike structures in paramecia and other unicellular organisms that aid in nutrition and locomotion.

classification A technique by which scientists sort, group, and name organisms for easier study.

cleavage A series of rapid mitotic divisions that increase cell number in a developing embryo without a corresponding increase in cell size.

climax community A stable, self-perpetuating community that results from an ecological succession.

cloning A technique of genetic investigation in which undifferentiated cells of an organism are used to produce new organisms with the same set of traits as the original cells.

closed-transport system A type of circulatory system in which the transport fluid is always enclosed within blood vessels (for example, earthworm, human).

clot A structure that forms as a result of enzyme-controlled reactions following the rupturing of a blood vessel and serves as a plug to prevent blood loss.

codominance A type of intermediate inheritance that results from the simultaneous expression of two dominant alleles with contrasting effects.

codon See **triplet codon**.

Coelenterata A phylum of the animal kingdom whose members (coelenterates) have bodies that resemble a sack (for example, hydra, jellyfish).

coenzyme A chemical substance or chemical subunit that functions to aid the action of a particular enzyme. (See **vitamin**.)

cohesion A force binding water molecules together that aids in the upward conduction of materials in the xylem.

commensalism A type of symbiosis in which one organism in the relationship benefits and the other is neither helped nor harmed.

common ancestry A concept central to the theory of evolution that postulates that all organisms share a common ancestry whose closeness varies with the degree of shared similarity.

community A level of biological organization that includes all of the species populations inhabiting a particular geographic area.

comparative anatomy The study of similarities in the anatomic structures of organisms, and their use as an indicator of common ancestry and as evidence of organic evolution.

comparative biochemistry The study of similarities in the biochemical make-ups of organisms, and their use as an indicator of common ancestry and as evidence of organic evolution.

comparative cytology The study of similarities in the cell structures of organisms, and their use as an indicator of

common ancestry and as evidence of organic evolution.

comparative embryology The study of similarities in the patterns of embryological development of organisms, and their use as an indicator of common ancestry and as evidence of organic evolution.

competition A condition that arises when different species in the same habitat attempt to use the same limited resources.

complete protein A protein that contains all eight essential amino acids.

compound A substance composed of two or more different kinds of atom (for example, water: H_2O).

compound light microscope A tool of biological study capable of producing a magnified image of a biological specimen by using a focused beam of light.

conditioned behavior A type of response that is learned but that becomes automatic with repetition.

conservation of resources The development and application of practices to protect valuable and irreplaceable soil and mineral resources—a positive aspect of human involvement with the environment.

constipation A disorder of the human digestive tract in which fecal matter solidifies and becomes difficult to egest.

consumer Any heterotrophic animal organism (for example, human being).

coronary artery An artery that branches off the aorta to feed the heart muscle.

coronary thrombosis A disorder of the human transport system in which the heart muscle becomes damaged as a result of blockage of the coronary artery.

corpus luteum A structure resulting from the hormone-controlled transformation of the ovarian follicle that produces the hormone progesterone.

corpus luteum stage A stage of the menstrual cycle in which the cells of the follicle are transformed into the corpus luteum under the influence of luteinizing hormone (LH).

cotyledon A portion of the plant embryo that serves as a source of nutrition for the young plant before photosynthesis begins.

cover cropping A proper agricultural practice in which a temporary planting (cover crop) is used to limit soil erosion between seasonal plantings of main crops.

crop A portion of the digestive tract of certain animals that stores food temporarily before digestion.

cross-pollination A type of pollination in which pollen from one flower pollinates flowers of a different plant of the same species.

crossing-over A pattern of inheritance in which linked genes may be separated during synapsis in the first meiotic division, when sections of homologous chromosomes may be exchanged.

cuticle A waxy coating that covers the upper epidermis of most leaves and acts to help the leaf retain water.

cutting A technique of plant propagation in which vegetative parts of the parent plant are cut and rooted to establish new plant organisms with identical characteristics.

cyclosis The circulation of the cell fluid (cytoplasm) within the cell interior.

cyton The cell body of the neuron, which generates the nerve impulse.

cytoplasm The watery fluid that provides a medium for the suspension of organelles within the cell.

cytoplasmic division The separation of daughter nuclei into two new daughter cells.

cytosine A nitrogenous base found in both DNA and RNA molecules.

daughter cell A cell that results from mitotic cell division.

daughter nucleus One of two nuclei that form as a result of mitosis.

deamination A process by which amino acids are broken down into their component parts for conversion into urea.

death The irreversible cessation of bodily functions and cellular activities.

deciduous A term relating to broad-leaved trees that shed their leaves in the fall.

decomposer Any saprophytic organism that derives its energy from the decay of plant and animal tissues (for example, bacteria of decay, fungus); the final stage of a food chain.

decomposition bacteria In the nitrogen cycle, bacteria that break down plant and animal protein and produce ammonia as a by-product.

dehydration synthesis A chemical process in which two organic molecules may be joined after removing the atoms needed to form a molecule of water as a by-product.

deletion A type of chromosome mutation in which a section of a chromosome is separated and lost.

dendrite A cytoplasmic extension of a neuron that serves to detect an environmental stimulus and carry an impulse to the cell body of the neuron.

denitrifying bacteria In the nitrogen cycle, bacteria that convert excess nitrate salts into gaseous nitrogen.

deoxygenated blood Blood that has released its transported oxygen to the body tissues.

deoxyribonucleic acid (DNA) A nucleic acid molecule known to be the chemically active agent of the gene; the fundamental hereditary material of living organisms.

deoxyribose A five-carbon sugar that is a component part of the nucleotide unit in DNA only.

desert A terrestrial biome characterized by sparse rainfall, extreme temperature variation, and a climax flora that includes cactus.

diabetes A disorder of the human regulatory system in which insufficient insulin production leads to elevated blood sugar concentrations.

diarrhea A disorder of the human digestive tract in which the large intestine fails to absorb water from the waste matter, resulting in watery feces.

diastole The lower pressure registered during blood pressure testing. (See **systole**.)

differentiation The process by which embryonic cells become specialized to perform the various tasks of particular tissues throughout the body.

diffusion A form of passive transport in which soluble substances are absorbed or released by cells.

digestion The process in which complex foods are broken down by mechanical or chemical means for use by the body.

dipeptide A chemical unit composed of two amino acid units linked by a peptide bond.

diploid chromosome number The number of chromosomes found characteristically in the cells (except gametes) of sexually reproducing species.

disaccharidase Any disaccharide-hydrolyzing enzyme.

disaccharide A type of carbohydrate known also as a double sugar; all disaccharides have the molecular formula $C_{12}H_{22}O_{11}$.

disjunction The separation of homologous chromosome pairs at the end of the first meiotic division.

disposal problems Problems, due to technological oversight, that result when commercial and technological activities produce solid and/or chemical wastes that must be disposed of.

dissecting microscope A tool of biological study that magnifies the image of a biological specimen up to 20 times normal size for purposes of gross dissection.

dominance A pattern of genetic inheritance in which the effects of a dominant allele mask those of a recessive allele.

dominant allele (gene) An allele (gene) whose effect masks that of its recessive allele.

double-stranded chromosome The two-stranded structure that results from chromosomal replication.

Down's syndrome In human beings, a condition characterized by mental and physical retardation that may be caused by the nondisjunction of chromosome number 21.

Drosophila The common fruit fly, an organism that has served as an object of genetic research in the development of the gene-chromosome theory.

ductless gland See **endocrine gland**.

ecology The science that studies the interactions of living things with each other and with the nonliving environment.

ecosystem The basic unit of study in ecology, including the plant and animal community in interaction with the nonliving environment.

ectoderm An embryonic tissue that differentiates into skin and nerve tissue in the adult animal.

effector An organ specialized to produce a response to an environmental stimulus; effectors may be muscles or glands.

egestion The process by which undigested food materials are eliminated from the body.

electron microscope A tool of biological study that uses a focused beam of electrons to produce an image of a biological specimen magnified up to 25,000 times its normal size.

element The simplest form of matter; an element is a substance (for example, nitrogen) made up of a single type of atom.

embryo An organism in the early stages of development following fertilization.

embryonic development A series of complex processes by which animal and plant embryos develop into adult organisms.

emphysema A disorder of the human respiratory system in which lung tissue deteriorates, leaving the lung with diminished capacity and efficiency.

emulsification A process by which fat globules are surrounded by bile to form fat droplets.

endocrine (ductless) gland A gland (for example, thyroid, pituitary) specialized for producing and secreting hormones directly into the bloodstream; such glands lack ducts.

endoderm An embryonic tissue that differentiates into the digestive and respiratory tract lining in the adult animal.

endoplasmic reticulum (ER) A cell organelle known to function in the transport of cell products from place to place within the cell.

environmental laws Federal, state, and local legislation enacted in an attempt to protect environmental resources—a positive aspect of human involvement with the environment.

enzymatic hydrolysis An enzyme-controlled reaction by which complex food molecules are broken down chemically into simpler subunits.

enzyme An organic catalyst that controls the rate of metabolism of a single type of substrate; enzymes are protein in nature.

enzyme-substrate complex A physical association between an enzyme molecule and its substrate within which the substrate is metabolized.

epicotyl A portion of the plant embryo that specializes to become the upper stem, leaves, and flowers of the adult plant.

epidermis The outermost cell layer in a plant or animal.

epiglottis In a human being, a flap of tissue that covers the upper end of the trachea during swallowing and prevents inhalation of food.

esophagus A structure in the upper portion of the digestive tract that conducts the food from the pharynx to the midgut.

essential amino acid An amino acid that cannot be synthesized by the human body but must be obtained by means of the diet.

estrogen A hormone, secreted by the ovary that regulates the production of female secondary sex characteristics.

evolution Any process of gradual change through time.

excretion The life function by which living things eliminate metabolic wastes from their cells.

exoskeleton A chitinous material that covers the outside of the bodies of most arthropods and provides protection for internal organs and anchorage for muscles.

exploitation of organisms Systematic removal of animals and plants with commercial value from their environments to sell them—a negative aspect of human involvement with the environment.

extensor A skeletal muscle that extends (opens) a joint.

external development Embryonic development that occurs outside the body of the female parent (for example, birds).

external fertilization Fertilization that occurs outside the body of the female parent (for example, fish).

extracellular digestion Digestion that occurs outside the cell.

fallopian tube See **oviduct**.

fatty acid An organic molecule that is a component of certain lipids.

fauna The animal species comprising an ecological community.

feces The semisolid material that results from the solidification of undigested foods in the large intestine.

fertilization The fusion of gametic nuclei in the process of sexual reproduction.

filament The portion of the stamen that supports the anther.

flagella Microscopic, whiplike structures found on certain cells that aid in locomotion and circulation.

flexor   A skeletal muscle that flexes (closes) a joint.

flora   The plant species comprising an ecological community.

flower   The portion of a flowering plant specialized for sexual reproduction.

fluid-mosaic model   A model of the structure of the cell membrane in which large protein molecules are thought to be embedded in a bilipid layer.

follicle   One of many areas within the ovary that serve as sites for the periodic maturation of ova.

follicle stage   The stage of the menstrual cycle in which an ovum reaches its final maturity under the influence of the hormone FSH.

follicle-stimulating hormone (FSH)   A pituitary hormone that regulates the maturation of and the secretion of estrogen by the ovarian follicle.

food chain   A series of nutritional relationships in which food energy is passed from producer to herbivore to carnivore to decomposer; a segment of a food web.

food web   A construct showing a series of interrelated food chains and illustrating the complex nutritional interrelationships that exist in an ecosystem.

fossil   The preserved direct or indirect remains of an organism that lived in the past, as found in the geologic record.

fraternal twins   In human beings, twin offspring that result from the simultaneous fertilization of two ova by two sperm; such twins are not genetically identical.

freshwater biome   An aquatic biome made up of many separate freshwater systems that vary in size and stability and may be closely associated with terrestrial biomes.

fruit   Any plant structure that contains seeds; a mechanism of seed dispersal.

Fungi   One of the five biological kingdoms; it includes organisms unable to manufacture their own organic foods (for example, mushroom).

gallbladder   An accessory organ that stores bile.

gallstones   A disorder of the human digestive tract in which deposits of hardened cholesterol lodge in the gallbladder.

gamete   A specialized reproductive cell produced by organisms of sexually reproducing species. (See **sperm**; **ovum**; **pollen**; **ovule**.)

gametogenesis   The process of cell division by which gametes are produced. (See **meiosis**; **spermatogenesis**; **oogenesis**.)

ganglion   An area of bunched nerve cells that acts as a switching point for nerve impulses traveling from receptors and to effectors.

garden pea   The research organism used by Mendel in his early scientific work in genetic inheritance.

gastric cecum   A gland in the grasshopper that secretes digestive enzymes.

gastrula   A stage of embryonic development in animals in which the embryo assumes a tube-within-a-tube structure and distinct embryonic tissues (ectoderm, mesoderm, endoderm) begin to differentiate.

gastrulation   The process by which a blastula becomes progressively more indented, forming a gastrula.

gene   A unit of heredity; a discrete portion of a chromosome thought to be responsible for the production of a single type of polypeptide; the factor responsible for the inheritance of a genetic trait.

gene frequency   The proportion (percentage) of each allele for a particular trait that is present in the gene pool of a population.

gene linkage   A pattern of inheritance in which genes located along the same chromosome are prevented from assorting independently but are linked together in their inheritance.

gene mutation   An alteration of the chemical nature of a gene that changes its ability to control the production of a polypeptide chain.

gene pool The sum total of all the inheritable genes for the traits in a given sexually reproducing population.

gene-chromosome theory A theory of genetic inheritance that is based on current understanding of the relationships between the biochemical control of traits and the process of cell division.

genetic counseling Clinical discussions concerning inheritance patterns that are designed to inform prospective parents of the potential for expression of a genetic disorder in their offspring.

genetic engineering The use of various techniques to move genes from one organism to another.

genetic screening A technique for the detection of human genetic disorders in which bodily fluids are analyzed for the presence of certain marker chemicals.

genome The total genetic makeup (DNA) of an organism.

genotype The particular combination of genes in an allele pair.

genus A level of biological classification that represents a subdivision of the phylum level; having fewer organisms with great similarity (for example, *Drosophila, Paramecium*).

geographic isolation The separation of species populations by geographic barriers, facilitating the evolutionary process.

geologic record A supporting item of evidence of organic evolution, supplied within the earth's rock and other geologic deposits.

germination The growth of the pollen tube from a pollen grain; the growth of the embryonic root and stem from a seed.

gestation The period of prenatal development of a placental mammal; human gestation requires approximately nine months.

gizzard A portion of the digestive tract of certain organisms, including the earthworm and the grasshopper in which food is ground into smaller fragments.

global warming A phenomenon characterized by a gradual increase in the average temperature of Earth that is thought to be the result of the addition of greenhouse gases to the atmosphere through industrialization and other human technologies.

glomerulus A capillary network lying within Bowman's capsule of the nephron.

glucagon A hormone, secreted by the islets of Langerhans, that regulates the release of blood sugar from stored glycogen.

glucose A monosaccharide produced commonly in photosynthesis and used by both plants and animals as a fuel in the process of respiration.

glycerol An organic compound that is a component of certain lipids.

glycogen A polysaccharide synthesized in animals as a means of storing glucose; glycogen is stored in the liver and in the muscles.

goiter A disorder of the human regulatory system in which the thyroid gland enlarges because of a deficiency of dietary iodine.

Golgi complex Cell organelles that package cell products and move them to the plasma membrane for secretion.

gonad An endocrine gland that produces the hormones responsible for the production of various secondary sex characteristics. (See **ovary**; **testis**.)

gout A disorder of the human excretory system in which uric acid accumulates in the joints, causing severe pain.

gradualism A theory of the time frame required for organic evolution that assumes that evolutionary change is slow, gradual, and continuous.

grafting A technique of plant propagation in which the stems of desirable plants are attached (grafted) to rootstocks of related varieties to produce new plants for commercial purposes.

grana The portion of the chloroplast within which chlorophyll molecules are concentrated.

grassland A terrestrial biome characterized by wide variation in temperature and a climax flora that includes grasses.

greenhouse gases Any of a number of gases in Earth's atmosphere that have the effect of trapping solar radiation in the form of heat (e.g., carbon dioxide, methane).

growth A process by which cells increase in number and size, resulting in an increase in size of the organism.

growth-stimulating hormone (GSH) A pituitary hormone regulating the elongation of the long bones of the body.

guanine A nitrogenous base found in both DNA and RNA molecules.

guard cell One of a pair of cells that surround the leaf stomate and regulate its size.

habitat The environment or set of ecological conditions within which an organism lives.

haploid chromosome number The number of chromosomes commonly found in the gametes of sexually reproducing species.

Hardy-Weinberg principle A hypothesis, advanced by G. H. Hardy and W. Weinberg that states that the gene pool of a population should remain stable as long as a set of ideal conditions is met.

heart In human beings, a four-chambered muscular pump that facilitates the movement of blood throughout the body.

helix Literally a spiral; a term used to describe the twisted ladder shape of the DNA molecule.

hemoglobin A type of protein specialized for the transport of respiratory oxygen in certain organisms, including earthworms and human beings.

herbivore A heterotrophic organism that consumes plant matter as its primary source of nutrition. (See **primary consumer**.)

hermaphrodite An animal organism that produces both male and female gametes.

heterotroph An organism that typically carries on heterotrophic nutrition.

heterotroph hypothesis A scientific hypothesis devised to explain the probable origin and early evolution of life on earth.

heterotrophic nutrition A type of nutrition in which organisms must obtain their foods from outside sources of organic nutrients.

heterozygous A term used to refer to an allele pair in which the alleles have different, contrasting effects (for example, *Aa, RW*).

high blood pressure A disorder of the human transport system in which systolic and diastolic pressures register higher than normal because of narrowing of the artery opening.

histamine A chemical product of the body that causes irritation and swelling of the mucous membranes.

homeostasis The condition of balance and dynamic stability that characterizes living systems under normal conditions.

homologous chromosomes A pair of chromosomes that carry corresponding genes for the same traits.

homologous structures Structures present within different species that can be shown to have had a common origin but that may or may not share a common function.

homozygous A term used to refer to an allele pair in which the alleles are identical in terms of effect (for example, *AA, aa*).

hormone A chemical product of an endocrine gland that has a regulatory effect on the cell's metabolism.

host The organism harmed in a parasitic relationship.

hybrid A term used to describe a heterozygous genotype. (See **heterozygous**.)

hybridization A technique of plant/animal breeding in which two varieties of the same species are crossbred in the hope of producing offspring with the favorable traits of both varieties.

hydrogen bond A weak electrostatic bond that holds together the twisted strands of DNA and RNA molecules.

hydrolysis The chemical process by which a complex food molecule is split into simpler components through the addition of a molecule of water to the bonds holding it together.

hypocotyl A portion of the plant embryo that specializes to become the root and lower stem of the adult plant.

hypothalamus An endocrine gland whose secretions affect the pituitary gland.

identical twins In human beings, twin offspring resulting from the separation of the embryonic cell mass of a single fertilization into two separate masses; such twins are genetically identical.

importation of organisms The introduction of nonactive plants and animals into new areas where they compete strongly with native species—a negative aspect of human involvement with the environment.

***in vitro* fertilization** A laboratory technique in which fertilization is accomplished outside the mother's body using mature ova and sperm extracted from the parents' bodies.

inbreeding A technique of plant/animal breeding in which a purebred variety is bred only with its own members so as to maintain a set of desired characteristics.

independent assortment A pattern of inheritance in which genes on different, nonhomologous chromosomes are free to be inherited randomly and regardless of the inheritance of the others.

ingestion The mechanism by which an organism takes in food from its environment.

inorganic compound A chemical compound that lacks the element carbon or hydrogen (for example, table salt: NaCl).

insulin A hormone, secreted by the islets of Langerhans, that regulates the storage of blood sugar as glycogen.

intercellular fluid (ICF) The fluid that bathes cells and fills intercellular spaces.

interferon A substance, important in the fight against human cancer, that may now be produced in large quantities with techniques of genetic engineering.

intermediate inheritance Any pattern of inheritance in which the offspring expresses a phenotype different from the phenotypes of its parents and usually representing a form intermediate between them.

internal development Embryonic development that occurs within the body of the female parent.

internal fertilization Fertilization that occurs inside the body of the female parent.

interneuron A type of neuron, located in the central nervous system, responsible for the interpretation of impulses received from sensory neurons.

intestine A portion of the digestive tract in which chemical digestion and absorption of digestive end products occur.

intracellular digestion A type of chemical digestion carried out within the cell.

iodine A chemical stain used in cell study; an indicator used to detect the presence of starch. (See **staining**.)

islets of Langerhans An endocrine gland, located within the pancreas, that produces the hormones insulin and glucagon.

karyotype An enlarged photograph of the paired homologous chromosomes of an individual cell that is used in the detection of certain genetic disorders involving chromosome mutation.

karyotyping A technique for the detection of human genetic disorders in which a karyotype is analyzed for abnormalities in chromosome structure or number.

kidney The excretory organ responsible for maintaining the chemical composition of the blood. (See **nephron**.)

kidney failure A disorder of the human excretory system in which there is a general breakdown of the kidney's ability to filter blood components.

kingdom A level of biological classification that includes a broad grouping of organisms displaying general structural similarity; five kingdoms have been named by scientists.

lacteal A small extension of the lymphatic system, found inside the villus, that absorbs fatty acids and glycerol resulting from lipid hydrolysis.

lactic acid fermentation A type of anaerobic respiration in which glucose is converted to two lactic acid molecules.

large intestine A portion of the digestive tract in which undigested foods are solidified by means of water absorption to form feces.

lateral meristem A plant growth region located under the epidermis or bark of a stem. (See **cambium**.)

Latin The language used in biological classification for naming organisms by means of binomial nomenclature.

lenticel A small pore in the stem surface that permits the absorption and release of respiratory gases within stem tissues.

leukemia A disorder of the human transport system in which the bone marrow produces large numbers of abnormal white blood cells. (See **cancer**.)

lichen A symbiosis of alga and fungus that frequently acts as a pioneer species on bare rock.

limiting factor Any abiotic or biotic condition that places limits on the survival of organisms and on the growth of species populations in the environment.

lipase Any lipid-hydrolyzing enzyme.

lipid An organic compound composed of carbon, hydrogen, and oxygen in which hydrogen and oxygen are *not* in a 2:1 ratio (for example, a wax, plant oil); many lipids are constructed of a glycerol and three fatty acids.

liver An accessory organ that stores glycogen, produces bile, destroys old red blood cells, deaminates amino acids, and produces urea.

lock-and-key model A theoretical model of enzyme action that attempts to explain the concept of enzyme specificity.

lung The major organ of respiratory gas exchange.

luteinizing hormone (LH) A pituitary hormone that regulates the conversion of the ovarian follicle into the corpus luteum.

lymph Intercellular fluid (ICF) that has passed into the lymph vessels.

lymph node One of a series of structures in the body that act as reservoirs of lymph and also contain white blood cells as part of the body's immune system.

lymph vessel One of a branching series of tubes that collect ICF from the tissues and redistribute it as lymph.

lymphatic circulation The movement of lymph throughout the body.

lymphocyte A type of white blood cell that produces antibodies.

lysosome A cell organelle that houses hydrolytic enzymes used by the cell in the process of chemical digestion.

Malpighian tubules In arthropods (for example, grasshopper), an organ specialized for the removal of metabolic wastes.

maltase A specific enzyme that catalyzes the hydrolysis (and dehydration synthesis) of maltose.

maltose A type of disaccharide; a maltose molecule is composed of two units of glucose joined together by dehydration synthesis.

marine biome An aquatic biome characterized by relatively stable conditions of moisture, salinity, and temperature.

marsupial mammal   See **nonplacental mammal**.

mechanical digestion   Any of the processes by which foods are broken apart physically into smaller particles.

medulla   The portion of the human brain responsible for regulating the automatic processes of the body.

meiosis   The process by which four monoploid nuclei are formed from a single diploid nucleus.

meningitis   A disorder of the human regulatory system in which the membranes of the brain or spinal cord become inflamed.

menstrual cycle   A hormone-controlled process responsible for the monthly release of mature ova.

menstruation   The stage of the menstrual cycle in which the lining of the uterus breaks down and is expelled from the body via the vaginal canal.

meristem   A plant tissue specialized for embryonic development. (See **apical meristem**; **lateral meristem**; **cambium**.)

mesoderm   An embryonic tissue that differentiates into muscle, bone, the excretory system, and most of the reproductive system in the adult animal.

messenger RNA (mRNA)   A type of RNA that carries the genetic code from the nuclear DNA to the ribosome for transcription.

metabolism   All of the chemical processes of life considered together; the sum total of all the cell's chemical activity.

methylene blue   A chemical stain used in cell study. (See **staining**.)

microdissection instruments   Tools of biological study used to remove certain cell organelles from within cells for examination.

micrometer (μm)   A unit of linear measurement equal in length to 0.001 millimeter (0.000001 meter), used for expressing the dimensions of cells and cell organelles.

mitochondrion   A cell organelle that contains the enzymes necessary for aerobic respiration.

mitosis   A precise duplication of the contents of a parent cell nucleus followed by an orderly separation of these contents into two new, identical daughter nuclei.

mitotic cell division   A type of cell division that results in the production of two daughter cells identical to each other and to the parent cell.

Monera   One of the five biological kingdoms; it includes simple unicellular forms lacking nuclear membranes (for example, bacteria).

monohybrid cross   A genetic cross between two organisms both heterozygous for a trait controlled by a single allele pair. The phenotypic ratio resulting is 3:1; the genotypic ratio is 1:2:1.

monosaccharide   A type of carbohydrate known also as a simple sugar; all monosaccharides have the molecular formula $C_6H_{12}O_6$.

motor neuron   A type of neuron that carries command impulses from the central nervous system to an effector organ.

mucus   A protein-rich mixture that bathes and moistens the respiratory surfaces.

multicellular   Having a body that consists of large groupings of specialized cells (for example, human being).

multiple alleles   A pattern of inheritance in which the existence of more than two alleles is hypothesized, only two of which are present in the genotype of any one individual.

muscle   A type of tissue specialized to produce movement of body parts.

mutagenic agent   Any environmental condition that initiates or accelerates genetic mutation.

mutation   Any alteration of the genetic material, either a chromosome or a gene, in an organism.

mutualism   A type of symbiosis beneficial to both organisms in the relationship.

nasal cavity A series of channels through which outside air is admitted to the body interior and is warmed and moistened before entering the lung.

natural selection A concept, central to Darwin's theory of evolution, to the effect that the individuals best adapted to their environment tend to survive and to pass their favorable traits on to the next generation

negative feedback A type of endocrine regulation in which the effects of one gland may inhibit its own secretory activity while stimulating the secretory activity of another gland.

nephridium An organ found in certain organisms, including the earthworm, specialized for the removal of metabolic wastes.

nephron The functional unit of the kidney. (See **glomerulus**; **Bowman's capsule**.)

nerve A structure formed from the bundling of neurons carrying sensory or motor impulses.

nerve impulse An electrochemical change in the surface of the nerve cell.

nerve net A network of nervelike cells in coelenterates such as the hydra.

neuron A cell specialized for the transmission of nerve impulses.

neurotransmitter A chemical substance secreted by a neuron that aids in the transmission of the nerve impulse to an adjacent neuron.

niche The role that an organism plays in its environment.

nitrifying bacteria In the nitrogen cycle, bacteria that absorb ammonia and convert it into nitrate salts.

nitrogen cycle The process by which nitrogen is recycled and made available for use by other organisms.

nitrogen-fixing bacteria A type of bacteria responsible for absorbing atmospheric nitrogen and converting it to nitrate salts in the soil.

nitrogenous base A chemical unit composed of carbon, hydrogen, and nitrogen that is a component part of the nucleotide unit.

nitrogenous waste Any of a number of nitrogen-rich compounds that result from the metabolism of proteins and amino acids in the cell. (See **ammonia**; **urea**; **uric acid**.)

nondisjunction A type of chromosome mutation in which the members of one or more pairs of homologous chromosomes fail to separate during the disjunction phase of the first meiotic division.

nonplacental mammal A type of mammal (marsupial) in which internal development is accomplished without the aid of a placental connection.

nucleic acid An organic compound composed of repeating units of nucleotide.

nucleolus A cell organelle located within the nucleus that functions in protein synthesis.

nucleotide The repeating unit making up the nucleic acid polymer (for example, DNA, RNA).

nucleus A cell organelle that contains the cell's genetic information in the form of chromosomes.

nutrition The life function by which living things obtain food and process it for their use.

omnivore A heterotrophic organism that consumes both plant and animal matter as sources of nutrition.

one gene–one polypeptide A scientific hypothesis concerning the role of the individual gene in protein synthesis.

oogenesis A type of meiotic cell division in which one ovum and three polar bodies are produced from each primary sex cell.

open transport system A type of circulatory system in which the transport fluid is *not* always enclosed within blood vessels (for example, grasshopper).

oral cavity In human beings, the organ used for the ingestion of foods.

oral groove   The ingestive organ of the paramecium.

organ transplant   An application of the study of immunity in which an organ or tissue of a donor is transplanted into a compatible recipient.

organelle   A small, functional part of a cell specialized to perform a specific life function (for example, nucleus, mitochondrion).

organic compound   a chemical compound that contains the elements carbon and hydrogen (for example, carbohydrate, protein).

organic evolution   The mechanism thought to govern the changes in living species over geologic time.

osmosis   A form of passive transport by which water is absorbed or released by cells.

ovary   A female gonad that secretes the hormone estrogen, which regulates female secondary sex characteristics; the ovary also produces ova, which are used in reproduction.

overcropping   A negative aspect of human involvement with the environment in which soil is overused for the production of crops, leading to exhaustion of soil nutrients.

overgrazing   The exposure of soil to erosion due to the loss of stabilizing grasses when it is overused by domestic animals—a negative aspect of human involvement with the environment.

overhunting   A negative aspect of human involvement with the environment in which certain species have been greatly reduced or made extinct by uncontrolled hunting practices.

oviduct   A tube that serves as a channel for conducting mature ova from the ovary to the uterus; the site of fertilization and the earliest stages of embryonic development.

ovulation   The stage of the menstrual cycle in which the mature ovum is released from the follicle into the oviduct.

ovule   A structure located within the flower ovary that contains a monoploid egg nucleus and serves as the site of fertilization.

ovum   A type of gamete produced as a result of oogenesis in female animals; the egg, the female sex cell.

oxygen 18   A radioactive isotope of oxygen that is used to trace the movement of this element in biochemical reaction sequences.

oxygenated blood   Blood that contains a high percentage of oxyhemoglobin.

oxyhemoglobin   Hemoglobin that is loosely bound to oxygen for purposes of oxygen transport.

pH   A chemical unit used to express the concentration of hydrogen ion (H^+), or the acidity, of a solution.

palisade layer   A cell layer found in most leaves that contains high concentrations of chloroplasts.

pancreas   An accessory organ that produces enzymes that complete the hydrolysis of foods to soluble end products; also the site of insulin and glucagon production.

parasitism   A type of symbiosis from which one organism in the relationship benefits, while the other (the host) is harmed, but not ordinarily killed.

parathormone   A hormone of the parathyroid gland that regulates the metabolism of calcium in the body.

parathyroid gland   An endocrine gland whose secretion, parathormone, regulates the metabolism of calcium in the body.

passive immunity   A temporary immunity produced as a result of the injection of preformed antibodies.

passive transport   Any process by which materials are absorbed into the cell interior from an area of high concentration to an area of low concentration without the expenditure of cell energy (for example, osmosis, diffusion).

penis   A structure that permits internal fertilization through direct implantation of sperm into the female reproductive tract.

peptide bond   A type of chemical bond that links the nitrogen atom of one amino acid with the terminal carbon atom of a second amino acid in the formation of a dipeptide.

peripheral nerves   Nerves in the earthworm and grasshopper that branch from the ventral nerve cord to other parts of the body.

peripheral nervous system   A major subdivision of the nervous system that consists of the part of the nervous system that is outside the central nervous system. (See **autonomic nervous system; somatic nervous system**.)

peristalsis   A wave of contractions of the smooth muscle lining; the digestive tract that causes ingested food to pass along the food tube.

petal   An accessory part of the flower that is thought to attract pollinating insects.

phagocyte   A type of white blood cell that engulfs and destroys bacteria.

phagocytosis   The process by which an ameba surrounds and ingests large food particles for intracellular digestion.

pharynx   The upper part of the digestive tube that temporarily stores food before digestion.

phenotype   The observable trait that results from the action of an allele pair.

phenylketonuria (PKU)   A genetically related human disorder in which the homozygous combination of a particular mutant gene prevents the normal metabolism of the amino acid phenylalanine.

phloem   A type of vascular tissue through which water and dissolved sugars are transported in plants from the leaf downward to the roots for storage.

phosphate group   A chemical group made up of phosphorus and oxygen and that is a component part of the nucleotide unit.

phosphoglyceraldehyde (PGAL)   An intermediate product formed during photosynthesis that acts as the precursor of glucose formation.

photochemical reactions   A set of biochemical reactions in photosynthesis in which light is absorbed and water molecules are split. (See **photolysis**.)

photolysis   The portion of the photochemical reactions in which water molecules are split into hydrogen atoms and made available to the carbon-fixation reactions.

photosynthesis   A type of autotrophic nutrition in which green plants use the energy of sunlight to convert carbon dioxide and water into glucose.

phylum   A level of biological classification that is a major subdivision of the kingdom level, containing fewer organisms with greater similarity (for example, Chordata).

pinocytosis   A special type of absorption by which liquids and particles too large to diffuse through the cell membrane may be taken in by vacuoles formed at the cell surface.

pioneer autotrophs   The organisms supposed by the heterotroph hypothesis to have been the first to evolve the ability to carry on autotrophic nutrition.

pioneer species   In an ecological succession, the first organisms to inhabit a barren environment.

pistil   The female sex organ of the flower. (See **stigma; style; ovary**.)

pituitary gland   An endocrine gland that produces hormones regulating the secretions of other endocrine glands; the master gland.

placenta   In placental mammals, a structure composed of both embryonic and maternal tissues that permits the diffusion of soluble substances to and from the fetus for nourishment and the elimination of fetal waste.

placental mammal   A mammal species in which embryonic development occurs internally with the aid of a placental connection to the female parent's body.

plant   One of the five biological kingdoms; it includes multicellular organisms whose cells are bounded by cell walls and that are capable of photosynthesis (for example, maple tree).

plasma The liquid fraction of blood, containing water and dissolved proteins.

plasma membrane A cell organelle that encloses the cytoplasm and other cell organelles and regulates the passage of materials into and out of the cell.

platelet A cell-like component of the blood that is important in clot formation.

polar body One of three nonfunctional cells produced during oogenesis that contain monoploid nuclei and disintegrate soon after completion of the process.

polio A disorder of the human regulatory system in which viral infection of the central nervous system may result in severe paralysis.

pollen The male gamete of the flowering plant.

pollen tube A structure produced by the germinating pollen grain that grows through the style to the ovary and carries the sperm nucleus to the ovule for fertilization.

pollination The transfer of pollen grains from anther to stigma.

pollution control The development of new procedures to reduce the incidence of air, water, and soil pollution—a positive aspect of human involvement with the environment.

polyploidy A type of chromosome mutation in which an entire set of homologous chromosomes fail to separate during the disjunction phase of the first meiotic division.

polysaccharide A type of carbohydrate composed of repeating units of monosaccharides that form a polymeric chain.

polyunsaturated fat A type of fat in which many bonding sites are unavailable for the addition of hydrogen atoms.

population All the members of a particular species in a given geographical location at a given time.

population control The use of various practices to slow the rapid growth in the human population—a positive aspect of human interaction with the environment.

population genetics A science that studies the genetic characteristics of a sexually reproducing species and the factors that affect its gene frequencies.

postnatal development The growth and maturation of an individual from birth, through aging, to death.

prenatal development The embryonic development that occurs before birth within the uterus. (See **gestation**.)

primary consumer Any herbivorous organism that receives food energy from the producer level (for example, mouse); the second stage of a food chain.

primary sex cell The diploid cell that undergoes meiotic cell division to produce monoploid gametes.

producer Any autotrophic organism capable of trapping light energy and converting it to the chemical bond energy of food (for example, green plants); the organisms forming the basis of the food chain.

progesterone A hormone produced by the corpus luteum and/or placenta that has the effect of maintaining the uterine lining and suppressing ovulation during gestation.

protease Any protein-hydrolyzing enzyme.

protein A complex organic compound composed of repeating units of amino acid.

Protista One of the five biological kingdoms; it includes simple unicellular forms whose nuclei are surrounded by nuclear membranes (for example, ameba, paramecium).

pseudopod A temporary, flowing extension of the cytoplasm of an ameba that is used in nutrition and locomotion.

pulmonary artery One of two arteries that carry blood from the heart to the lungs for reoxygenation.

pulmonary circulation Circulation of blood from the heart through the lungs and back to the heart.

pulmonary vein One of four veins that carry oxygenated blood from the lungs to the heart.

pulse Rhythmic contractions of the artery walls that help to push the blood fluid through the capillary networks of the body.

punctuated equilibrium A theory of the time frame required for evolution that assumes that evolutionary change occurs in bursts with long periods of relative stability intervening.

pyramid of biomass A construct used to illustrate the fact that the total biomass available in each stage of a food chain diminishes from producer level to consumer level.

pyramid of energy A construct used to illustrate the fact that energy is lost at each trophic level in a food chain and is most abundant at the producer level.

pyruvic acid An intermediate product in the aerobic or anaerobic respiration of glucose.

receptor An organ specialized to receive a particular type of environmental stimulus.

recessive allele (gene) An allele (gene) whose effect is masked by that of its dominant allele.

recombinant DNA DNA molecules that have been moved from one cell to another in order to give the recipient cell a genetic characteristic of the donor cell.

recombination The process by which the members of segregated allele pairs are randomly recombined in the zygote as a result of fertilization.

rectum The portion of the digestive tract in which digestive wastes are stored until they can be released to the environment.

red blood cell Small, nonnucleated cells in the blood that contain hemoglobin and carry oxygen to body tissues.

reduction division See **meiosis**.

reflex A simple, inborn, involuntary response to an environmental stimulus.

reflex arc The complete path involving a series of three neurons (sensory, interneuron, and motor) working together in a reflex action.

regeneration A type of asexual reproduction in which new organisms are produced from the severed parts of a single parent organism; the replacement of loss or damaged tissues.

regulation The life process by which living things respond to changes within and around them, and by which all life processes are coordinated.

replication An exact self-duplication of the chromosome during the early stages of cell division; the exact self-duplication of a molecule of DNA.

reproduction The life process by which new cells arise from preexisting cells by cell division.

reproductive isolation The inability of species varieties to interbreed and produce fertile offspring because of variations in behavior or chromosome structure.

respiration The life function by which living things convert the energy of organic foods into a form more easily used by the cell.

response The reaction of an organism to an environmental stimulus.

restriction enzyme Any of a number of specific enzymes capable of "snipping" DNA molecules at particular base sequences.

rhizoid A rootlike fiber produced by fungi that secrete hydrolytic enzymes and absorb digested nutrients.

ribonucleic acid (RNA) A type of nucleic acid that operates in various ways to facilitate protein synthesis.

ribose A five-carbon sugar found as a component part of the nucleotides of RNA molecules only.

ribosomal RNA (rRNA) The type of RNA that makes up the ribosome.

ribosome A cell organelle that serves as the site of protein synthesis in the cell.

root A plant organ specialized to absorb water and dissolved substances from the soil as well as to anchor the plant to the soil.

root hair A small projection of the growing root that serves to increase the surface area of the root for absorption.

roughage A variety of undigestible carbohydrates that add bulk to the diet and facilitate the movement of foods through the intestine.

runner A type of vegetative propagation in which an above-ground stem (runner) produces roots and leaves and establishes new organisms with identical characteristics.

saliva A fluid secreted by salivary glands that contains hydrolytic enzymes specific to the digestion of starches.

salivary gland The gland that secretes saliva, which is important in the chemical digestion of certain foods.

salt A chemical composed of a metal and a nonmetal joined by means of an ionic bond (for example, sodium chloride).

saprophyte A heterotrophic organism that obtains its nutrition from the decomposing remains of dead plant and animal tissues (for example, fungi, bacteria).

saturated fat A type of fat molecule in which all available bonding sites on the hydrocarbon chains are taken up with hydrogen atoms.

scrotum A pouch extending from the wall of the lower abdomen that houses the testes at a temperature optimum for sperm production.

secondary consumer Any carnivorous animal that derives its food energy from the primary consumer level (for example, a snake); the third level of a food chain.

secondary sex characteristics The physical features, different in males and females, that appear with the onset of sexual maturity.

seed A structure that develops from the fertilized ovule of the flower and germinates to produce a new plant.

seed dispersal Any mechanism by which seeds are distributed in the environment so as to widen the range of a plant species. (See **fruit**.)

segregation The random separation of the members of allele pairs that occurs during meiotic cell division.

self-pollination A type of pollination in which the pollen of a flower pollinates another flower located on the same plant organism.

sensory neuron A type of neuron specialized for receiving environmental stimuli, which are detected by receptor organs.

sepal An accessory part of the flower that functions to protect the bud during development.

sessile A term that relates to the unmoving state of certain organisms, including the hydra.

seta One of several small, chitinous structures (setae) that aid the earthworm in its locomotor function.

sex chromosomes A pair of homologous chromosomes carrying genes that determine the sex of an individual; these chromosomes are designated as X and Y.

sex determination A pattern of inheritance in which the conditions of maleness and femaleness are determined by the inheritance of a pair of sex chromosomes (XX = female; XY = male).

sex linkage A pattern of inheritance in which certain nonsex genes are located on the X sex chromosome but have no corresponding alleles on the Y sex chromosome.

sex-linked trait A genetic trait whose inheritance is controlled by the genetic pattern of sex linkage (for example, color blindness).

sexual reproduction A type of reproduction in which new organisms form as a result of the fusion of gametes from two parent organisms.

shell An adaptation for embryonic development in many terrestrial, externally developing species that protects the developing embryo from drying and physical damage (for example, birds).

sickle-cell anemia A genetically related human disorder in which the homozygous combination of a mutant gene leads to the production of abnormal hemoglobin and crescent-shaped red blood cells.

skeletal muscle A type of muscle tissue associated with the voluntary movements of skeletal levers in locomotion.

small intestine In human beings, the longest portion of the food tube, in which final digestion and absorption of soluble end products occur.

smooth muscle See **visceral muscle**.

somatic nervous system A subdivision of the peripheral nervous system that is made up of nerves associated with voluntary actions.

speciation The process by which new species are thought to arise from previously existing species.

species A biological grouping of organisms so closely related that they are capable of interbreeding and producing fertile offspring (for example, human being).

species presentation The establishment of game lands and wildlife refuges that have permitted the recovery of certain endangered species—a positive aspect of human involvement with the environment.

sperm A type of gamete produced as a result of spermatogenesis in male animals; the male reproductive cell.

spermatogenesis A type of meiotic cell division in which four sperm cells are produced from each primary sex cell.

spinal cord The part of the central nervous system responsible for reflex action as well as impulse conduction between the peripheral nervous system and the brain.

spindle apparatus A network of fibers that form during cell division and to which centromeres attach during the separation of chromosomes.

spiracle One of several small pores in arthropods, including the grasshopper, that serve as points of entry of respiratory gases from the atmosphere to the tracheal tubes.

spongy layer A cell layer found in most leaves that is loosely packed and contains many air spaces to aid in gas exchange.

spore A specialized asexual reproductive cell produced by certain plants.

sporulation A type of asexual reproduction in which spores released from special spore cases on the parent plant germinate and grow into new adult organisms of the species.

staining A technique of cell study in which chemical stains are used to make cell parts more visible for microscopic study.

stamen The male reproductive structure in a flower. (See **anther**; **filament**.)

starch A type of polysaccharide produced and stored by plants.

stem A plant organ specialized to support the leaves and flowers of a plant as well as to conduct materials between the roots and the leaves.

stem cell An undifferentiated cell that resides in the tissues of the body and can give rise to new, specialized cells of any kind.

stigma The sticky upper portion of the pistil, which serves to receive pollen.

stimulus Any change in the environment to which an organism responds.

stomach A muscular organ that acts to liquefy food and that produces gastric protease for the hydrolysis of protein.

stomate A small opening that penetrates the lower epidermis of a leaf and through which respiratory and photosynthetic gases diffuse.

strata The layers of sedimentary rock that contain fossils, whose ages may be determined by studying the patterns of sedimentation.

stroke   A disorder of the human regulatory system in which brain function is impaired because of oxygen starvation of brain centers.

stroma   An area of the chloroplast within which the carbon-fixation reactions occur; each stroma lies between pairs of grana.

style   The portion of the pistil that connects the stigma to the ovary.

substrate   A chemical metabolized by the action of a specific enzyme.

succession   A situation in which an established ecological community is gradually replaced by another until a climax community is established.

survival of the fittest   The concept, frequently associated with Darwin's theory of evolution, that in the intraspecies competition among naturally occurring species, the organisms best adapted to the particular environment will survive.

sweat glands   In human beings, the glands responsible for the production of perspiration.

symbiosis   A term that refers to a variety of biotic relationships in which organisms of different species live together in close physical association.

synapse   The gap that separates the terminal branches of one neuron from the dendrites of an adjacent neuron.

synapsis   The intimate, highly specific pairing of homologous chromosomes that occurs in the first meiotic division, forming tetrads.

synthesis   The life function by which living things manufacture the complex compounds required to sustain life.

systemic circulation   The circulation of blood from the heart through the body tissues (except the lungs) and back to the heart.

systole   The higher pressure registered during blood pressure testing. (See **diastole**.)

taiga   A terrestrial biome characterized by long, severe winters and climax flora that includes coniferous trees.

Tay-Sachs disease   A genetically related human disorder in which fatty deposits in the cells, particularly of the brain, inhibit proper functioning of the nervous system.

technological oversight   A term relating to human activities that adversely affect environmental quality due to failure to assess the environmental impact of a technological development adequately.

teeth   Structures located in the mouth that are specialized to aid in the mechanical digestion of foods.

temperate deciduous forest   A terrestrial biome characterized by moderate climatic conditions and climax flora that includes deciduous trees.

template   A pattern or design provided by the DNA molecule for the synthesis of protein molecules.

tendon   A type of connective tissue that attaches a skeletal muscle to a bone.

tendonitis   A disorder of the human locomotor system in which the junction between a tendon and a bone becomes irritated and inflamed.

tentacle   A grasping structure in certain organisms, including the hydra, that contains stinging cells and is used for capturing prey.

terminal branch   A cytoplasmic extension of the neuron that transmits a nerve impulse to adjacent neurons via the secretion of neurotransmitters.

terrestrial biome   A biome that comprises primarily land ecosystems, the characteristics of which are determined by the major climate zone of the earth.

testcross   A genetic cross to determine the genotype of an organism expressing a dominant phenotype; the unknown is crossed with a homozygous recessive.

testis   A gonad in human males that secretes the hormone testosterone, which regulates male secondary sex characteristics; the testis also produces sperm cells for reproduction.

testosterone   A hormone secreted by the testis that regulates the production of male secondary sex characteristics.

tetrad A grouping of four chromatids that results from synapsis.

thymine A nitrogenous base found only in DNA.

thyroid gland An endocrine gland that regulates the body's general rate of metabolism through secretion of the hormone thyroxin.

thyroid-stimulating hormone (TSH) A pituitary hormone that regulates the secretions of the thyroid gland.

thyroxin A thyroid hormone that regulates the body's general metabolic rate.

tongue A structure that aids in the mechanical digestion of foods.

trachea A cartilage-ringed tube that conducts air from the mouth to the bronchi.

tracheal tube An adaptation in arthropods (for example, grasshopper) that functions to conduct respiratory gases from the environment to the moist internal tissues.

Tracheophyta A phylum of the plant kingdom whose members (tracheophytes) contain vascular tissues and true roots, stems, and leaves (for example, geranium, fern, bean, maple tree, corn).

transfer RNA (tRNA) A type of RNA that transports specific amino acids from the cytoplasm to the ribosome for protein synthesis.

translocation A type of chromosome mutation in which a section of a chromosome is transferred to a nonhomologous chromosome.

transpiration The evaporation of water from leaf stomates.

transpiration pull A force that aids the upward conduction of materials in the xylem by means of the evaporation of water (transpiration) from leaf surfaces.

transport The life function by which substances are absorbed, circulated, and released by living things.

triplet codon A group of three nitrogenous bases that provide information for the placement of amino acids in the synthesis of proteins.

tropical forest A terrestrial biome characterized by a warm, moist climate and a climax flora that includes many species of broad-leaved trees.

tropism A plant growth response to an environmental stimulus.

tuber A type of vegetative propagation in which an underground stem (tuber) produces new tubers, each of which is capable of producing new organisms with identical characteristics.

tundra A terrestrial biome characterized by permanently frozen soil and climax flora that includes lichens and mosses.

tympanum A receptor organ in arthropods (for example, grasshopper) that is specialized to detect vibrational stimuli.

ulcer A disorder of the human digestive tract in which a portion of its lining erodes and becomes irritated.

ultracentrifuge A tool of biological study that uses very high speeds of centrifugation to separate cell parts for examination.

umbilical cord In placental mammals, a structure containing blood vessels that connects the placenta to the embryo.

unicellular Having a body that consists of a single cell (for example, paramecium).

uracil A nitrogenous base that is a component part of the nucleotides of RNA molecules only.

urea A type of nitrogenous waste with moderate solubility and moderate toxicity.

ureter In human beings, a tube that conducts urine from the kidney to the urinary bladder.

urethra In human beings, a tube that conducts urine from the urinary bladder to the exterior of the body. The urethra also conducts semen to the outside of the body.

uric acid A type of nitrogenous waste with low solubility and low toxicity.

urinary bladder An organ responsible for the temporary storage of urine.

urine A mixture of water, salts, and urea excreted from the kidney.

use and disuse A term associated with the evolutionary theory of Lamarck, since proved incorrect.

uterus In female placental mammals, the organ within which embryonic development occurs.

vaccination An inoculation of dead or weakened disease organisms that stimulates the body's immune system to produce active immunity.

vacuole A cell organelle that contains storage materials (for example, starch, water) housed inside the cell.

vagina In female placental mammals, the portion of the reproductive tract into which sperm are implanted during sexual intercourse and through which the baby passes during birth.

variation A concept, central to Darwin's theory of evolution, that refers to the range of adaptation that can be observed in all species.

vascular tissues Tubelike plant tissues specialized for the conduction of water and dissolved materials within the plant. (See **xylem**; **phloem**.)

vegetative propagation A type of asexual reproduction in which new plant organisms are produced from the vegetative (nonfloral) parts of the parent plant.

vein (human) A relatively thin-walled blood vessel that carries blood from capillary networks back toward the heart.

vein (plant) An area of vascular tissues located in the leaf that aid the upward transport of water and minerals through the leaf and the transport of dissolved sugars to the stem and roots.

vena cava One of two major arteries that return blood to the heart from the body tissues.

ventral nerve cord The main pathway for nerve impulses between the brain and peripheral nerves of the grasshopper and earthworm.

ventricle One of two thick-walled, muscular chambers of the heart that pump blood out to the lungs and body.

villi Microscopic projections of the lining of the small intestine that absorb the soluble end products of digestion. (See **lacteal**.)

visceral muscle A type of muscle tissue associated with the involuntary movements of internal organs (for example, peristalsis in the small intestine).

vitamin a type of nutrient that acts as a coenzyme in various enzyme-controlled reactions.

water cycle The mechanism by which water is made available to living things in the environment through the processes of precipitation, evaporation, runoff, and percolation.

water pollution A type of technological oversight that involves the addition of some unwanted factor (for example, sewage, heavy metals, heat, toxic chemicals) to our water resources.

Watson-Crick model A model of DNA structure devised by J. Watson and F. Crick that hypothesizes a twisted ladder arrangement for the DNA molecule also known as a "double helix."

white blood cell A type of blood cell that functions in disease control. (See **phagocyte**; **lymphocyte**.)

xylem A type of vascular tissue through which water and dissolved minerals are transported upward through a plant from the root to the stems and leaves.

yolk A food substance, rich in protein and lipid, found in the eggs of many animal species.

yolk sac The membrane that surrounds the yolk food supply of the embryos of many animal species.

zygote The single diploid cell that results from the fusion of gametes in sexual reproduction; a fertilized egg.

Examination June 2019

Living Environment

PART A

Answer all questions in this part. [30]

Directions **(1–30): For** *each* **statement or question, record in the space provided the** *number* **of the word or expression that, of those given, best completes the statement or answers the question.**

1 Which activity is an example of a decomposer recycling organic compounds back into the environment?

(1) A tree synthesizes starch from simpler molecules.

(2) A bacterial cell performs photosynthesis.

(3) A bird digests proteins from its food.

(4) A fungus breaks down the body of a dead animal. 1 _____

2 Itching and other skin problems are signs that a cat or dog may have fleas. Fleas are parasites known for their biting and blood-sucking abilities. When they bite, flea saliva enters the pet's circulatory system, sometimes causing an allergic response commonly seen as a "hot spot" on the pet's neck or the base of its tail.

Source: https://www.planetnatural.com/
pest-problem-solver/household-pests/flea-control/

These observations are best explained by the fact that

(1) flea saliva may stimulate an immune response in cats and dogs

(2) fleas are microbes whose bites cause a decreased blood flow

(3) flea saliva is a toxic substance that is released when fleas prey on cats and dogs

(4) fleas are host organisms whose saliva digests cat and dog fur, leaving "hot spots"

2 _____

3 A German measles (rubella) epidemic during the years 1963 to 1965 resulted in approximately 30,000 babies being born with birth defects. The specific cause of these birth defects was most likely

(1) the development of rubella virus infections in embryos

(2) the failure of zygotes infected with rubella to develop

(3) mutations in the nerve cells of pregnant females at the time of the rubella epidemic

(4) an increase in the amount of time needed for healthy embryonic development

3 _____

4 Placenta previa is a medical condition that occurs in some pregnant women. Women with this condition are often placed on bed rest, which prohibits them from any strenuous activity that may cause the blood vessels in the placenta to rupture. If not diagnosed, placenta previa can be a very dangerous condition because the placenta is

(1) the primary source of oxygen for the mother
(2) where the fetus obtains milk from the mother
(3) where nutrients and wastes are exchanged
(4) the primary source of estrogen and progesterone in the mother

4 _____

5 Over time, a tree that once had a total mass of 300 g increased in mass to 3000 kg. This increase in mass comes mostly from

(1) carbon dioxide that enters through the leaf openings
(2) oxygen that enters through the leaf openings
(3) soil that all plants need to grow
(4) chloroplasts that enter the roots and move to the leaves

5 _____

6 Recently, a type of genetically modified fish has been approved for sale for human consumption. The modified fish contain a growth hormone gene from a different fish species. As a result, the modified fish grow rapidly and are ready to sell in almost half the time it normally would take. The modified fish are able to produce the new growth hormone because

(1) each of their cells contains the new gene to produce growth hormone
(2) each gene contains the code to synthesize carbohydrates
(3) the altered gene directs the mitochondria to synthesize the hormone
(4) the modified body cells are able to reproduce by meiosis

6 _____

7 Melanoma is a type of skin cancer that can spread to vital organs in the body. Doctors believe that exposure to ultraviolet (UV) radiation from the Sun is a leading cause of melanoma. One practical way governments can help prevent the harmful effects of UV radiation is to

(1) require everyone to remain indoors during daylight hours
(2) regulate the production and release of gases that damage the ozone shield
(3) encourage the building of a greater number of cancer treatment centers
(4) prohibit the use of solar panels on homes and businesses   7 _____

8 Some birds have recently modified their migratory behavior. Instead of flying to warmer climates during the winter months, the birds are remaining in northern areas where they can consume discarded food that is abundant in landfills. As a result of this change in migratory behavior, many insect populations that the birds normally feed on in the warmer climate areas are now increasing. This is an example of human activity

(1) interfering with ecological succession
(2) increasing competition for infinite resources
(3) disrupting the homeostasis of organisms
(4) altering the equilibrium of ecosystems   8 _____

9 New York State charges consumers a fee when purchasing beverages sold in aluminum cans and plastic bottles. This money is returned to purchasers when they return these items for recycling. Programs such as these are an attempt to

(1) encourage people to spend more money on their beverages
(2) conserve the resources these containers are made from
(3) reduce the amount of carbon dioxide produced by deforestation
(4) totally eliminate the use of reusable containers   9 _____

10 Recently, a human trachea (a respiratory organ) was produced by using a patient's own stem cells. The benefit of using the patient's own cells to produce a trachea instead of receiving one from a donor is that

(1) there will be more enzymes produced to help maintain homeostasis in the trachea
(2) there will be an increase in the quantity of antibodies that the patient produces in response to the new trachea
(3) there is less of a chance that the patient's immune system will attack the trachea
(4) there will be a greater response to any infectious agent that may enter the body   10

11 The diagram below represents the organization of structures within an organism.

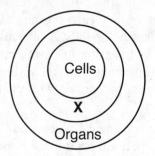

Which term best indicates the structures represented by the circle labeled X?

(1) organelles   (3) organ systems
(2) chromosomes   (4) tissues   11 _____

12 The chart below shows a comparison of the blood sugar levels for two individuals who took part in a scientific study.

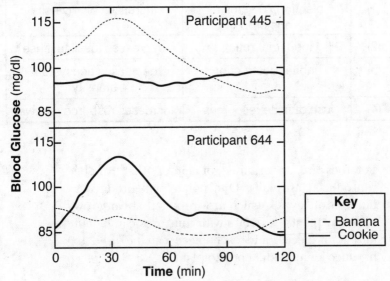

Source: Science Daily 11/19/15

Scientists have observed that blood sugar levels rose by different amounts in the two individuals even though they were given identical portions of bananas and cookies. These results were obtained because

(1) glucose is too large a molecule to be absorbed into the blood, so the researchers were only measuring the amount of glucose already present
(2) participant 445 didn't like bananas, and his body absorbed more of the food that he likes
(3) individuals have genetic differences that alter their responses to environmental factors
(4) two different foods were used; the scientists should have had only one experimental variable   12 _____

13 Which row in the chart below correctly matches the human activity with its effect?

Row	Human Activity	Effect
(1)	planting 20 acres of one crop	increases biodiversity
(2)	industrialization	decreases fossil fuel use
(3)	habitat destruction	decreases ecosystem stability
(4)	use of finite resources	increases resource renewal

13 _____

14 Potatoes are an example of a crop that can be reproduced asexually. One potato will produce a number of "eyes," which are sprouts that can grow into new plants. A potato with four eyes can be cut into four pieces, and each piece can be used to produce an individual potato plant.

"Eyes"

Source: https://www.quickcrop.ie/
blog/2014/02/growing-potatoes/

A gardener could produce a small crop of potatoes by planting the eyes from a single potato in her garden. Some of the potatoes grown in this way could be used to obtain eyes for the next season's crop.

One likely *disadvantage* of growing potatoes cloned in this way, year after year, would be that

(1) after a few years, the potatoes would stop producing eyes altogether, so no potatoes could be grown in the garden
(2) the potatoes produced each succeeding year would get larger and larger, eventually being too big for use as food
(3) the cost for growing your own potatoes in the garden would be greatly reduced
(4) a potato plant could become infected with a disease, and it could easily spread to the entire crop, killing all of the plants   14 _____

15 The back of the Namib Desert darkling beetle, shown in the photograph below, is covered in little bumps that collect water from the air. When it tilts forward, the water runs off its back into its mouth.

Source: http://myinforms.com

These specialized structures on the beetle's back allow it to

(1) locate food within the harsh desert environment
(2) obtain a substance that is required for survival
(3) reproduce asexually if mates are not available in the area
(4) increase the chances of survival by producing organic raw materials   15 _____

16 An increase in human population puts a stress on resources that can be renewed, such as

(1) trees and coal   (3) oil and natural gas

(2) water and gasoline   (4) water and trees   16 _____

17 Mitochondria provide ribosomes with

(1) ATP for protein synthesis

(2) amino acids for protein synthesis

(3) oxygen for respiration

(4) carbon dioxide for the production of sugars   17 _____

18 Mutations are most directly caused by changes in the

(1) cell organelles of tissues

(2) genes of chromosomes

(3) ribosomes in gametes

(4) receptors on membranes   18 _____

19 Animals and green plants are similar in that they

(1) both carry out heterotrophic nutrition

(2) all produce offspring by asexual reproduction

(3) both use DNA to transmit hereditary information to offspring

(4) all require oxygen to carry out photosynthesis   19 _____

20 Two organisms of different species are *not* likely to compete for the same

(1) food   (3) space

(2) mate   (4) water   20 _____

21 Some salmon have been genetically modified to grow bigger and faster than wild salmon. They are grown in fish-farming facilities. These genetically modified fish should *not* be introduced into a natural habitat because

(1) the salmon would recycle nutrients at a rapid rate
(2) their rapid growth rate could cause them to outcompete native salmon
(3) they would not have enough oxygen for survival
(4) they would reproduce asexually once they were released

21 _____

22 The diagram below represents a portion of a cell membrane.

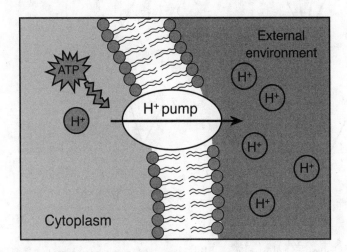

The arrow indicates that the cell membrane is carrying out the process of

(1) respiration   (3) diffusion
(2) cell recognition   (4) active transport

22 _____

23 The expression of a trait is directly dependent on the

(1) arrangement of amino acids in the protein synthesized
(2) shape of the subunits in the DNA molecule
(3) number of chromosomes present in the nucleus
(4) sequence of bases coded for by the ribosome

23 _____

24 Global warming is most closely associated with

    (1) increased use of solar panels
    (2) increased industrialization
    (3) reducing the rate of species extinction
    (4) removal of environmental wastes   24 _____

25 Which diagram below indicates that species *D* is more closely related to *C* than it is to either *A* or *B*?

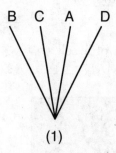

(1)

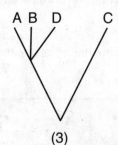

(3)

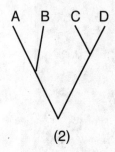

(2)

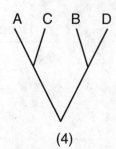

(4)   25 _____

26 As climate changes, which type of reproduction would most likely result in a greater chance of survival for a species?

    (1) sexual reproduction, with a short reproductive cycle
    (2) sexual reproduction, with a long reproductive cycle
    (3) asexual reproduction, with a short reproductive cycle
    (4) asexual reproduction, with a long reproductive cycle   26 _____

27 Adults of the *Aedes* mosquito genus are responsible for transmitting the viral diseases Zika and Dengue. Scientists have produced a modified form of male *Aedes* mosquitoes. The offspring of these male mosquitoes die before reaching adulthood. This method of reducing the spread of disease is dependent on

(1) vaccines stimulating the immune system of infected people
(2) providing medication to reduce the symptoms of disease
(3) the use of natural selection to modify the viruses so they are no longer pathogenic
(4) the use of genetic engineering to reduce the population of mosquitoes that carry the virus   27 _____

28 Humans deplete the most resources when

(1) using wind energy as a power source
(2) generating power by using fossil fuels
(3) using water power to generate electricity
(4) recycling glass and plastics   28 _____

29 The diagram below does *not* represent a sustainable energy pyramid in an ecosystem because

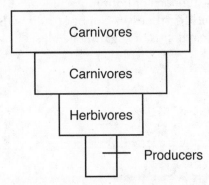

(1) energy is never transferred between levels in ecosystems
(2) ecosystems never have more than three levels of energy transfer
(3) more energy must be available in the producer level than in the consumer levels
(4) producers feed on herbivores in most ecosystems   29 _____

375

30 The two diagrams below represent a sugar molecule and a fat molecule that are used by living organisms.

Sugar molecule

Fat molecule

Which statement best describes these two molecules?

(1) Sugar molecules are inorganic and fat molecules are organic.

(2) Sugar molecules are organic and fat molecules are inorganic.

(3) Energy for life processes can be stored within the chemical bonds of both molecules.

(4) Energy for life processes can be stored within the chemical bonds of sugar molecules, only.

30 _____

PART B–1

Answer all questions in this part. [13]

Directions (31–43): For *each* statement or question, record in the space provided the *number* of the word or expression that, of those given, best completes the statement or answers the question.

31 A scientist analyzed a segment of DNA from a human chromosome and found that the percentage of thymine molecular bases (T) was 35%. Which row in the chart below contains the correct percentages of the other molecular bases in the DNA segment?

Row	Guanine (G)	Cytosine (C)	Adenine (A)
(1)	15%	25%	25%
(2)	25%	25%	15%
(3)	15%	15%	35%
(4)	35%	15%	15%

31 _____

32 The graph below shows changes in the populations of hares and lynx in a Canadian ecosystem.

Changes in Hare and Lynx Populations

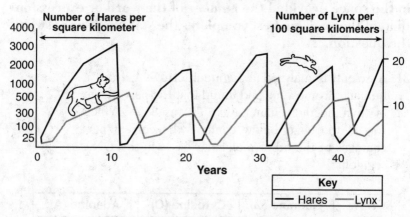

Source: Adapted from http://lbyiene-jardin-wikispaces.com

Which statement about the hares and lynx can be supported with information from the graph?

(1) The hare is the predator of the lynx because it is a larger animal.
(2) The lynx population begins to drop after the hare population drops.
(3) Both populations go through cycles due to the succession of plant species.
(4) Both populations have a carrying capacity of 3000 per square kilometer.

32 _____

33 The diagram below represents a cell in the human body.

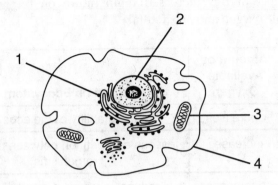

Which statement concerning the structures within this cell is accurate?

(1) Structure 1 is a chloroplast that carries out photosynthesis.
(2) Structure 2 is a vacuole that contains DNA.
(3) Structure 3 is a mitochondrion, where respiration takes place.
(4) Structure 4 is the cell membrane, which provides rigid support for the cell.

33 _____

Base your answers to questions 34 and 35 on the data table below and on your knowledge of biology. The table below indicates the amount of oxygen present at various water temperatures in a pond.

Amount of Available Oxygen in Water at Various Temperatures

Temperature (°F)	Dissolved Oxygen (ppm)
68.0	9.2
71.6	8.8
78.8	8.2
82.4	7.9
86.0	7.6

34 An aquatic ecosystem experiences an increase in temperature. Which row in the chart below shows the effect of this increased temperature on the available oxygen and ecosystem?

Row	Amount of Available Oxygen	Effect on Ecosystem
(1)	decreases	greater stability of the ecosystem
(2)	increases	lessens competition between predatory organisms
(3)	decreases	reduces carrying capacity for fish
(4)	increases	increases genetic mutations in bacteria

34 _____

35 Which process performed by organisms produces oxygen for the aquatic ecosystem?

(1) respiration   (3) active transport
(2) replication   (4) autotrophic nutrition   35 _____

Base your answers to questions 36 and 37 on the diagram below and on your knowledge of biology. The diagram represents a food web illustrating some relationships in a tidal marsh ecosystem.

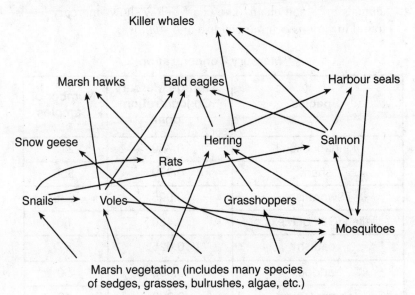

Adapted from: http://www.physicalgeography.net/fundamental/9o.html

36 Examples of autotrophs in this food web are

(1) killer whales and grasses
(2) sedges and bulrushes
(3) mosquitoes and grasshoppers
(4) snails and seals

36 _____

37 In addition to grasshoppers, herring may also get energy from

(1) algae   (3) snails
(2) bald eagles   (4) voles

37 _____

Base your answers to questions 38 and 39 on the information below and on your knowledge of biology.

Mercury is a toxic chemical that accumulates in the tissues of animals in a food chain. The chart below shows mercury levels found in various commercial fish and shellfish.

Mercury Concentration

Species	Average Mercury Concentration (ppm)	Number of Samples
king mackerel	0.730	213
shark	0.979	356
swordfish	0.995	636
tilefish (Gulf of Mexico)	1.450	60
catfish	0.025	57
haddock	0.055	50
lobster (spiny)	0.093	13

Source: www.fda.gov/food/foodborneillnesscontaminants/metals/ucm115644.html

38 Each species listed is a predator. If the prey organisms that each predator consumes were tested, they would most likely contain

(1) the same amount of mercury as the predator species
(2) less mercury than the predator species
(3) more mercury than the predator species
(4) no mercury, since the predators probably get it from the polluted water

38 _____

39 Which statement is best supported by the data in the chart?

(1) Any fish caught in the Gulf of Mexico would have low levels of mercury.
(2) Eating catfish or haddock would be most likely to cause deadly mercury poisoning.
(3) Spiny lobsters may have more or less mercury than indicated because only a few were sampled.
(4) Tilefish are the most nutritious of all the species listed.

39 _____

40 The diagram below represents a laboratory process.

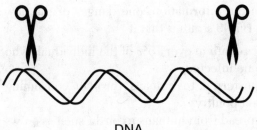

DNA

The substance represented by the scissors shown cutting the DNA is

(1) an enzyme   (3) a carbohydrate
(2) a starch molecule (4) a fat molecule

40 _____

41 The human body has many cells that are deep inside the body. For this reason, the human body requires

(1) a transport system and other organs
(2) carbon dioxide from the air
(3) the synthesis of many inorganic compounds
(4) the breakdown of glucose by the digestive system

41 _____

Base your answers to questions 42 and 43 on the information below and on your knowledge of biology.

Bird Flu

Researchers are not sure when the H7N9 virus, referred to as bird flu, hit the China poultry markets. In February of 2012, the virus was found to have spread from birds to humans. All cases resulted from direct contact with infected poultry.

The bird flu can cause severe respiratory illness in humans. Since flu viruses constantly mutate, it would be difficult to develop a vaccine ahead of time. Scientists are worried that the virus could spread easily among people, causing a worldwide outbreak of the disease.

42 Based on the information, one danger of the new Bird Flu H7N9 strain is that it

(1) causes death in over 75% of the individuals who become infected
(2) is transferred to humans through consuming cooked poultry
(3) can spread from humans to birds, such as crows and pigeons
(4) mutates rapidly, making it hard to produce an effective vaccine   42 _____

43 The fact that the H7N9 virus has only recently infected humans helps explain why

(1) it is highly transmissible through both the air and water
(2) it is found only in the U.S.
(3) humans have little or no immunity to the virus
(4) the human population has formed antibodies against the virus   43 _____

PART B–2

Answer all questions in this part. [12]

Directions (44–55): For those questions that are multiple choice, record your answer in the space provided. For all other questions in this part, record your answer in accordance with the directions.

Base your answers to questions 44 through 47 on the information below and on your knowledge of biology.

As part of an experiment, a bacterial culture was grown in a lab for two days. No additional nutrients were added to the culture after the initial set-up. As the bacteria reproduced asexually, the population of the culture was measured every six hours. Some of the data related to the bacterial growth are shown in the data table below.

Bacterial Growth

Time (hrs)	Population (millions)
0	2.0
6	4.5
18	16.0
30	28.0
48	37.0

Directions (44–45): **Using the information in the data table, construct a line graph on the grid below, following the directions below.**

44 Mark an appropriate scale, without any breaks in the data, on each labeled axis. [1]

45 Plot the data on the grid provided. Connect the points and surround each point with a small circle. [1]

Example: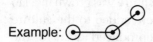

Bacterial Growth

Population (millions)

Time (hrs)

46 If data for the growth of this bacterial population continued to be recorded, would the data point at 60 hours be above or below 37 million? Support your answer. [1]

_____

_____

Note: The answer to question 47 should be recorded in the space provided.

47 One likely reason bacteria would be grown in laboratory cultures would be to

(1) increase the number of antibiotics produced by human cells
(2) eliminate the cloning of cells that can fight disease
(3) increase the production of specialized proteins by using genetic engineering
(4) decrease the amount of bacteria naturally present in organisms

47 _____

Base your answers to questions 48 and 49 on the information and diagram below and on your knowledge of biology. The diagram represents a biological process.

Fossil evidence has demonstrated that birds evolved from a group of small carnivorous dinosaurs. Scientists have hypothesized that some evolved into birds as they filled available niches.

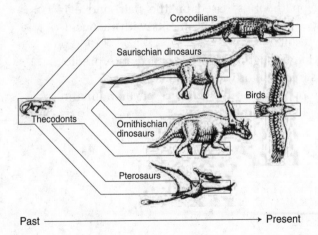

Thecodonts

Crocodilians

Saurischian dinosaurs

Birds

Ornithischian dinosaurs

Pterosaurs

Past ————————————————→ Present

48 Identify *two* groups of organisms from the diagram that still exist on Earth today. Describe how they may have been able to survive to the present. [1]

Organisms: _____ and _____

_____

_____

_____

Note: The answer to question 49 should be recorded in the space provided.

49 The most recent fossil discoveries have filled in many of the gaps in the evolution of birds from dinosaurs. Before the latest fossils were found, there were some scientists who questioned this idea that birds evolved from dinosaurs. In general, scientists constantly work to

(1) clarify scientific explanations so they can be made into a law that never changes
(2) develop theories based on the data and evidence from a few experiments with inconclusive results
(3) provide enough evidence and accurate predictions to allow for widespread acceptance
(4) develop explanations that are permanent and do not change over time

49 _____

Base your answers to questions 50 through 52 on the diagram below and on your knowledge of biology. The diagram indicates some parts of the human female reproductive system.

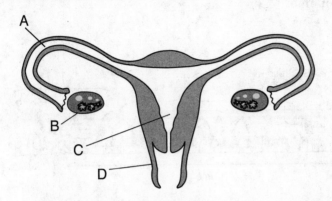

Note: The answer to question 50 should be recorded in the space provided.

50 The structure in which fertilization normally takes place is

(1) *A*   (3) *C*

(2) *B*   (4) *D*   50 _____

51 State *one* function of organ *B*.   [1]

_____

52 State *one* advantage of internal development for the human embryo.   [1]

_____

_____

_____

Base your answers to questions 53 through 55 on the information below and on your knowledge of biology. The diagram represents an ecological process that occurs in New York State over a long period of time.

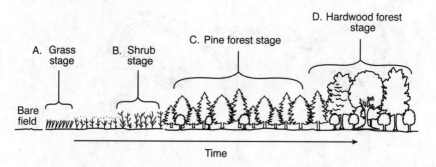

53 Identify the ecological process that is represented from stage A through stage D, and explain why each stage is important to the stage that follows it. [1]

Process: _____

_____

_____

54 Identify *two* abiotic factors that can determine which types of organisms can inhabit an ecosystem. [1]

_____ and _____

55 Identify the short-term effect that a forest fire during stage D would have on the biodiversity of the area. [1]

_____

_____

PART C

Answer all questions in this part. [17]

Directions (56–72): Record your answers in the spaces provided.

Base your answers to questions 56 and 57 on the information below and on your knowledge of biology.

Turtle Cells and Human Skin

New research has demonstrated that turtles and humans may have had a common ancestor 310 million years ago. A recent study looked at the genes responsible for the skin layers of turtle shells compared to the genes for human skin. The findings of the study suggest that about 250 million years ago, when turtle evolution split from other reptiles, a mutation in a specific group of genes occurred. The basic organization of this group of genes is similar in turtles and humans, and they produce the important skin proteins that produce shells in turtles and protect against infection in the skin of humans.

56 Identify the molecule that contains the hereditary material and the organelle in which it is found in turtle cells. [1]

Molecule: _____

Organelle in turtle cells: _____

57 Describe how the mutation in the genes of a turtle ancestor turned out to be a beneficial evolutionary adaptation. [1]

_____

_____

Base your answers to questions 58 through 60 on the illustration and information below and on your knowledge of biology.

The Little Brown Bat

Source: http://knatolee.blogspot.com/2011/09/
not-ducklings.html

The illustration is of a species commonly called the little brown bat. It has 38 teeth and usually lives near bodies of water. The animal is considered beneficial by many people because it eats mosquitoes and many types of garden pests. They feed at night, detecting their prey by echolocation—a form of sonar similar to what is used on ships. They can determine the location and size of their prey by listening to the return echo.

58 The little brown bat eats mainly mosquitoes and night-flying insects. State *one* way in which the animal is adapted to prey on these organisms. [1]

_____

_____

_____

59 If a mutation occurs in some of these bats, it may result in a new inheritable trait that makes them better able to catch insects than other bats in the population. Describe what will most likely happen to the frequency of the *original* trait in the population. Support your answer.   [1]

_____

_____

_____

60 Coevolution occurs when the evolution of an adaptation by one species affects the evolution of an adaptation in a second species. Some species of moths have evolved the ability to emit high frequency sounds that can block the little brown bat's echolocation. Based on the information provided, explain how this relationship between moths and bats is an example of coevolution.   [1]

_____

_____

_____

_____

Base your answers to questions 61 through 64 on the information below and on your knowledge of biology.

Kaolin as a Spray to Control a Bean Pest

Spraying kaolin, a clay-like material, on the leaves of plants has been effective in reducing insect damage to plants that grow in temperate regions, but has not been tried in tropical areas.

Researchers in the tropical Andean region of South America have recently conducted experiments to see if kaolin can be used there to control the greenhouse whitefly, a significant pest of the region's bean crops.

In the study, four groups of bean plants were used with the following treatments:

Group	Treatment	Whiteflies Killed (%)
1 (control)	No insecticide or other substance applied to the plants	0
2	Synthetic chemical insecticide applied to leaves	90
3	Leaves treated with 2.5% concentration of kaolin spray	80
4*	Leaves treated with 5% concentration of kaolin spray	80

* Note: In group 4 the plants lost 40% less water and showed a 45% increase in chlorophyll content in the leaves.

61 State *one* likely effect of the whiteflies on the bean plants in the control group (group 1) by the end of the study. Support your answer.   [1]

_____

_____

62 Should the group 3 kaolin treatments be considered as an acceptable alternative control method to the group 2 insecticide treatment for whiteflies? Support your answer with data from the chart.   [1]

_____

_____

63 Based on the results of groups 3 and 4, identify the kaolin treatment that would be best for bean plants grown in areas where low rainfall is a common occurrence. Support your answer.   [1]

_____

_____

64 State *one* reason why the scientists are interested in reducing whitefly populations in the Andean region.   [1]

_____

_____

Base your answers to questions 65 through 68 on the passage below and on your knowledge of biology.

Medical Mystery

Recently, a man went to a hospital. He felt tired and was coughing and dehydrated. At first, the doctor thought he had pneumonia, but an x ray showed a spot on his lung. Because the man was a smoker, the doctor expected to find a tumor.

Instead, the surgeon discovered a pea seed growing inside the man's lung. When the pea seedling was removed, the patient quickly regained his health.

65 When he first arrived at the hospital, the man reported feeling unusually tired. Explain why damage to the man's lung caused fatigue. [1]

_____

_____

66 In this case, the pea seed entered into the man's lung, but the immune system was not able to defend against it. Describe *one* specific way the cells of the immune system usually protect the body against certain molecules or microbes that are breathed into the lungs. [1]

_____

_____

67 Identify *two* environmental factors inside a human lung that would help the pea begin to germinate. [1]

_____

68 State whether the pea seedling could have continued to grow and develop in the lung over a long period of time. Support your answer. [1]

_____

_____

Base your answers to questions 69 and 70 on the information below and on your knowledge of biology.

Scientists Reprogram Plants for Drought Tolerance

Source: Lancaster Farming 2/21/15/AAAS

Arabidopsis plants respond to drought conditions by producing a stress hormone called ABA. This hormone slows down plant growth and leads to a decrease in the plant's use of water.

ABA binds to specific receptors in the plant that cause the guard cells on the leaf surfaces to close the stomatal openings through which water vapor can normally pass. This reduces water loss during the drought conditions.

Although it has been suggested that spraying plants with ABA during a drought could be beneficial, it is not practical. The chemical is expensive to produce and quickly loses its ability to bind to cell receptors in the plant cells.

Recently, however, scientists have found a way to modify the ABA receptors in *Arabidopsis* plants so they can be activated by another chemical that is both stable and inexpensive.

69 Describe how the shape of molecules, such as the hormone ABA, is critical to their function in the *Arabidopsis* plant.   [1]

_____

_____

_____

70 Explain how the response of the guard cells to a drought is part of a feedback mechanism.   [1]

_____

_____

_____

Base your answers to questions 71 and 72 on the passage and graph below and on your knowledge of biology.

Atmospheric Carbon Dioxide

Records from polar ice cores show that the natural range of atmospheric carbon dioxide (CO_2) over the past 800,000 years was 170 to 300 parts per million (ppm) by volume. In the early 20th century, scientists began to suspect that CO_2 in the atmosphere might be increasing beyond this range due to human activities, but there were no clear measurements of this trend. In 1958, Charles David Keeling began measuring atmospheric CO_2 at the Mauna Loa observatory on the big island of Hawaii.

Carbon Dioxide Concentration at Mauna Loa Observatory

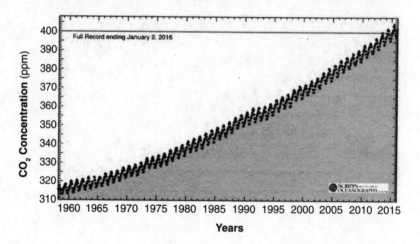

71 Record the approximate concentration of carbon dioxide at the start of the study and describe how it compares to the concentration in 2015.   [1]

_____ ppm CO_2

Description: _____

_____

72 Identify *one* likely reason for the overall change in CO_2 concentration observed between 1958 and 2015.   [1]

_____

_____

PART D

Answer all questions in this part. [13]

Directions (73–85): **For those questions that are multiple choice, record your answer in the space provided. For all other questions in this part, record your answer in accordance with the directions.**

Note: The answer to question 73 should be recorded in the space provided.

73 Which group of materials would be most useful to a student planning to separate a mixture of leaf pigments using paper chromatography?

(1) filter paper, dropper, solvent, beaker
(2) enzymes, beaker, goggles, compound microscope
(3) compound microscope, filter paper, coverslip, glass slide
(4) meterstick, thermometer, solvent, enzymes   73 _____

Note: The answer to question 74 should be recorded in the space provided.

74 In many parts of the world, plants are used as a source of medicine. Many of these plants are in danger of becoming extinct. It is therefore important for researchers to

(1) collect and dry all the medicinal plants to preserve them for future use
(2) search for other plant species that could be used as a new source of that medicine
(3) use the plants now while we still have them
(4) apply fertilizer to reduce the numbers of the plants that grow in the wild   74 _____

Note: The answer to question 75 should be recorded in the space provided.

75 In the lab activity *Making Connections*, an experiment was designed to test the effect of exercise on the ability to squeeze a clothespin. The number of times the clothespin was squeezed served as the

(1) independent variable   (3) hypothesis

(2) dependent variable   (4) control   75 _____

Base your answer to question 76 on the Universal Genetic Code Chart below and on your knowledge of biology.

Universal Genetic Code Chart

		SECOND BASE			
		U	**C**	**A**	**G**	
FIRST BASE	**U**	UUU } PHE, UUC; UUA } LEU, UUG	UCU, UCC, UCA, UCG } SER	UAU } TYR, UAC; UAA } STOP, UAG	UGU } CYS, UGC; UGA } STOP; UGG } TRP	U C A G
	C	CUU, CUC, CUA, CUG } LEU	CCU, CCC, CCA, CCG } PRO	CAU } HIS, CAC; CAA } GLN, CAG	CGU, CGC, CGA, CGG } ARG	U C A G
	A	AUU, AUC, AUA } ILE; AUG } MET or START	ACU, ACC, ACA, ACG } THR	AAU } ASN, AAC; AAA } LYS, AAG	AGU } SER, AGC; AGA } ARG, AGG	U C A G
	G	GUU, GUC, GUA, GUG } VAL	GCU, GCC, GCA, GCG } ALA	GAU } ASP, GAC; GAA } GLU, GAG	GGU, GGC, GGA, GGG } GLY	U C A G

THIRD BASE

Note: The answer to question 76 should be recorded in the space provided.

76 When provided with a sequence of bases in one segment of mRNA, the Universal Genetic Code Chart is used to

(1) directly identify the DNA from an animal cell

(2) determine the sequence of amino acids in a protein

(3) change the RNA sequence of a protein into DNA

(4) identify the specific mutations in the genetic material in a cell

76 _____

77 A student was setting up beakers that contained different solutions in order to conduct a laboratory investigation, but the next day he could not tell which beaker contained the starch and water mixture. In order to find out which beaker contained starch, he took a small sample from each of the beakers and conducted a test for starch on each of them.

Describe the test for starch that the student should use and the result that would indicate the presence of starch. [1]

_____

_____

78 In order to survive in its environment, a single-celled organism uses a contractile vacuole to remove excess water that diffuses into its cell. Another species, a hydra, also excretes excess water. Both processes involve the use of energy.

Based on this information, state whether these two organisms live in fresh water or salt water. Support your answer. [1]

_____

_____

79 The diagram below represents two types of carbohydrate molecules, glucose and sucrose.

Glucose

Sucrose

State *one* reason why a glucose molecule is more likely
than a sucrose molecule to diffuse through an artificial
membrane.   [1]

_____

Base your answers to questions 80 through 82 on the information below and on your knowledge of biology. The diagram represents some of the various types of giant tortoises that live on the
Galapagos Islands. The chart provides information about some
individual island environments.

Giant Tortoises of the Galapagos Islands

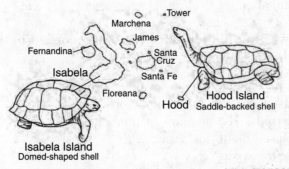

Isabela Island
Domed-shaped shell

Hood Island
Saddle-backed shell

Tower
Marchena
James
Fernandina
Santa Cruz
Isabela
Santa Fe
Floreana
Hood

Source: Adapted from http://slideplayer.com/slide/7372273

Environmental Conditions on
Certain Galapagos Islands

Galapagos Island	Island Characteristics
Hood Island	sparse vegetation located high off of the ground; hot, dry, arid
Isabela Island	rich variety of vegetation located low to the ground; much rainfall; humid

80 Explain why specific Galapagos tortoise species are able to live only on certain islands.   [1]

_____

_____

Note: The answer to question 81 should be recorded in the space provided.

81 The role that the environment plays in determining which species survive is referred to as

(1) a trade-off   (3) an ecological niche
(2) a gene mutation   (4) a selecting agent   81 _____

Note: The answer to question 82 should be recorded in the space provided.

82 Over the years, human activity introduced organisms such as goats and other herbivores to the Galapagos Islands. The addition of these invasive organisms caused the tortoise species to be threatened because there was

(1) an increase in competition for food sources
(2) a decrease in ecological succession
(3) an increase in the availability of vegetation
(4) a decrease in direct harvesting   82 _____

83 As fish are frozen for storage, the water in the cells expands as it cools from 4°C to 0°C and may cause cells to burst. This lowers the quality of the fish. Explain how soaking the fish briefly in salt water before freezing them might prevent this damage to the cells. [1]

_____

_____

Base your answers to questions 84 and 85 on the diagram below and on your knowledge of biology.

Variations in Beaks of Galapagos Islands Finches

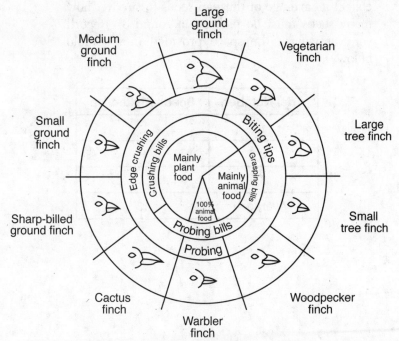

Source: *Galapagos: A Natural History Guide*

84 Identify *one* finch population that would be *negatively* affected if the birth rate of small tree finches increased significantly. Support your answer. [1]

Finch: _____

Support: _____

_____

_____

85 A student completed two trials of the *Beaks of Finches* lab, each time picking up eleven seeds, as shown in the table below. If the student needs to collect an average of thirteen seeds to survive, how many seeds must he pick up in round 3? Record your answer in the space provided in the table below. [1]

Trial Number	Seeds Picked Up
1	11
2	11
3	....
Average	13

Answers
June 2019

Living Environment

Answer Key

PART A

1. 4	6. 1	11. 4	16. 4	21. 2	26. 1
2. 1	7. 2	12. 3	17. 1	22. 4	27. 4
3. 1	8. 4	13. 3	18. 2	23. 1	28. 2
4. 3	9. 2	14. 4	19. 3	24. 2	29. 3
5. 1	10. 3	15. 2	20. 2	25. 2	30. 3

PART B–1

31. 3	34. 3	36. 2	38. 2	40. 1	42. 4
32. 2	35. 4	37. 1	39. 3	41. 1	43. 3
33. 3					

PART B–2

47. 3
49. 3
50. 1

PART D

73. 1
74. 2
75. 2
76. 2
81. 4
82. 1

In **PARTS B–2**, **C**, and **D**, for any answers not listed here, you are required to show your work and explain your answers. For these responses, see *Barron's Regents Exams and Answers* book for Living Environment.

Index